AGING OF ORGANISMS

BIOLOGY OF AGING AND ITS MODULATION

VOLUME 4

AGING OF ORGANISMS

Edited by

HEINZ D. OSIEWACZ

J.W. Goethe University
Frankfurt
Germany

KLUWER ACADEMIC PUBLISHERS
DORDRECHT / BOSTON / LONDON

Library of Congress Cataloging-in-Publication Data is available.

ISBN 1-4020-1406-6

Published by Kluwer Academic Publishers,
PO Box 17, 3300 AA Dordrecht, The Netherlands.

Sold and distributed in North, Central and South America
by Kluwer Academic Publishers,
101 Philip Drive, Norwell, MA 02018, USA

In all other countries, sold and distributed
by Kluwer Academic Publishers, Distribution Center,
PO Box 322, 3300 AH Dordrecht, The Netherlands

Printed on acid-free paper

Printed and bound in Great Britain by Antony Rowe Limited.

Contents

About the series “Biology of aging and its modulation”

During the last 40 years, the study of the biological basis of aging has progressed tremendously, and it has now become an independent and respectable field of study and research. Several universities, medical institutes and research centers throughout the world now offer full-fledged courses on biogerontology. The interest of students taking such courses, followed by undertaking research projects for MSc and PhD studies, has also increased significantly. Cosmetic, cosmeceutical and pharmaceutical industry’s ever increasing interest in aging research and therapy is also obvious. Moreover, increased financial support by the national and international financial agencies to biogerontological research has given much impetus to its further development.

This five-volume series titled “Biology of Aging and its Modulation” fulfills the demand for books on the biology of aging, which can provide critical and comprehensive overview of the wide range of topics, including the descriptive, conceptual and interventive aspects of biogerontology. The titles of the books in this series and the names of their respective editors are:

1. Aging at the molecular level (Thomas von Zglinicki, UK)
2. Aging of cells in and outside the body (S. Kaul and R. Wadhwa, Japan)
3. Aging of organs and systems (R. Aspinall, UK)
4. Aging of organisms (H. D. Osiewacz, Germany)
5. Modulating aging and longevity (S. Rattan, Denmark)

The target readership is both the undergraduate and graduate students in the universities, medical and nursing colleges, and the post-graduates taking up research projects on different aspects of biogerontology. We hope that these books will be an important series for the college, university and state libraries maintaining a good database in biology, medical and biomedical sciences. Furthermore, these books will also be of much interest to pharmaceutical, cosmaceutical, nutraceutical and health-care industry for an easy access to accurate and reliable information in the field of aging research and intervention.

Suresh I.S. Rattan, Ph.D., D.Sc.
Series Editor and Editor-in-Chief, Biogerontology
Danish Centre for Molecular Gerontology, Department of Molecular Biology,
University of Aarhus, Denmark

Preface

Biological aging as the time-depending general decline of biological systems associated with a progressively increasing mortality risk is a general phenomenom of great significance. The underlying processes are very complex and depending on genetic and environment factors. These factors encode or affect a network of interconnected cellular pathways. In no system this network has been deciphered in greater detail. However, the strategy of studying various biological systems has let to the identification of pathways and specific modules and makes it obvious that aging is the result of different overlapping mechanisms and pathways. Some of these appear to be conserved ("public") among species, others are specific or "private" and only of significance in one or a few organisms.

This volume in the series on "Biology of aging and its modulation" specifically focuses on organismic aging. The book covers research on organisms from lower to higher complexity representing examples from very diverse taxa like photosynthetic plants, fungi, sponges, nematodes, flies, birds and mammals. Such a broad treatise of this complex topic provides a comprehensive "flavor" about the current issues dealt with in this rapidly growing scientific discipline.

Heinz D. Osiewacz
Frankfurt, Juli 2003

Yeast Longevity and Aging

S. Michal Jazwinski
Department of Biochemistry and Molecular Biology, LSU Health Sciences Center, 1901 Perdido Street, New Orleans, LA 70112, USA

Definitions of yeast aging

The possibility that the yeast *Saccharomyces cerevisiae* may be prone to age changes in its reproductive potential was first pointed out by Barton [1]. This suggestion was based on the observation that cell division is associated with the deposition of a "bud scar" on the surface of the mother cell. This fundamental aspect of reproduction was subsequently affirmed and expanded upon [2–4], and its cumulative features were directly demonstrated [5].

The finite budding or reproductive/replicative capacity of individual yeast cells was clearly shown by Mortimer and Johnston [6]. These authors reasoned that non-disjunction or recessive lethal genetic changes, other deleterious nuclear events or random depletion of essential autonomous cytoplasmic constituents are not likely to explain the finite replicative life span, because the original population is already at equilibrium for such events. This led to the conclusion that genetic mutation theories are not likely to explain the limitations on yeast life span [7]. However, this rationale does not encompass the cumulative effects of damage and functional losses within individual yeast cells as they progress through their replicative life spans. Progeny cells may be shielded from such damage, which may make its presence known primarily in parental cells. If excessive, functional losses of this sort would be selected against in the population. In any event, they would not impact the fitness of the population as explained later.

A life span limited by the number of divisions available to the cell has recently been demonstrated in the fission yeast *Schizosaccharomyces pombe* [8]. This discovery required a cunning analysis of the development of age asymmetry in lineages of cells dividing by fission. Thus far, this experimental system has not been exploited. We will not refer to this yeast any further in this article.

H.D. Osiewacz (ed.), Aging of Organisms, 1–30.

Yeasts gradually lose viability when they are maintained in stationary phase [9]. Under these conditions of nutrient depletion, the cells exhibit markedly reduced metabolic activity compared to exponentially growing cells [9, 10]. However, metabolic time is still running [11, 12], and it can clearly impact the subsequent replicative life span of the cells [12, 13]. The loss of viability in stationary phase is used to define the chronological life span of yeast cells [14].

Dividing yeast cells display an exponential increase in mortality rate with age [15, 16]. The mortality rate, however, plateaus for the last ten percent or so of an aging cohort [17]. On the other hand, mortality rate fluctuates wildly during the yeast chronological life span, exhibiting increases, decreases and plateaus [18].

Phenomenology of yeast aging

During the quarter of a century following the discovery of the limited replicative capacity of yeast cells, most of the attention was focused on a description of the changes that occur in these cells as they progress through their life spans. Some of these studies were ingenious in employing manipulations of these cells to provide insights into the mechanisms underlying this phenomenon. The laboratory of Ilse Müller played the leading role during this period. Of utmost significance was the finding that the number of divisions available to yeast cells was not altered by holding them in a non-dividing state through nutrient elimination [19]. The cells essentially picked up where they left off in completing their remaining life spans. There is no discrepancy between this finding and the loss of viability or curtailment of life span of cells held in stationary phase, mentioned earlier, because the issue in this case is related to the number of cell divisions remaining for completion of the life span and not those available to cells held in stationary phase to begin with.

Since Johnston [20] determined that daughter cells have the same life span as their mothers had, it has been assumed that cell division involves the establishment of age asymmetry between mother and daughter. In other words, regardless of the nature of the "clock" measuring life span, it is reset in the daughter. Indeed, this principle is maintained throughout the bulk of the life span [11]. However, it breaks down for daughter cells produced from mothers late in their life spans [A. Högel and I. Müller cited in ref. 11]. Daughters of old cells have a shorter life span than daughters of young cells. This finding has been affirmed [21]. Owing to the limited number of cells examined, Johnston [20] would not have recognized this fate of daughters of old mothers.

Johnston [20] also determined that cells that cease dividing early in the life span of the cohort frequently fail to release their daughters in a viable form, while cells displaying longer life spans do this rarely. This suggests that it may be possible to distinguish a phenomenon of premature death. However, such premature demise may be peculiar to the particular yeast strain. Recently, McVey *et al.* [22] have confirmed the finding that normal yeasts cease dividing as unbudded cells [23]. They also showed that certain mutations that shorten life span lead to two distinct populations of cells. The first ceases division as large-budded cells and has short life spans, while the second arrests as unbudded cells and has longer life spans. These morphologies were used to interpret the shapes of the survival curves, which differed for the mutants and the wild type. However, the shapes of the survival curves were likely over-interpreted, given the relatively small number of cells in the cohorts examined. It is not clear how these observations relate to those of Johnston [20] concerning differences in the terminal arrest phenotype of cells. Survival curve shapes are frequently encountered that do not correlate with any overt population stratification in both wild type and mutant yeast.

Early death has been interpreted as the result of premature aging in some studies of yeast replicative life span [24–27]. Several criteria have been used to define premature aging. Interestingly, they appear to be all related to the accumulation of extrachromosomal ribosomal DNA circles (see below) and include sterility and nucleolar fragmentation, all of which coincide with the migration of silencing complexes to the nucleolus [28]. Thus, they are really markers of a single event and cannot be used as an adequate barometer of the global and varied changes associated with the finite replicative life span.

Ploidy does not appear to affect the life span of yeasts [11, 29], at least up to the tetraploid. This conclusion has been confirmed recently with the caveat that it is correct in most but not all genetic backgrounds [30]. The lack of an effect of ploidy on yeast replicative longevity clearly argues against a role for genetic mutation as a cause of limited life span. In fact, mutations do not appear to accumulate during the yeast replicative life span [29].

Yeasts undergo a variety of changes during their replicative life spans, in addition to those listed above (Table 1). The most universally recognized is an increase in generation time; that is the time between consecutive buddings [23]. The increase in generation time becomes acute during the last two to three cell divisions [6, 23]. The age changes observed in yeasts are various, and they vary from one individual cell to the next. They can be characterized as morphological or physiological, and some of them represent decline. Thus, it is proper to refer to an aging process in yeasts.

At the end of their life spans, yeasts stop dividing. Some cells promptly lyse or at the least lose refractility, signifying the loss of differential permeability of the cell membrane. However, others linger. The possibility of a post-mitotic life span cannot be excluded. In one case, it may be possible to associate it with a terminal differentiation process akin to filamentous growth [31], which suggests that the aged cells are starved. The decline in protein synthesis with age [32] suggests reduction and eventual cessation of biosynthetic activity. Studies on the effects of estradiol on generation time and ATP levels in old yeast cells suggest that deficits in energy metabolism, which appear to be related to glycolysis, accumulate with age [33]. Perhaps, these deficits become as acute as the increase in generation time with age, ultimately causing cell death. Recently, evidence has been presented that yeasts undergo a process akin to apoptosis at the end of the replicative life span [34]. It will be of interest to see how widespread this apoptosis-like process is and how closely it resembles apoptosis in mammalian cells.

The study of yeast aging began with the realization that chitin-containing bud scars mark the surface of the mother cell with each division, as indicated earlier. The effect of these bud scars as a causal factor in limiting yeast life span cannot be

***Table 1**. Age-dependent changes in yeast morphology and physiology*

Phenotype	Change	References
Cell size	Increase	2, 5, 6, 117
Cell shape	Altered	JB Chen, SM Jazwinski, unpublished
Granular appearance	Develops	6
Surface wrinkles	Develop	6, 29
Loss of turgor	Develops	29
Cell fragility (prior to death)	None	5
Cell lysis	Occurs	6
Loss of refractility	Occurs	5
Bud scar number	Increase	1–3, 5
Cell wall chitin	Increase	5
Vacuole size	Increase	5
Generation (cell cycle) time	Increase	6, 23
Response to pheromones (haploids)	None/decrease	51 (none), 59 (decrease)
Mating ability (haploids)	Decrease	51
Sporulation ability (diploids)	Increase	118
Cessation of division at G_1/S boundary of the cell cycle (putative)	Occurs	23
Senescence factor	Appears	23
Mutability of mtDNA	Decrease	119
UV resistance	Increase then decrease	71
Resistance to methylating agents	Decrease	71
Ty1 transposon mobility	None	C Pinswasdi, SM Jazwinski, unpublished
Telomere length	None	53
Random budding	Increase	17
Specific gene expression	Altered	41
rRNA levels	Increase	SP Kale, SM Jazwinski, unpublished; 32
rDNA circles	Increase	55
Cellular rRNA concentration	Decrease	SP Kale, SM Jazwinski, unpublished; 32
Protein synthesis	Decrease	32
Ribosome activity, polysome recruitment	Decrease	32
Transcriptional silencing	Decrease	58, 59
Nucleolar fragmentation	Appears	28
Migration of silencing complexes to nucleolus	Appears	28
NAD levels	Increase	25
ATP levels	Decrease	33

dismissed out of hand. Bud scars may exert their effect by limiting sites for budding or by reduction of metabolically active cell surface in relation to volume. There are many arguments against such a role for bud scars, and these were first gathered together exhaustively by Jazwinski [11]. They are tabulated here (Table 2). The most significant of these are that replicative life span can readily be extended by genetic manipulation, while bud scars still accumulate, and that forced deposition of chitin in the cell wall does not curtail life span. There may well be some absolute limit to yeast longevity due to cell wall changes (or to other factors); however, we are currently far from this limit it appears.

Genetics of yeast aging

The genetic analysis of the finite replicative life span of yeast began with an analysis of the yeast *RAS* genes. The v-Ha-*RAS* oncogene was used as a surrogate for yeast *RAS2* initially [35]. This was an informed choice, because this gene can replace *RAS2* in yeast, and the yeast gene can substitute for c-H-*RAS* in animal cells. The activating mutation in the oncogene is not detrimental unless it is coupled to the presence of the unique C-terminal domain of yeast Ras2p [36]. Expression of v-Ha-*RAS* resulted in a marked extension of mean and maximum life span, but only up to a point [35]. Too much expression abrogated the life extension. Moderation seems to be a recurrent theme when aging and longevity are concerned, and such bimodal or nonlinear relationships are often observed. The roles of the two yeast *RAS* genes, *RAS1* and *RAS2*, in determining life span were found to be opposed; *RAS1* curtails life span while *RAS2* extends it [37]. These genes function as a homeostatic device in yeast longevity [12] by modulating, but not directly participating in, several cellular processes that impact life span [17, 38–40].

The reverse genetics approach used to examine *RAS* was next applied to the study of yeast longevity in a more general way. Genes that are differentially expressed during the yeast life span were cloned [41], in the first such effort in any species. One of these genes called *LAG1* was found to play a role in determining yeast life span [42]. The Lag1p is a component of ceramide synthase [43, 44]. Ceramide is a signaling molecule apart from being a component of sphingolipids. It signals stress and alters several responses in eukaryotic cells, including growth, differentiation, proliferation, and apoptosis, by activating or potentiating several signaling pathways [45]. Significantly, one of these is the insulin/IGF-1 pathway, which plays such an important role in *Caenorhabditis elegans* and *Drosophila melanogaster* longevity [46]. In yeast, ceramide affects exocytosis, endocytosis, endoplasmic reticulum to Golgi transport, and heat stress responses, and it interacts with the TOR (target of rapamycin) nutrient signaling pathway to affect cell cycling [47]. The bulk of yeast longevity genes have been identified using reverse genetics (Table 3).

Forward genetics has been used to isolate mutants that affect yeast life span. Kennedy *et al.* [48] isolated mutants that affect yeast replicative life span by using the surrogate phenotypes of resistance to cold and starvation. One of the four uth mutants isolated was determined to be an unusual gain of function mutant in the *SIR4* gene, which is a component of the gene silencing apparatus that operates at the

***Table 2**. Bud scars do not determine yeast longevity*

Evidence/rationale	References
Less than 50% of possible sites for scars are ever occupied	2, 6
Cell wall expands more than necessary to oblige a bud scar at each cell division	3, 117
Bud scars can overlap	2
Life spans of individual cells of a given strain vary; mean and maximum life spans of different strains also vary	7, 52
Overexpression of certain genes extends longevity without affecting bud scars	11, 35, 37
Adaptive increases in surface-to-volume ratio are not the rule during evolution over about 300 generations in a chemostat	120
Isogenic strains of increasing ploidy (and size) up to tetraploid do not differ in life span	29; DS Franklin, SM Jazwinski, unpublished
The large daughters of older mothers (up to middle age) do not differ in life expectancy from the small daughters of younger mothers	C Pinswasdi, SM Jazwinski unpublished; 20, 23
Ethanol increases yeast life span without affecting bud scars	19
Induced deposition of chitin, the major component of the bud scar, does not curtail life span	23
Yeasts display the Lansing effect, which cannot be mediated by the bud scar	A Högel, I Müller cited in 11; also 21
Transmission of the mother cell effect on generation time within a lineage cannot be mediated by the bud scar	23
Life span of zygote reflects that of the older parent, even though the number of bud scars per unit of cell surface is smaller than in that parent	51
Overexpression of v-Ha-*RAS* at moderate levels increases life span irrespective of the increase in cell size	35
LAG1 mutant that displays increased longevity does not differ in size from its isogenic parent	42
Induced thermotolerance extends life span without reducing bud scar number	39
Caloric restriction extends life span without reducing bud scar number	68, 103

In the case of yeasts, the Lansing effect refers to the extinction of a lineage through the continuous selection of the final daughters produced during the life span.

silent mating type locus, telomeres, and as realized subsequently ribosomal DNA (rDNA) locus. This line of investigation ultimately pointed to events at the rDNA locus as important in determining life span [28]. An unbiased hunt for replicative life span mutants based on the primary phenotype of longevity has not been attempted at this time. Thus, the picture we have of yeast replicative life span may be very limited. In addition, it is not clear how this picture translates to aging of wild isolates of yeast in their natural environment.

Forward genetics has also been used to identify genes that contribute to the chronological life span of yeast [49]. Among them is *SCH9*, which encodes a protein kinase B homologue, tying yeast chronological aging into a pathway possessing some similarity to the insulin/IGF-1 pathway mentioned above. Clearly, resistance to stress plays an important role in chronological aging, as the involvement of the *MSN2/MSN4* encoded transcription factor which drives expression from the stress response regulatory element (STRE) would indicate (Table 3). The connection to stress resistance joins replicative and chronological aging.

A cursory examination of the biochemical functions encoded by longevity genes indicates the panoply of cellular pathways and processes involved in determining life span. This array is likely to be even broader, given the biased nature of the search for longevity genes in yeast. Nevertheless, it was already possible several years ago to propose that the pathways and processes determining yeast longevity are circumscribed by four broad physiological principles: metabolic control, stress resistance, gene dysregulation, and genome stability [50]. These same principles were proposed to encompass aging in other organisms as well. Those features portend the marshaling of resources evident in dispersal forms and in diapause [50]. The clarity of this description has withstood the test of time [46].

Mechanisms of yeast aging

Senescence factor

The generation time of individual yeast cells increases as they progress through their replicative life spans [23]. It was determined that daughter cells mimic the generation time of the mothers that produced them, such that the daughters of older mothers had a longer generation time. However, these daughters recovered, and their generation time became similar to that of young cells after they had divided a few times. The lengthened generation time described here is a dominant feature that is transmitted from mother to daughter then to grand-daughter and so on within a pedigree, but it ultimately disappears in this lineage in an all-or-none rather than gradual fashion [23]. This manifestation of aging was interpreted in terms of the operation of a cytoplasmic, diffusible senescence factor that appears with age in the mother cell. This was the first clear description of such an entity in yeast aging. However, earlier studies [51] can also be interpreted in those terms. In the latter studies, it was shown that the life span of a zygote obtained by mating an old with a younger cell was characteristic of the remaining life span of the older mating partner. The changes in the aging cell that constitute the senescence factor cannot be readily reversed, especially as that cell becomes older and older. It should be noted that the

Table 3. *Yeast longevity genes*

Gene	Function/pathway/process	References
LAG1	Ceramide synthesis	42
LAC1	Homologue of *LAG1*; ceramide synthesis	121
RAS1	GTP-binding (G-) protein; signal transduction	37
RAS2	G-protein; signal transduction	37
CDC35	Adenylate cyclase	37
BCY1	Regulatory subunit of protein kinase A	37
PHB1	Membrane-bound mitochondrial protein chaperone	50, 122, 123
PHB2	Homologue of *PHB1*; membrane-bound mitochondrial protein chaperone	122, 123
CDC7	Protein kinase; cell cycle control	15
BUD1	G-protein; cell polarity	17
RTG2	Unknown; retrograde response	40, 103
RTG3	Basic helix-loop-helix/leucine zipper transcription factor; retrograde response	103
RPD3	Histone deacetylase; chromatin-dependent gene regulation	60
HDA1	Histone deacetylase; chromatin-dependent gene regulation	60
SIR2	Histone deacetylase; chromatin-dependent transcriptional silencing	60, 65
SIR4	Transcriptional silencing	48
UTH1	Mitochondrial biogenesis	48
UTH4	Unknown	48
YGL023	Unknown; homologue of *UTH4*	28
SGS1	DNA helicase; DNA recombination	24, 57
RAD52	DNA repair	124
FOB1	Replication block	56
CDC25	GDP-GTP exchange factor for Ras	68
TPK2	Protein kinase A catalytic subunit	68
GPR1	Glucose-binding protein	68
GPA2	G-protein	68
ZDS1	Transcriptional silencing	125
ZDS2	Transcriptional silencing	125
NMT1	N-myristoyltransferase	25
SIP2	Unknown; N-myristoylprotein	25
SNF1	AMP-activated protein kinase homologue	25
SNF4	Regulatory subunit of Snf1complex	25
SOD1	Cu, Zn-superoxide dismutase	74, 75, 126
SOD2	Mn-superoxide dismutase	74
CTA1	Peroxisomal catalase	76, 77
CTT1	Cytoplasmic catalase	76, 77
NPT1	Nicotinate phosphoribosyltransferase; NAD salvage	68, 69
*SCH9**	Protein kinase B (Akt) homolog	49
*CYR1**	Adenylate cyclase (same as *CDC35*)	49
*RIM15**	Protein kinase; stress resistance	49
*MSN2**	Transcription factor; stress resistance	49
*MSN4**	Transcription factor; stress resistance	49

References cite articles in which longevity gene function was established. Asterisks (*) denote genes involved in determining chronological life span, as opposed to the remainder that determine replicative life span.

two studies reviewed here examined the senescence factor from two distinct vantage points. In the first, its effects on the daughter cell were analyzed, while in the second its effects on the mother cell were the object of analysis. Additional support for the operation of a senescence factor comes from the studies showing that the life spans of daughters of old mother cells are shorter than those of young mothers [21; A. Högel and I. Müller cited in ref. 11].

Epigenetic inheritance of different regulatory states of chromatin

A theoretical analysis of the yeast replicative life span led to the proposal that the transcriptional states of chromatin may play a role in determining yeast longevity [52]. This analysis postulated a "molecular memory" of cell cycles completed by an individual yeast cell. Molecular memory is established at each cell division in a probabilistic fashion by the gene silencing machinery, although mechanisms at the post-transcriptional level can also be entertained. The operation of this machinery at multiple sites throughout the genome could lead to cumulative effects.

The discovery of telomere shortening as a function of population doubling levels in mammalian cells prompted the notion that a similar process might lead in yeasts to a loss of silencing of subtelomeric genes. However, it was established that telomeres do not shorten as a function of replicative age in yeast [53]. This finding did not put an end to epigenetic inheritance as a mechanism of aging in yeast. Instead, it prompted a refinement of experimental approaches, as seen below.

The molecular memory referred to above is not meant to imply a genetic aging program, in which a series of events each dependent on completion of the former in the series results in cell senescence. There is no need to postulate such a program to explain yeast aging. Thus, aging is a passive process. Furthermore, there is no natural selection for aging in yeast, because old yeasts constitute a minute fraction of the population, which contributes little to its reproductive fitness [11, 50]. (The fraction of the cells of a particular replicative age decreases exponentially with age in a logarithmically growing yeast population, because one-half of the population is always newborn due to asymmetric division.)

Extrachromosomal ribosomal DNA circles (ERCs)

The gain of function mutant in the *SIR4* gene, mentioned earlier, that displayed extended life span was also deficient in mating [48]. This was interpreted as the result of the loss of transcriptional silencing at the silent mating type loci. The *SIR4* mutation generated a phenocopy of a diploid cell, due to the simultaneous expression of both ***a*** and α mating type information. It was suggested that in effect the silencing complexes that were normally present at the silent mating type loci migrated elsewhere. Their new location was termed the AGE locus.

The identity of this AGE locus became apparent gradually. Further analyses showed that the silencing complexes could be co-localized with nuclear structures called nucleoli, which are the location of rDNA. The *SIR4* mutant induced the migration of silencing complexes to nucleoli, and this was associated with extended life span [28]. At the same time, this migration also occurred as a normal consequence of aging. This prompted the suggestion that the transcriptional complexes were responding to changes in nucleoli as a function of cell age and at the same time counteracting these changes. In the *SIR4* mutant, this resulted in delay of these changes. The changes that were observed were enlargement and fragmentation of the nucleoli during yeast aging. The focus was thus damage at the rDNA locus.

Nucleolar enlargement and fragmentation were also found in the *sgs1* mutant in yeast, which results in a shortened life span [24]. *SGS1* is a homologue of the *WRN* gene, which encodes a DNA helicase that is defective in Werner's syndrome patients [54]. Werner's is a premature aging syndrome. Thus, the connection was made between *SGS1*, nucleolar events, and premature aging. The molecular basis for these effects of *sgs1* was deduced to be the generation of ERCs from the tandem repeats of rDNA in the yeast genome [55]. Indeed, it was shown that ERCs were present at high concentration in old cells and that the artificial induction of ERCs led to the early demise of yeast cells. Prevention of ERC formation by the *fob1* mutation extended the life span of yeasts [56]. *FOB1* causes stalling of replication forks. Such stalling would favor recombination events in the rDNA resulting in the production of ERCs, although Fob1p is also likely to act at replication forks throughout the genome making it difficult to restrict any conclusions to events at the rDNA. Indeed, replication deficits in general appear to lead to yeast aging [27]. Nevertheless, ERCs were deemed to be the mechanism of aging in yeast, and it was suggested that they also cause aging in the cells of Werner's syndrome patients. ERCs were concluded to represent the senescence factor [55], whose operation had been described earlier [23].

Several difficulties with the ERC theory began gradually to appear. ERCs have not been to this day detected in animal cells, despite gallant efforts. More critically, it was shown that ERCs do not accumulate in *sgs1* mutant cells [57]. Surprisingly, *sgs1* was complemented for longevity by another human DNA helicase gene *BLM1* and not by *WRN* [57]. In addition, it was shown that an increase in ERC levels does not curtail life span under all conditions [40], as discussed further below. An attempt has been made to mitigate these discrepancies [22], by postulating two separate effects of *sgs1* mutation on yeast longevity. However, it remains difficult to reconcile the divergent findings concerning ERCs.

Silencing and gene dysregulation

Despite the lack of telomere shortening during yeast aging [53], the possibility was explored that other elements of the silencing machinery might be affected leading to loss of silencing with age. Loss of silencing with age of subtelomeric genes, which was dependent on distance from the telomere, was demonstrated, and it was shown that individual telomeres differed in their response to aging [58]. This finding indicated that telomere shortening as occurs during cell cycling in somatic cells of higher eukaryotes is not essential for loss of silencing, and it suggested that such an epigenetic change could take place in non-dividing cells. Subsequently, it was shown that there is also a loss of silencing with age at the silent mating type loci as well [59].

The loss of silencing during the life span was suggested to result in inappropriate gene expression and ultimately gene dysregulation and to contribute to aging in yeast and in other organisms [50]. Two genes, *CDC7* [15] and *SIR4* [48] that are important in transcriptional silencing had been implicated in yeast longevity. Thus, the question that needed to be addressed was which of the three known heterochromatic regions of the yeast genome, telomeres, silent mating type loci, and/or rDNA, are associated with yeast aging. Yeast genes known to affect silencing at these genomic regions were used as tools in this query [60].

The histone deacetylases Rpd3p and Hda1p are present in distinct histone deacetylase complexes that regulate silencing at telomeres and transcription [61]. Sir2p is not only involved in silencing at telomeres and the silent mating type loci but also at the rDNA locus [62]. The effects of *RPD3*, *HDA1*, and *SIR2* on yeast longevity and on silencing at telomeres, silent mating type loci, and rDNA were examined [60]. All three genes were found to play a role in yeast longevity. Their effects on silencing at the three heterochromatic regions of the yeast genome varied. However, their effects on silencing of rDNA were most closely associated with longevity. This prompted the conclusion that loss of silencing of rDNA is an important factor in yeast aging. The effect at the rDNA locus did not coincide with ERC formation [60]. This study firmly shifted the emphasis from ERC production to gene silencing in yeast aging.

The new role of *RPD3* and *HDA1* in rDNA silencing identified in the analysis of yeast longevity was supported by other studies [63]. However, these histone deacetylases affect the transcription of many genes, and not only those at the three known heterochromatic regions of the yeast genome. Thus, the possibility was raised that their effects on life span are elicited not only through silencing but also transcription of genes throughout the yeast genome [60]. *SIR2* does not appear to be among these many genes [64] (J.C. Jiang and S.M. Jazwinski, unpublished). Thus, the silencing effects of the Rpd3p and Hda1p histone deacetylases at the rDNA locus are not mediated indirectly through Sir2p. The suggestion was made that the silencing effects of Rpd3p and Hda1p are the result of differential effects of the acetylation status of various lysine residues on histones [60]. As will be seen below, limits have been recently placed on the suites of genes regulated by Rpd3p and Hda1p that are involved in determining life span. The role of Sir2p in determining life span has received strong support [30], and it has been shown to be an NAD-dependent histone deacetylase [65–67]. In fact, the pivotal role of NAD in yeast longevity appears to be certain, based on studies with genes encoding enzymes of NAD metabolism [68, 69]. These studies suggest the involvement of an NAD salvage pathway in the nucleus, which provides the cofactor for Sir2p [69]. However, NAD plays an essential role in a multitude of biochemical reactions in the cell, and it is easy to be misled.

Environmental factors and stress resistance

Most of the emphasis in yeast aging research has been on genetic factors. However, some studies have highlighted the significance of the environment and stress. The key role of these latter factors in chronological aging has already been emphasized. This role extends to the involvement of superoxide dismutase for protection from oxidative stress [70]. The possibility that resistance to cold and starvation may be important for the replicative life span was suggested by the utility of these stresses as surrogate phenotypes in the isolation of life span mutants [48]. However, this association has not been exploited further, and, in fact, it has been decidedly lost in the examination of the so-called AGE locus, ERC production, and their sequel.

The association of resistance to ultraviolet radiation with yeast replicative life span was shown by Kale and Jazwinski [71]. *RAS2* was implicated as a positive element in this phenomenon. In some ways this was surprising, because of the repressive effect of Ras2p on stress responses, which is mediated through the Msn2p-Msn4p transcription factor acting at the STRE [72]. However, the response to ultraviolet radiation requires Ras2p function acting through a distinct pathway [73]. *RAS2* can have a negative effect on replicative life span, as indicated earlier, and this effect may be a by-product of its influence on resistance to oxidative stress [72]. The life-extending effect of *RAS2* is mediated through a cyclic AMP-independent pathway, while the curtailment of life span occurs through a cAMP-dependent pathway [37]. It is the latter pathway that inhibits stress resistance through the STRE. Thus, it is essential to consider whether or not the yeasts are undergoing stress, internal as well as external. Clearly, resistance to oxidative stress is important in maintaining replicative life span, as the involvement of *SOD1* and *SOD2* [74, 75] and *CTA1* and *CTT1* indicate [76, 77].

Resistance to stress and longevity show a complex relationship. This is best illustrated by the role of heat stress and the *RAS* genes in yeast replicative life span. Lethal heat shock curtails life span by definition. What about sublethal heat stress? Yeasts subjected to chronic bouts of sublethal heat stress, the sort of situation that organisms frequently encounter in nature, display a shortened life span [38]. This life span curtailment is exacerbated by the deletion of the *RAS2* gene but not *RAS1*. There was a marked delay in the resumption of growth and division following the heat stress in cells lacking Ras2p. There was no difference between *ras1*, *ras2*, and wild-type cells in the activation of stress genes or the suppression of growth-promoting genes. However, the downregulation of the stress genes and upregulation of the growth-promoting genes was dramatically delayed in the *ras2* mutant cells after the heat stress was alleviated. These deficits are due to lack of adequate control at the STRE [72]. They were completely overcome by overexpression of *RAS2*, to the extent that the life span of the cells went well beyond that of unstressed cells [38]. However, both the cAMP-dependent and cAMP-independent pathways were required during chronic sublethal heat stress, whereas in the absence of stress the latter was necessary for life extension, while the former had a negative effect on longevity [37]. This highlights the importance of context and the nonlinear effects of *RAS2* on longevity. There appears to be a trade-off at play: the cost of the capability to respond to chronic stress is a shorter life span in the absence of overt stress.

Despite certain similarities, there are also marked distinctions between the resistance to stress in dividing cells (replicative life span) and non-dividing cells (chronological life span). In non-dividing cells, a premium is placed on the mounting of an effective response to lethal stress, because the survival of the clone depends on it [78]. Under these conditions, the deletion of *RAS2* increases viability [36]. A different strategy is adopted by dividing cells. In this case, expansion of the population guarantees survival due to the generation of an overwhelming number of cells, some of which can survive. This solution is supported by *RAS2* activity [78]. Generally speaking, cells can divide asymmetrically, in which case the damage due to stress, constituting the senescence factor, can be sequestered in one of the mitotic partners [79]. In cells that do not display asymmetric division, undamaged clones that arise randomly may be selected. It has been proposed more generally that asymmetric reproduction is the foundation of aging and longevity [11].

Transient, sublethal heat stress has a very different effect than chronic stress on yeast replicative life span [39, 80]. Delivered early in life, such stress increases longevity by a large depression in mortality rate, which is persistent but not permanent. This constitutes a heritable epigenetic effect. *RAS2* is required for this increase in life span through its effect on the timely recovery from the stress at the level of gene expression, as is the case for chronic stress. In contrast to chronic stress, *RAS1* is also required for the life extension, as is the major heat shock gene *HSP104* that is responsible for induced thermotolerance. Functioning mitochondria are also necessary. This suggests that this mechanism of life extension does not operate concurrently with the retrograde response (see below).

The *RAS1* gene appears to be uniquely involved in the long-term response to transient, sublethal heat stress. *RAS1* is not likely to play this role by its weak stimulation of adenylate cyclase, because this gene cannot substitute for *RAS2* in the response to chronic heat stress. The involvement of *RAS1* may instead stem from its stimulation of inositol phospholipids turnover [81, 82]. Inositol sphingolipids are required for resistance to a variety of stresses in yeast, including heat stress [83]. It has been suggested that the Ras1p can produce a persistent change in inositol phospholipid pools in response to heat stress [84]. Ceramide, a component of these phospholipids, signals in heat stress responses mediated by the Msn2p-Msn4p transcription factor [85]. The longevity gene *LAG1* operates ceramide synthesis in yeast and promotes changes in ceramide levels similar to those seen during heat stress [45]. It is possible that *LAG1* and the *RAS* genes interact in determining yeast life span.

RAS2 is required for recovery of yeasts from glucose starvation [36], similar to its role in survival under chronic heat stress. This suggests a relationship between metabolism and stress resistance. The *SNF1* protein kinase gene is required for emergence of yeasts from glucose repression on starvation for glucose [86]. It is required for survival in stationary phase and in the activation of protective systems against other forms of stress by expression of genes that are repressed by glucose [87]. This provides additional support for the interdependence of metabolism and stress. The involvement of *RAS2* in pseudohyphal growth, which is elicited by nutrient deprivation, is yet another example of a juxtaposition of metabolism and stress, and it has prompted the notion that replicative life span and filamentous growth may partially overlap [11]. Both are potentiated by the action of Ras2p operating through a cAMP-independent pathway [37, 88, 89].

Retrograde response

The importance of metabolism in determining yeast replicative life span is given a priori by the fact that life span is measured by the number of daughter cells produced, a process that requires significant biosynthetic activity. However, the first direct demonstration of the relationship between metabolic control and life span came with the elucidation of the role of the retrograde response in determining yeast longevity [40]. The retrograde response was first discovered as a curious induction of the expression of certain nuclear genes in petite yeast cells, which lack fully functional mitochondria due to partial or complete loss of mitochondrial DNA (mtDNA) [90]. The induced genes appear to compensate for the loss of mitochondrial function by shifting metabolism from the Krebs cycle to the glyoxylate pathway [reviewed in ref. 91]. This provides biosynthetic intermediates in the face of metabolic duress, with the use of acetate as a net source of carbon. The glyoxylate pathway conserves the two carbon atoms of acetate, in contrast to their fate in the Krebs cycle. In fact, the metabolic changes associated with the retrograde response are even more profound, encompassing enhanced expression of genes encoding proteins involved in transport of small molecules, peroxisomal biogenesis and function, stress responses, and anaplerotic reactions [92, 93]. Indeed, the retrograde response entails an increase in stress resistance [40, 92].

The retrograde response was a phenomenon in search of a role until it was found that its induction results in extension of the replicative life span [40]. This is a robust effect. It has been demonstrated in four different genetic backgrounds. The issue is not whether the strain is a petite, but whether the retrograde response is actually induced. It is also not important whether the petite is generated through a loss of mtDNA sequences or through the deletion of nuclear genes that encode components of the respiratory chain. The life extension through induction of the retrograde response can also be elicited by environmental manipulation. The induction of the retrograde response not only extends life span, but it also postpones the development of the senescent phenotype. The retrograde response depends on the action of *RTG2*, which is necessary for the activation of the Rtg1p-Rtg3p transcription factor and its translocation to the nucleus [94]. The life extension observed in strains in which the retrograde response is induced is suppressed by deletion of *RTG2* [40], demonstrating the essential role of the retrograde response. Control of Krebs cycle genes shifts from the HAP transcription complex to the *RTG* genes under metabolic duress [95], and certain aspects of the retrograde response are modulated by the TOR pathway [96], which modulates the response of cells to nutrients. These observations further support the notion of the metabolic remodeling of the cells. The retrograde response is potentiated by *RAS2*, which also facilitates the life extension observed on induction of this response [40]. This is yet another example of the pleiotropic effects of *RAS2* on yeast longevity.

The retrograde response performs as a rheostat rather than a simple on-off switch in extending yeast life span [79]. Greater induction of the retrograde response results in greater life span extension. This suggested that the retrograde response may function as a signal-strength sensitive device that sustains life span in the face of accumulating mitochondrial dysfunction during aging [91]. It has now been shown that mitochondrial dysfunction in fact accumulates during yeast aging [97], perhaps due to an increase in oxidative stress with age [34]. Furthermore, the retrograde response is induced concomitantly in aging yeasts (C. Borghouts and S.M. Jazwinski, unpublished). Thus, the retrograde response has a function during normal aging in

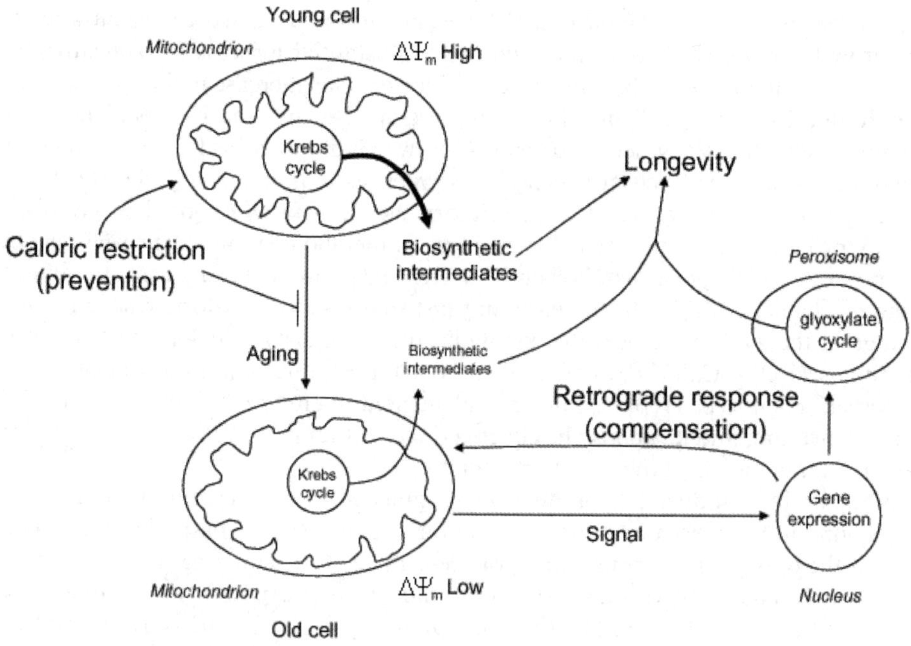

Figure 1. Compensatory and preventive mechanisms of aging converge on mitochondrial function. This model emphasizes the role of mitochondrial membrane potential ($\Delta\Psi_m$) and biosynthetic intermediates provided by the Krebs and glyoxylate cycles for longevity. The retrograde response induces expression of activities targeted to the cytoplasm, in addition to those imported into mitochondria and peroxisomes. See text for further discussion.

yeast (Figure 1). It acts as a compensatory mechanism for the increasing mitochondrial dysfunction during aging and thus may allow yeasts to live as long as they do [47]. This molecular mechanism of aging may well apply to mammals, because mammalian cells also display mitochondrial decay with age [98]. Mitochondrial dysfunction is likely to be compensated in a mouse model of mitochondrial cardiomyopathy [99]. There also appears to be an interaction between the mitochondrial genome and *HRAS1* in human longevity reminiscent of the interaction of *RAS2* and the retrograde response in yeast [100]. Thus, the compensatory mechanism of the retrograde response may play a role in aging in mammals.

Caloric restriction

The retrograde response is a metabolic mechanism in yeast longevity. It is not, however, the only metabolic response that determines yeast life span. Based on studies showing a decline in blood glucose on caloric restriction in mammals [101], we hypothesized that a reduction in glucose availability to yeast cells would extend their replicative life span, because caloric restriction extends rodent life span [102].

This indeed was found to be the case [103]. Reduction of the glucose concentration in the growth medium led to a more and more substantial increase in replicative life span, as the glucose levels became lower and lower. This increase in life span was not simply due to a release from glucose repression, because the life span increased continuously when the glucose concentration was reduced by at least two orders of magnitude below the levels at which glucose repression is completely abolished. The issue appears to be the extent of respiratory metabolism as opposed to glycolytic energy production, or it is related to changes in metabolism that occur well beyond the point at which yeast metabolism is entirely respiratory. In any case, the underlying mechanism of this phenomenon in yeast that resembles caloric restriction is a change in the way that glucose is metabolized in similarity to caloric restriction in mammals [102]. "Caloric restriction" in yeast postpones the appearance of the sensescent phenotype [103]. It can be implemented in either broth or in chemically-defined medium, and it can also be effected by reduction in the concentration of non-essential amino acids in the growth medium.

Whether it is induced by reduction in glucose or non-essential amino acid availability in the medium, caloric restriction operates through a pathway independent of the retrograde response. This was determined by combining the limitation of these nutrients with deletion of the retrograde response pathway signaling genes *RTG2* and *RTG3* and examining the effect on life span [103]. Caloric restriction did not induce expression of the *CIT2* gene, a diagnostic of the retrograde response. However, the studies showed that although the retrograde response and caloric restriction operate by separate pathways, the longevity effectors they control may overlap to some extent. The operation of the two pathways at the same time does not appear to be favored. However, caloric restriction seems to overcome defective retrograde signaling.

In another study, the identity of the signaling events in caloric restriction in yeast was addressed [68]. This study addressed only the effect of lowering glucose concentrations and by only 4-fold. Two different pathways were implicated. However, there was no demonstration that these pathways were actually mediating the effect on life span of lowering glucose levels. Rather, the effect on life span of manipulation of these pathways was evaluated on its own. The negative effect on life span of adenylate cyclase and its target protein kinase A shown by Lin *et al.* [68] confirmed the results of earlier work [37]. Adenylate cyclase is a component of one of the Ras2p signaling pathways. The connection between the Ras2p-cAMP pathway and caloric restriction that was drawn by Lin *et al.* [68] is based on questionable involvement of Ras2p in glucose signaling [104]. The second pathway examined was the Gpr1p/Gpa2p glucose sensor/effector, which is a plausible candidate for the caloric restriction signaling pathway [68]. However, again no direct evidence for its involvement in the life extension obtained on lowering glucose levels was presented. Interestingly, the same pathway is involved in the induction of pseudohyphal growth [105], suggesting some relationship between caloric restriction and the induction by starvation of additional cell divisions during pseudohyphal growth. No signs of filamentous growth were, however, observed during caloric restriction [103] or the retrograde response [40].

The life span effects observed by Lin *et al.* [68] could have been due simply to the release from glucose repression and not to caloric restriction as such. The life extension seen on deletion of *HXK2* [68] supports the notion that release from glucose repression was the culprit, as this gene is known to be required for glucose repression. The concern that glucose repression is involved in the life span effects seen by Lin *et al.* [68] is significant, because the retrograde response is activated and extends life span in some yeast strains on release from glucose repression [40]. At present, it is still not certain whether caloric restriction in yeast is the result of nutrient signaling events, or whether it is caused by changes in metabolism that occur due to lower substrate levels.

An interesting mechanism by which reduction of glucose levels might extend life span has gained support in studies involving *SIR2* [68]. Manipulation of the Ras2p-cAMP pathway was used as a surrogate for caloric restriction in this study. Deletion of *SIR2* suppressed life extension achieved by elimination of an upstream effector of Ras2p, and it was argued that this indicates the participation of Sir2p in a pathway of glucose signaling. Ras2p affects life span through several pathways and processes [12], and one of these pathways impacts chromatin-dependent transcriptional silencing in which Sir2p is involved [17]. No direct analysis of the effect of *SIR2* deletion on life extension afforded by lowering glucose levels was presented [68]. This aspect has become more complicated recently (see below).

The connection of Sir2p to caloric restriction and to metabolism more generally was based on its function as an NAD-dependent histone deacetylase [65–67]. The slowing of metabolism suggested to occur on lowering glucose levels was proposed to result in a higher NAD/NADH ratio, favoring Sir2p activity and thus transcriptional silencing [68]. NAD is certainly important for maintaining yeast life span, as the studies involving manipulation of NAD biosynthetic genes mentioned earlier indicate [68, 69]. However, NAD levels increase with age in yeast [25], rendering loss of Sir2p histone deacetylase activity as a normal cause of aging not very likely. The *NPT1* gene, which encodes an enzyme of NAD salvage, increases yeast life span and may cooperate with *SIR2* in this regard [69]. However, increased expression of *NPT1* did not raise NAD levels or the NAD/NADH ratio in the cell, complicating the interpretation of the *NPT1* effect as a consequence of increased NAD-dependent Sir2p activity. The Npt1p is located in the cell nucleus [69]. Thus, the relevant effects on NAD metabolism may be limited to the nucleus. In this case, the NAD may be protected, however, from metabolic processes occurring in other parts of the cell, and this would diminish its potential role as a sensor of cellular energy metabolism.

Another potential glucose signaling pathway in caloric restriction could be the Snf1 protein kinase pathway, which is essential for emergence of yeasts from glucose repression allowing the cells to use alternative carbon sources [86]. Components upstream of Snf1p in this pathway are not known. Downregulation of this protein kinase by deletion of the *SNF4* activator gene extends life span, while deletion of the inhibitor gene *SIP2* curtails it [25]. Global transcriptional profiling indicated that replicative aging is associated with gluconeogenesis and glucose storage, which are accelerated in *sip2Δ* strains and forestalled by both *SNF4* deletion and a 4-fold reduction in glucose levels in the growth medium [106]. These results were confirmed by analysis of enzyme activities and metabolites in the cells. This analysis provided information on metabolic changes during yeast aging and on the possible effectors of caloric restriction in yeast. However, the method used to isolate young and old cells yielded fairly heterogeneous preparations, and the old cells obtained were considerably younger than the average life span of the yeast strain. It is curious that the Snf1 protein kinase, which is required for glucose de-repression, promotes aging in yeast, while lowering of glucose to levels at which glucose repression is relieved enhances life span and that the metabolic changes in the cell are always consistent with an association of short life span with gluconeogenesis and glucose storage, irrespectively.

Further light has been shed on the effectors of life extension by 4-fold reduction of glucose levels through comparison of global profiles of gene expression in this regime with those obtained by deletion of *HXK2* or overexpression of *HAP4* [107]. Each of these manipulations increased life span. The conclusion was reached that the major effects involve enhanced respiration, and indeed this was demonstrated directly by measurement of oxygen consumption. Despite the 2-fold increase in this parameter over basal levels, the actual contribution of respiration to energy production versus fermentation is likely to still be minimal, given the low baseline levels of about ten percent [108]. The life extensions obtained were suppressed by deletion of *CYT1*, confirming the role of respiration. However, the cells would of necessity rely on respiration in part for their energy needs under these conditions, and thus curtailment of life span could ensue without suppression of the processes that are actually involved in life extension by lowering glucose 4-fold or overexpressing *HAP4*. In other words, respiration might simply have a permissive effect for life extension. A shift from fermentation to respiration as such clearly does not increase life span [5].

Experiments were also performed to address the question of whether life extension was due to enhancement of resistance to oxidative stress [107]. The conclusion was that it was not, based on the lack of increase in resistance to paraquat or hydrogen peroxide under conditions in which life extension is observed. However, it is possible that any enhancement of antioxidant defenses was only sufficient to counter the internal oxidative stress without leaving any excess defense capacity to cope with the external oxidative stress.

The role of *SIR2* in caloric restriction was revisited. Deletion of *SIR2* suppressed the life extension caused by 4-fold reduction of glucose or *HAP4* overexpression, indicating that this gene is required [107]. However, the cells also were deleted for *FOB1*, which complicates the interpretation. It is significant that there was no induction of *SIR2* expression under the conditions in which life extension was observed by Lin *et al.* [107].

It is interesting to note that the metabolic strategies employed by the yeast cells during life extension by *SNF4* deletion or 4-fold reduction of glucose levels involved the maintenance of active glycolysis in the study of Lin *et al.* [106]. Under conditions of shortened life span, the gluconeogenic activator *HAP4* was induced, and this was associated with a shift to gluconeogenesis and glucose storage. This contrasts with the metabolic shift away from glycolysis and the lack of induction of gluconeogenesis on 4-fold glucose reduction, *HXK2* deletion, and even *HAP4* overexpression in the study of Lin *et al.* [107]. On the other hand, *HAP4* was not induced in this latter study on deletion of *HXK2* or 4-fold reduction of glucose.

A different approach to the identification of the longevity effectors of caloric restriction has been taken by examining the effects on life span of reduction of glucose or amino acid concentrations in concert with genetic manipulations that alter global patterns of gene expression and also affect yeast longevity [109]. The reasoning behind this design was that caloric restriction alters the expression of many genes, so an overlap with the suites of genes affected by the genetic manipulations could be inferred from the effects on life span of combining them with caloric restriction. The histone deacetylase genes *RPD3*, *HDA1*, and *SIR2* were chosen for this analysis, because the global patterns of gene expression [64, 110, 111] caused by deletion of these genes and the attendant effects on longevity [60] were known. Deletion of *RPD3* extended life span, and there was no additional effect of caloric restriction. The deletion of *HDA1* did not affect life span on its own, but it had a synergistic effect on the life extension afforded by caloric restriction resulting in a further increase. *SIR2* deletion curtailed life span, but it did not prevent life extension by caloric restriction. The results suggested a model in which Rpd3p affects processes that are required and those that synergize with caloric restriction, while Hda1p and Sir2p do not affect the obligate longevity effectors of caloric restriction, but instead affect processes that synergize with them. However, this synergy acts in opposite directions for Hda1p and Sir2p. By comparing the profiles of gene expression for the three gene deletions based on the effects of the genes on extension of life span by caloric restriction, the conclusion was reached that the major longevity effectors of caloric restriction are carbohydrate/energy metabolism and mitochondrial function.

Jiang *et al.* [109] report lack of suppression of the life span-extending effect of caloric restriction by deletion of *SIR2*. This contrasts with the results reported by Lin *et al.* [107]. There may be several explanations for this discrepancy. As noted earlier, Lin *et al.* [107] combined deletion of *SIR2* with that of *FOB1*, complicating the analysis. There may also be differences dictated by the strain used and the experimental conditions. In particular, the 4-fold reductions in glucose concentration utilized by Lin *et al.* [107] may simply effect release from glucose repression without the maximum conversion to respiratory metabolism that results from further reduction of glucose levels.

The retrograde response compensates for the mitochondrial dysfunction that occurs during aging (see above). Caloric restriction shares some of the same longevity effectors. In particular, it results in mitochondrial function remodeling (Figure 1). In this way, it may prevent damage and dysfunction, especially to this organelle, and thus it may delay aging and alleviate its symptoms. The retrograde response and caloric restriction represent compensatory and preventive mechanisms of aging, respectively.

The role of chance in yeast aging

Chance is considered at play when we are dealing with random or stochastic events. Chance is not deterministic, although deterministic rules can describe its action and effects. Environmental factors can have a stochastic quality. However, we are not dealing with the responses, often programmed, to environmental factors as such in this section. Instead, we are considering the net effects of random events, whether external or internal to the yeast cell, deriving from their very unpredictability. The study of this facet of aging began with the description of the epigenetic stratification of yeast populations due to individual change [17]. It was postulated that change is not simply an effect of aging, but rather its cause. A model was constructed based on scale invariant and increasing change in the form of a simple difference equation to describe the nonlinear dynamics of such an aging system. Unexpectedly, the solution to this equation, regardless of the initial state of the system, indicates that the augmentation of aging processes can result in a lack of aging. Experimental evidence at the population, cellular, and molecular levels supports the model [17]. In fact, the equation correctly predicted that stochastic change in individual yeast cells results in stratification of the population into rapidly aging cells and cells that display no increase in mortality rate. Thus, mortality rate was shown to plateau for cells with the longest life spans [17].

The nonlinearity of yeast aging has been further documented [112]. In addition, evidence for stochastic change in the individual aging yeast cell was gathered [112]. This evidence pointed to the random changes in cell polarity experienced by yeast cells during the replicative life span and to the unique character of the profiles of these changes for individual cells.

Chance can have profound effects at the level of individual molecules, as well as the level of the complex system of the yeast cell. If a particular molecule such as an enzyme is present in small numbers in the cell, there is always the finite possibility that all of the molecules can end up in one or another particular state, resulting in dramatic departures from equilibrium. This is an intrinsically random process, which can be affected by outside factors. At the level of the cell at the other extreme, there are many interactions supported by individual cellular components. A change in any one interaction can therefore have profound effects on the system as a whole. Epigenetic stratification due to chance alone readily explains the different ages at death of individuals in a population of genetically identical individuals maintained under constant conditions. It also explains the presence of multiple and independent limiting factors for longevity in such a population.

Is there information that can be extracted from stochastic change during aging? The difference equation alluded to above predicts that it can. A model of the budding profiles of individual yeast cells was constructed to describe the changes in cell polarity with each cell division during the life span [113]. This model was a linear equation with seven independent variables, each of which described a particular graphic feature of the budding profile. The solution of this equation in seven-dimensional space is a value termed the bud print. Examination of numerous bud prints confirmed the unique character of age changes in individual yeast cells in an aging population. Furthermore, the bud prints were shown to be accurate predictors of longevity. The analysis of the variables in the equation predicted that random budding would be associated with longer life span of individual cells, and this was indeed verified experimentally. Perhaps, some degree of flexibility allows the cell to accommodate the various exigencies encountered as it ages and thus facilitates a longer life span.

Age asymmetry between mother and daughter cells

There are two fundamental questions that must be answered to understand the limited replicative life span of yeast cells. The first concerns the nature of the processes that lead to the ultimate demise of the mother cell. The second question is what determines that a daughter is born young, allowing the perpetuation of the yeast clone or strain. As we have seen above, there are numerous genetically and environmentally based determinants of longevity. On the other hand, the factors that establish age asymmetry have remained obscure until recently. There is no fundamental requirement that a relationship, let alone identity, exists between determinants of the life span of mother cells and of age asymmetry between mother and daughter cells, although this would be esthetically pleasing.

Telomere shortening was raised as a potential cause of aging of mother cells [53], based on a report that chromatids of similar replicative age segregate together in yeast [114]. This co-segregation was shown not to be the case, and it was demonstrated that telomeres do not shorten during the replicative life span in either mother or daughter cells, eliminating telomere shortening as the source of age asymmetry [53]. The induction of telomere shortening by mutation causes the extinction of a yeast clone (clonal senescence) during the course of approximately one hundred cell generations [115]. However, this is not the result of loss of age asymmetry.

ERCs accumulate during the replicative life span of yeast cells, and they preferentially segregate to mother cells during cell division [55]. They have many of the features of the senescence factor referred to earlier, and their increased numbers in old mother cells could result in spill over into daughters, although this has never been demonstrated. The artificial acceleration of ERC production shortens replicative life span of mother cells. However, there is no evidence that this results in the loss of age asymmetry, and ERCs clearly do not cause clonal senescence, even under conditions when their accumulation is dramatically enhanced [55]. In fact, increased ERC levels are associated with longer replicative life spans under certain circumstances [40, 116].

In an effort to identify the key features of age asymmetry, temperature-sensitive clonal senescence mutants were isolated, and one of these was analyzed in detail [97]. It was demonstrated that the clonal senescence was in fact due to loss of age asymmetry, such that daughters were born possessing the replicative age of their mothers at the time of birth. This phenotype was the result of a point mutation in the *ATP2* gene, which encodes the β-subunit of mitochondrial F_1, F_0-ATPase. The mutation did not affect growth on non-fermentable carbon sources; its effect was subtle and manifested itself in the loss of division potential over a period of many cell generations, resulting in clonal senescence. The clonal senescence was reversed on shift-down to permissive temperature. The clonal senescence phenotype was only observed on a fermentable carbon source, and the same temperature-sensitive phenotype was elicited by deletion of *ATP2*. These last two properties may have a common basis. Mitochondrial function is required for life extension by transient, sublethal heat stress [39], which may be connected to the function of F_1-ATPase in the generation of mitochondrial membrane potential ($\Delta\Psi_m$). This activity of the ATPase is important during growth on fermentable carbon sources, because of the greatly reduced activity of the mitochondrial electron transport chain.

The clonal senescence phenotype elicited by mutation of *ATP2* was not dependent on telomere shortening or accumulation of ERCs [97]. Given the localization of the defect in the age asymmetry mutant to the mitochondrion, the functional status of this organelle was examined by measurement of $\Delta\Psi_m$ [97]. It was shown by flow cytometry that there was a progressive decline in $\Delta\Psi_m$ with onset and development of clonal senescence, and that this was followed by loss of mitochondrial mass. It was the young cells in the population that displayed a paucity of active mitochondria. Microscopic examination showed a concomitant change in mitochondrial morphology and distribution in the cell. This resulted in a deficit in the segregation of active mitochondria to daughter cells. The net effect was the generation of cells totally lacking mitochondria, and these were the dominant species as yeast clones became extinct.

The studies with the *atp2* mutant demonstrate that defects in this gene can result in the loss of age asymmetry and the development of clonal senescence. Thus, the normal function of this gene is essential for maintenance of age asymmetry. The question arises whether some of the deficits associated with *atp2* are found during the course of normal aging. In fact, there is a five-fold decline in $\Delta\Psi_m$ as yeasts progress through the replicative life span [97]. This may be due to oxidative damage [34]. Importantly, older yeast cells tend to segregate dysfunctional mitochondria to their daughters [97]. Thus, mitochondrial dysfunction could readily be a cause of aging in yeast. Deletion of *ATP2*, in fact, curtails replicative life span (P.A. Kirchman and S.M. Jazwinski, unpublished). Dysfunctional mitochondria possess the properties of the senescence factor, described above. Not only may they cause aging in mother cells, but they are segregated to daughters of old mother cells, which shortens the replicative life span of these daughters as the maternal-age effect of *atp2* on age asymmetry between mothers and daughters demonstrates [97]. The reversibility of the clonal senescence phenotype also corresponds to one of the properties of the senescence factor; that is, the ability of the great grand-daughters of a senescing mother cell to recover from its effects [23].

We generalize to state that the proper partition of active and undamaged cellular components to daughter cells is required for the maintenance of age asymmetry. There may exist "filters" that guarantee the accuracy of this process, and these filters may breakdown with age. The operation of such filters requires communication between subcellular compartments and organelles, and there may be checkpoint controls that monitor the activities of these filters. Decreased reliability of these controls could result in loss of connectivity with aging. Stem cell lineages, just as yeast, may be dependent on these filters and controls.

The studies on age asymmetry bring mitochondria to the fore in aging (Figure 1). The significance of mitochondria for life span is also visible in the operation of the retrograde response and caloric restriction. In the latter, life extension is obtained through the remodeling of mitochondrial function, while in the former this is achieved by compensating for mitochondrial deficits through activation of alternate pathways. In both cases, the focal point appears to be the provision of biosynthetic intermediates by either the Krebs cycle or the glyoxylate cycle. These pathways of intermediary metabolism depend on communication between subcellular compartments and on transport of metabolites across the mitochondrial inner membrane, which requires the maintenance of $\Delta\Psi_m$. The maintenance of $\Delta\Psi_m$ therefore appears crucial for replicative life span and for the establishment of age asymmetry. Provision of biosynthetic intermediates determines replicative life span, in the sense that these metabolites are necessary for production of daughter cells whose quantity is the measure of life span. Thus, biosynthetic intermediates from the Krebs cycle and $\Delta\Psi_m$ may be more significant than mitochondrial electron transport, as such, for longevity.

Conclusions and future directions

The study of aging in yeast has been devoted primarily to the examination of the limited reproductive capacity of *S. cerevisiae*. This effort is now a half-century old, but it has intensified during the past fifteen years with the application of genetic and molecular methods. Some thirty genes, which encode a wide variety of biochemical activities, have been implicated in yeast longevity. However, this diversity belies the simplicity of the grouping of these genes into four broad physiologic categories, encompassing metabolic control, stress resistance, gene silencing, and genome stability. There are only a handful of mechanisms governing yeast longevity that have been enumerated. These can be categorized as compensatory and preventive mechanisms of aging. Chance, in addition to genes and environment, has recently become the subject of analysis in yeast aging studies. The phenomenological description of a senescence factor in yeast aging has now received a molecular identity in the form of dysfunctional mitochondria, whose inappropriate segregation to daughter cells results in the loss of age asymmetry between mother and daughter. The decline in the efficiency of the "filter" that sequesters damaged organelles in the mother cell may be the proximal cause of this loss of age asymmetry. Together with the analysis of age asymmetry, studies of the retrograde response and caloric restriction have brought metabolism and mitochondria to the fore of yeast aging.

Much remains to be done. An unbiased screen for replicative life span mutants will be necessary to round out the picture we have currently. It appears that exploration of age asymmetry will be a particularly fruitful avenue of future research. It will also be important to continue the examination of chronological aging, including aging of yeasts in colonies. The issue of whether the genes and processes uncovered using laboratory strains will pertain to life span of wild strains in their natural environments remains to be addressed. Finally, the relevance of the information gained in studying yeast aging to aging in higher organisms must still be evaluated. There are already several areas of controversy in yeast aging research that will be resolved more easily when the number of laboratories studying this phenomenon increases.

Acknowledgments

Research in the author's laboratory is supported by grants from the National Institute on Aging of the National Institutes of Health (U.S.P.H.S.) and from the Louisiana Board of Regents through the Millenium Trust Health Excellence Fund [HEF (2001-06)-02].

References

1. Barton AA (1950). Some aspects of cell division in *Saccharomyces cerevisiae. J Gen Microbiol.* 4: 84–7.
2. Bartholomew JW, Mittwer T (1953). Demonstration of yeast bud scars with the electron microscope. *J Bacteriol.* 65: 272–5.
3. Beran K, Malek E, Streiblova E, Lieblova J (1967). The distribution of the relative age of cells in yeast populations In: Powell EO, Evans CGT, Strange RE, Tempest DW, eds. *Microbial Physiology and Continuous Culture.* London: Her Majesty's Stationery Office, pp. 57–67.
4. Cabib E, Ulane R, Bowers B (1974). A molecular model for morphogenesis: the primary septum of yeast. *Curr Top Cell Regul.* 8: 1–32.
5. Egilmez NK, Chen JB, Jazwinski SM (1990). Preparation and partial characterization of old yeast cells. *J Gerontol.* 45: B9–17.
6. Mortimer RK, Johnston JR (1959). Life span of individual yeast cells. *Nature* 183: 1751–2.
7. Jazwinski SM (1990). Aging and senescence of the budding yeast *Saccharomyces cerevisiae. Mol Microbiol.* 4: 337–43.
8. Barker MG, Walmsley RM (1999). Replicative ageing in the fission yeast *Schizosaccharomyces pombe. Yeast* 15: 1511–18.
9. Werner-Washburne M, Braun E, Johnston GC, Singer RA (1993). Stationary phase in the yeast *Saccharomyces cerevisiae. Microbiol Rev.* 57: 383–401.
10. Lagunas R, Ruiz E (1988). Balance of production and consumption of ATP in ammonium-starved *Saccharomyces cerevisiae. J Gen Microbiol.* 134: 2507–11.
11. Jazwinski SM (1993). The genetics of aging in the yeast *Saccharomyces cerevisiae. Genetica* 91: 35–51.
12. Jazwinski SM (1999). The *RAS* genes: a homeostatic device in *Saccharomyces cerevisiae* longevity. *Neurobiol Aging* 20: 471–8.
13. Ashrafi K, Sinclair D, Gordon JI, Guarente L (1999). Passage through stationary phase advances replicative aging in *Saccharomyces cerevisiae. Proc Natl Acad Sci USA* 96: 9100–5.
14. Longo VD (1999). Mutations in signal transduction proteins increase stress resistance and longevity in yeast, nematodes, fruit flies, and mammalian neuronal cells. *Neurobiol Aging* 20: 479–86.
15. Jazwinski SM, Egilmez NK, Chen JB (1989). Replication control and cellular life span. *Exp Gerontol.* 24: 423–36.

16. Pohley HJ (1987). A formal mortality analysis for populations of unicellular organisms (*Saccharomyces cerevisiae*). *Mech Ageing Dev.* 38: 231–43.
17. Jazwinski SM, Kim S, Lai CY, Benguria A (1998). Epigenetic stratification: the role of individual change in the biological aging process. *Exp Gerontol.* 33: 571–80.
18. Vaupel JW, Carey JR, Christensen K, *et al.* (1998). Biodemographic trajectories of longevity. *Science* 280: 855–60.
19. Müller I, Zimmermann M, Becker D, Flömer M (1980). Calendar life span versus budding life span of *Saccharomyces cerevisiae. Mech Ageing Dev.* 12: 47–52.
20. Johnston JR (1966). Reproductive capacity and mode of death of yeast cells. *Antonie Van Leeuwenhoek* 32: 94–8.
21. Kennedy BK, Austriaco NR, Guarente L (1994). Daughter cells of *Saccharomyces cerevisiae* from old mothers display a reduced life span. *J Cell Biol.* 127: 1985–993.
22. McVey M, Kaeberlein M, Tissenbaum HA, Guarente L (2001). The short life span of *Saccharomyces cerevisiae sgs1* and *srs2* mutants is a composite of normal aging processes and mitotic arrest due to defective recombination. *Genetics* 157: 1531–42.
23. Egilmez NK, Jazwinski SM (1989). Evidence for the involvement of a cytoplasmic factor in the aging of the yeast *Saccharomyces cerevisiae. J Bacteriol.* 171: 37–42.
24. Sinclair DA, Mills K, Guarente L (1997). Accelerated aging and nucleolar fragmentation in yeast *sgs1* mutants. *Science* 277: 1313–16.
25. Ashrafi K, Lin SS, Manchester JK, Gordon JI (2000). Sip2p and its partner Snf1p kinase affect aging in *S. cerevisiae. Genes Dev.* 14: 1872–85.
26. Merker RJ, Klein HL (2002). *hpr1Δ* affects ribosomal DNA recombination and cell life span in *Saccharomyces cerevisiae. Mol Cell Biol.* 22: 421–9.
27. Hoopes LL, Budd M, Choe W, Weitao T, Campbell JL (2002). Mutations in DNA replication genes reduce yeast life span. *Mol Cell Biol.* 22: 4136–46.
28. Kennedy BK, Gotta M, Sinclair DA, *et al.* (1997). Redistribution of silencing proteins from telomeres to the nucleolus is associated with extension of life span in S. cerevisiae. *Cell* 89: 381–91.
29. Müller I (1971). Experiments on ageing in single cells of *Saccharomyces cerevisiae. Arch Mikrobiol.* 77: 20–5.
30. Kaeberlein M, McVey M, Guarente L (1999). The SIR2/3/4 complex and SIR2 alone promote longevity in *Saccharomyces cerevisiae* by two different mechanisms. *Genes Dev.* 13: 2570–80.
31. Woldringh CL, Fluiter K, Huls PG (1995). Production of senescent cells of *Saccharomyces cerevisiae* by centrifugal elutriation. *Yeast* 11: 361–9.
32. Motizuki M, Tsurugi K (1992). The effect of aging on protein synthesis in the yeast *Saccharomyces cerevisiae. Mech Ageing Dev.* 64: 235–45.
33. Motizuki M, Tsurugi K (1992). Effect of 17β-estradiol on the generation time of old cells of the yeast *Saccharomyces cerevisiae. Biochem Biophys Res Commun.* 183: 1191–6.
34. Laun P, Pichova A, Madeo F, *et al.* (2001). Aged mother cells of *Saccharomyces cerevisiae* show markers of oxidative stress and apoptosis. *Mol Microbiol.* 39: 1166–73.
35. Chen JB, Sun J, Jazwinski SM (1990). Prolongation of the yeast life span by the v-Ha-*RAS* oncogene. *Mol Microbiol.* 4: 2081–6.
36. Tatchell K (1993). *RAS* genes in the budding yeast *Saccharomyces cerevisiae* In: Kurjan T, ed. *Signal Transduction: Prokaryotic and Simple Eukaryotic Systems.* San Diego: Academic Press, pp. 147–88.
37. Sun J, Kale SP, Childress AM, Pinswasdi C, Jazwinski SM (1994). Divergent roles of *RAS1* and *RAS2* in yeast longevity. *J Biol Chem.* 269: 18638–45.

38. Shama S, Kirchman PA, Jiang JC, Jazwinski SM (1998). Role of *RAS2* in recovery from chronic stress: effect on yeast life span. *Exp Cell Res.* 245: 368–78.
39. Shama S, Lai CY, Antoniazzi JM, Jiang JC, Jazwinski SM (1998). Heat stress-induced life span extension in yeast. *Exp Cell Res.* 245: 379–88.
40. Kirchman PA, Kim S, Lai CY, Jazwinski SM (1999). Interorganelle signaling is a determinant of longevity in *Saccharomyces cerevisiae. Genetics* 152: 179–90.
41. Egilmez NK, Chen JB, Jazwinski SM (1989). Specific alterations in transcript prevalence during the yeast life span. *J Biol Chem.* 264: 14312–17.
42. D'mello NP, Childress AM, Franklin DS, Kale SP, Pinswasdi C, Jazwinski SM (1994). Cloning and characterization of *LAG1*, a longevity-assurance gene in yeast. *J Biol Chem.* 269: 15451–9.
43. Guillas I, Kirchman PA, Chuard R, *et al.* (2001). C26-CoA-dependent ceramide synthesis of *Saccharomyces cerevisiae* is operated by Lag1p and Lac1p. *EMBO J.* 20: 2655–65.
44. Schorling S, Vallee B, Barz WP, Riezman H, Oesterhelt D (2001). Lag1p and Lac1p are essential for the acyl-CoA-dependent ceramide synthase reaction in *Saccharomyces cerevisae. Mol Biol Cell* 12: 3417–27.
45. Jazwinski SM, Conzelmann A (2002). *LAG1* puts the focus on ceramide signaling. *Int J Biochem Cell Biol.* 34: 1491–5.
46. Kenyon C (2001). A conserved regulatory system for aging. *Cell* 105: 165–8.
47. Jazwinski SM (2002). Growing old: metabolic control and yeast aging. *Annu Rev Microbiol.* 56: 769–92.
48. Kennedy BK, Austriaco NR, Zhang J, Guarente L (1995). Mutation in the silencing gene *SIR4* can delay aging in S. cerevisiae. *Cell* 80: 485–96.
49. Fabrizio P, Pozza F, Pletcher SD, Gendron CM, Longo VD (2001). Regulation of longevity and stress resistance by Sch9 in yeast. *Science* 292: 288–90.
50. Jazwinski SM (1996). Longevity, genes and aging. *Science* 273: 54–9.
51. Müller I (1985). Parental age and the life-span of zygotes of *Saccharomyces cerevisiae. Antonie Van Leeuwenhoek* 51: 1–10.
52. Jazwinski SM (1990). An experimental system for the molecular analysis of the aging process: the budding yeast *Saccharomyces cerevisiae. J Gerontol.* 45: B68–74.
53. D'mello NP, Jazwinski SM (1991). Telomere length constancy during aging of *Saccharomyces cerevisiae. J Bacteriol.* 173: 6709–13.
54. Yu CE, Oshima J, Fu YH, *et al.* (1996). Positional cloning of the Werner's syndrome gene. *Science* 272: 258–62.
55. Sinclair DA, Mills K, Guarente L (1998). Molecular mechanisms of yeast aging. *Trends Biochem Sci.* 23: 131–4.
56. Defossez PA, Prusty R, Kaeberlein M, *et al.* (1999). Elimination of replication block protein Fob1 extends the life span of yeast mother cells. *Mol Cell* 3: 447–55.
57. Heo SJ, Tatebayashi K, Ohsugi I, Shimamoto A, Furuichi Y, Ikeda H (1999). Bloom's syndrome gene suppresses premature ageing caused by Sgs1 deficiency in yeast. *Genes Cells* 4: 619–25.
58. Kim S, Villeponteau B, Jazwinski SM (1996). Effect of replicative age on transcriptional silencing near telomeres in *Saccharomyces cerevisiae. Biochem Biophys Res Commun.* 219: 370–6.
59. Smeal T, Claus J, Kennedy B, Cole F, Guarente L (1996). Loss of transcriptional silencing causes sterility in old mother cells of *S. cerevisiae. Cell* 84: 633–42.
60. Kim S, Benguria A, Lai CY, Jazwinski SM (1999). Modulation of life-span by histone deacetylase genes in *Saccharomyces cerevisiae. Mol Biol Cell* 10: 3125–36.

61. Rundlett SE, Carmen AA, Kobayashi R, Bavykin S, Turner BM, Grunstein M (1996). *HDA1* and *RPD3* are members of distinct yeast histone deacetylase complexes that regulate silencing and transcription. *Proc Natl Acad Sci USA* 93: 14503–8.
62. Smith JS, Brachmann CB, Pillus L, Boeke JD (1998). Distribution of a limited Sir2 protein pool regulates the strength of yeast rDNA silencing and is modulated by Sir4p. *Genetics* 149: 1205–19.
63. Smith JS, Caputo E, Boeke JD (1999). A genetic screen for ribosomal DNA silencing defects identifies multiple DNA replication and chromatin-modulating factors. *Mol Cell Biol.* 19: 3184–97.
64. Bernstein BE, Tong JK, Schreiber SL (2000). Genomewide studies of histone deacetylase function in yeast. *Proc Natl Acad Sci USA* 97: 13708–13.
65. Imai S, Armstrong CM, Kaeberlein M, Guarente L (2000). Transcriptional silencing and longevity protein Sir2 is an NAD-dependent histone deacetylase. *Nature* 403: 795–800.
66. Landry J, Slama JT, Sternglanz R (2000). Role of NAD(+) in the deacetylase activity of the SIR2-like proteins. *Biochem Biophys Res Commun.* 278: 685–90.
67. Smith JS, Brachmann CB, Celic I, *et al.* (2000). A phylogenetically conserved NAD+-dependent protein deacetylase activity in the Sir2 protein family. *Proc Natl Acad Sci USA* 97: 6658–63.
68. Lin SJ, Defossez PA, Guarente L (2000). Requirement of NAD and *SIR2* for life-span extension by calorie restriction in *Saccharomyces cerevisiae. Science* 289: 2126–8.
69. Anderson RM, Bitterman KJ, Wood JG, *et al.* (2002). Manipulation of a nuclear NAD+ salvage pathway delays aging without altering steady-state NAD+ levels. *J Biol Chem.* 277: 18881–90.
70. Longo VD, Gralla EB, Valentine JS (1996). Superoxide dismutase activity is essential for stationary phase survival in *Saccharomyces cerevisiae.* Mitochondrial production of toxic oxygen species *in vivo. J Biol Chem.* 271: 12275–80.
71. Kale SP, Jazwinski SM (1996). Differential response to UV stress and DNA damage during the yeast replicative life span. *Dev Genet.* 18: 154–60.
72. Marchler G, Schuller C, Adam G, Ruis H (1993). A *Saccharomyces cerevisiae* UAS element controlled by protein kinase A activates transcription in response to a variety of stress conditions. *EMBO J.* 12: 1997–2003.
73. Engelberg D, Klein C, Martinetto H, Struhl K, Karin M (1994). The UV response involving the Ras signaling pathway and AP-1 transcription factors is conserved between yeast and mammals. *Cell* 77: 381–90.
74. Wawryn J, Krzepilko A, Myszka A, Bilinski T (1999). Deficiency in superoxide dismutases shortens life span of yeast cells. *Acta Biochim Pol.* 46: 249–53.
75. Barker MG, Brimage LJ, Smart KA (1999). Effect of Cu,Zn superoxide dismutase disruption mutation on replicative senescence in *Saccharomyces cerevisiae. FEMS Microbiol Lett.* 177: 199–204.
76. Nestelbacher R, Laun P, Vondrakova D, Pichova A, Schuller C, Breitenbach M (2000). The influence of oxygen toxicity on yeast mother cell-specific aging. *Exp Gerontol.* 35: 63–70.
77. Van Zandycke SM, Sohier PJ, Smart KA (2002). The impact of catalase expression on the replicative lifespan of *Saccharomyces cerevisiae. Mech Ageing Dev.* 123: 365–73.
78. Jazwinski SM (2000). Coordination of metabolic activity and stress resistance in yeast longevity. *Results Probl Cell Differ.* 29: 21–44.
79. Jazwinski SM (2000). Metabolic control and gene dysregulation in yeast aging. *Ann NY Acad Sci.* 908: 21–30.

80. Swiecilo A, Krawiec Z, Wawryn J, Bartosz G, Bilinski T (2000). Effect of stress on the life span of the yeast *Saccharomyces cerevisiae*. *Acta Biochim Pol.* 47: 355–64.
81. Kaibuchi K, Miyajima A, Arai K, Matsumoto K (1986). Possible involvement of *RAS*-encoded proteins in glucose-induced inositolphospholipid turnover in *Saccharomyces cerevisiae*. *Proc Natl Acad Sci USA* 83: 8172-8176.
82. Hawkins PT, Stephens LR, Piggott JR (1993). Analysis of inositol metabolites produced by *Saccharomyces cerevisiae* in response to glucose stimulation. *J Biol Chem.* 268: 3374–83.
83. Patton JL, Srinivasan B, Dickson RC, Lester RL (1992). Phenotypes of sphingolipid-dependent strains of *Saccharomyces cerevisiae*. *J Bacteriol.* 174: 7180–4.
84. Jazwinski SM (1999). Molecular mechanisms of yeast longevity. *Trends Microbiol.* 7: 247–52.
85. Dickson RC, Nagiec EE, Skrzypek M, Tillman P, Wells GB, Lester RL (1997). Sphingolipids are potential heat stress signals in *Saccharomyces*. *J Biol Chem.* 272: 30196–200.
86. Johnston M (1999). Feasting, fasting and fermenting. Glucose sensing in yeast and other cells. *Trends Genet.* 15: 29–33.
87. Alepuz PM, Cunningham KW, Estruch F (1997). Glucose repression affects ion homeostasis in yeast through the regulation of the stress-activated *ENA1* gene. *Mol Microbiol.* 26: 91–8.
88. Gimeno CJ, Ljungdahl PO, Styles CA, Fink GR (1992). Unipolar cell divisions in the yeast *S. cerevisiae* lead to filamentous growth: regulation by starvation and *RAS*. *Cell* 68: 1077–90.
89. Mosch HU, Roberts RL, Fink GR (1996). Ras2 signals via the Cdc42/Ste20/mitogen-activated protein kinase module to induce filamentous growth in *Saccharomyces cerevisiae*. *Proc Natl Acad Sci USA* 93: 5352–6.
90. Parikh VS, Morgan MM, Scott R, Clements LS, Butow RA (1987). The mitochondrial genotype can influence nuclear gene expression in yeast. *Science* 235: 576–80.
91. Jazwinski SM (2000). Metabolic control and ageing. *Trends Genet.* 16: 506–11.
92. Traven A, Wong JM, Xu D, Sopta M, Ingles CJ (2000). Inter-organellar communication: altered nuclear gene expression profiles in a yeast mitochondrial DNA mutant. *J Biol Chem.* 276: 4020–7.
93. Epstein CB, Waddle JA, Hale W,*et al.* (2001). Genome-wide responses to mitochondrial dysfunction. *Mol Biol Cell.* 12: 297–308.
94. Sekito T, Thornton J, Butow RA (2000). Mitochondria-to-nuclear signaling is regulated by the subcellular localization of the transcription factors Rtg1p and Rtg3p. *Mol Biol Cell.* 11: 2103–15.
95. Liu Z, Butow RA (1999). A transcriptional switch in the expression of yeast tricarboxylic acid cycle genes in response to a reduction or loss of respiratory function. *Mol Cell Biol.* 19: 6720–8.
96. Komeili A, Wedaman KP, O'Shea EK, Powers T (2000). Mechanism of metabolic control. Target of rapamycin signaling links nitrogen quality to the activity of the Rtg1 and Rtg3 transcription factors. *J Cell Biol.* 151: 863–78.
97. Lai CY, Jaruga E, Borghouts C, Jazwinski SM (2002). A mutation in the *ATP2* gene abrogates the age asymmetry between mother and daughter cells of the yeast *Saccharomyces cerevisiae*. *Genetics* 162: 73–87.
98. Shigenaga MK, Hagen TM, Ames BN (1994). Oxidative damage and mitochondrial decay in aging. *Proc Natl Acad Sci USA* 91: 10771–8.

99. Li H, Wang J, Wilhelmsson H, *et al.* (2000). Genetic modification of survival in tissue-specific knockout mice with mitochondrial cardiomyopathy. *Proc Natl Acad Sci USA* 97: 3467–72.
100. Bonafè M, Barbi C, Olivieri F, *et al.* (2002). An allele of *HRAS1* 3′variable number of tandem repeats is a frailty allele: implication for an evolutionarily-conserved pathway involved in longevity. *Gene* 286: 121–6.
101. Cartee GD, Dean DJ (1994). Glucose transport with brief dietary restriction: heterogenous responses in muscles. *Am J Physiol.* 266: E946–52.
102. Masoro EJ (1995). Dietary restriction. *Exp Gerontol.* 30: 291–8.
103. Jiang JC, Jaruga E, Repnevskaya MV, Jazwinski SM (2000). An intervention resembling caloric restriction prolongs life span and retards aging in yeast. *FASEB J.* 14: 2135–7.
104. Thevelein JM, de Winde JH (1999). Novel sensing mechanisms and targets for the cAMP-protein kinase A pathway in the yeast *Saccharomyces cerevisiae. Mol Microbiol.* 33: 904–18.
105. Lorenz MC, Pan X, Harashima T, *et al.* (2000). The G protein-coupled receptor gpr1 is a nutrient sensor that regulates pseudohyphal differentiation in *Saccharomyces cerevisiae. Genetics* 154: 609–22.
106. Lin SS, Manchester JK, Gordon JI (2001). Enhanced gluconeogenesis and increased energy storage as hallmarks of aging in *Saccharomyces cerevisiae. J Biol Chem.* 276: 36000–7.
107. Lin SJ, Kaeberlein M, Andalis AA, *et al.* (2002). Calorie restriction extends *Saccharomyces cerevisiae* lifespan by increasing respiration. *Nature* 418: 344–8.
108. Gancedo C, Serrano, R. (1989). Energy-yielding metabolism. In: Rose AH, Harrison JS, eds. *The Yeasts*, 2nd edn, vol 3. San Diego: Academic Press, pp. 205–59.
109. Jiang JC, Wawryn J, Shantha Kumara HM, Jazwinski SM (2002). Distinct roles of processes modulated by histone deacetylases Rpd3p, Hda1p, and Sir2p in life extension by caloric restriction in yeast. *Exp Gerontol.* 37: 1023–30.
110. Wyrick JJ, Holstege FC, Jennings EG, *et al.* (1999). Chromosomal landscape of nucleosome-dependent gene expression and silencing in yeast. *Nature* 402: 418–21.
111. Hughes TR, Marton MJ, Jones AR, *et al.* (2000). Functional discovery via a compendium of expression profiles. *Cell* 102: 109–26.
112. Jazwinski SM (1999). Nonlinearity of the aging process revealed in studies with yeast In: Bohr VA, Clark, BFC, Stevnsner T, eds. *Molecular Biology of Aging*. Copenhagen: Munksgaard, pp. 35–44.
113. Jazwinski SM, Wawryn J (2001). Profiles of random change during aging contain hidden information about longevity and the aging process. *J Theor Biol.* 213: 599–608.
114. Williamson DH, Fennell DJ (1981). Non-random assortment of sister chromatids in yeast mitosis In: von Wettstein D, Stenderup A, Kielland-Brandt M, Friis J, eds. *Molecular Genetics in Yeast*. Copenhagen: Munksgaard, pp. 89–102.
115. Lundblad V, Szostak JW (1989). A mutant with a defect in telomere elongation leads to senescence in yeast. *Cell* 57: 633–43.
116. Conrad-Webb H, Butow RA (1995). A polymerase switch in the synthesis of rRNA in *Saccharomyces cerevisiae. Mol Cell Biol.* 15: 2420–8.
117. Johnson BF, Lu C (1975). Morphometric analysis of yeast cells. IV. Increase of the cylindrical diameter of *Schizosaccharomyces pombe* during the cell cycle. *Exp Cell Res.* 95: 154–8.
118. Sando N, Maeda M, Endo T, Oka R, Hayashibe M (1973). Induction of meiosis and sporulation in differently aged cells of *Saccharomyces cerevisiae. J Gen Appl Microbiol.* 19: 359–73.

119. James AP, Johnson BF, Inhaber ER, Gridgeman NT (1975). A kinetic analysis of spontaneous *rho*- mutations in yeast. *Mutat Res.* 30: 199–208.
120. Adams J, Paquin C, Oeller PW, Lee LW (1985). Physiological characterization of adaptive clones in evolving populations of the yeast, *Saccharomyces cerevisiae*. *Genetics* 110: 173–85.
121. Jiang JC, Kirchman PA, Zagulski M, Hunt J, Jazwinski SM (1998). Homologs of the yeast longevity gene *LAG1* in *Caenorhabditis elegans* and human. *Genome Res.* 8: 1259–72.
122. Coates PJ, Jamieson DJ, Smart K, Prescott AR, Hall PA (1997). The prohibitin family of mitochondrial proteins regulate replicative lifespan. *Curr Biol.* 7: 607–10.
123. Berger KH, Yaffe MP (1998). Prohibitin family members interact genetically with mitochondrial inheritance components in *Saccharomyces cerevisiae*. *Mol Cell Biol.* 18: 4043–52.
124. Park PU, Defossez PA, Guarente L (1999). Effects of mutations in DNA repair genes on formation of ribosomal DNA circles and life span in *Saccharomyces cerevisiae*. *Mol Cell Biol.* 19: 3848–56.
125. Roy N, Runge KW (2000). Two paralogs involved in transcriptional silencing that antagonistically control yeast life span. *Curr Biol.* 10: 111–14.
126. Wawryn J, Swiecilo A, Bartosz G, Bilinski T (2002). Effect of superoxide dismutase deficiency on the life span of the yeast *Saccharomyces cerevisiae*. An oxygen-independent role of Cu,Zn-superoxide dismutase. *Biochim Biophys Acta* 1570: 199–202.

Aging and Longevity in the Filamentous Fungus *Podospora anserina*

Heinz D. Osiewacz
J.W. Goethe University, Marie-Curie Str. 9, 60439 Frankfurt, Germany

Introduction

The fungi, a diverse group of heterotrophic eukaryotes

The fungi represent a diverse assembly of heterotrophic eukaryotes. In principal, they can be divided into two groups, the yeasts and the mycelial (filamentous) fungi. In yeasts, a single cell predominantly represents the individuum. In contrast, the vegetation body of mycelial fungi is represented by the so-called mycelium, a thallus consisting of a network of long filamentous "cells," termed hyphae. Mycelial growth proceeds via the elongation of the peripheral hyphae [1]. From time to time, these hyphae branch at their tips. If two independent mycelia come into close contact they may fuse by hyphal anastomoses giving rise to the formation of a heterokaryon, a mycelium containing the nuclei of the two fusion partners. This process is genetically controlled [2].

Hyphae of fungi from different taxa are characterized by a distinct anatomy. In one group, no cross-walls are formed, in others, so-called septae are dividing the hyphae and lead to the appearance of a "cellular" organization of the mycelium. However, since septae are incomplete cross-walls containing a central opening that allows exchange of organelles, cytoplasm and molecules, the thallus of mycelial fungi represents a coenocyte (syncytium) containing numerous nuclei and thus is poly-energetic.

The distinction between yeasts and mycelial fungi is artificial and does not reflect a natural and systematic relationship. In fact, different yeasts may belong to different fungal taxa, including ascomycetes (e.g., the budding yeast *Saccharomyces cerevisiae* or fission yeast *Schizosaccaromyces pombe*), basidiomycetes (e.g., *Cryptococcus neoformans*), or deuteromycetes (e.g., *Candida albicans*). Moreover, under certain growth conditions, yeasts can switch to pseudohyphal or even hyphal growth [3]. Vice versa typical mycelial fungi can grow as single cells, a phenomenon referred to as mycelial-yeast dimorphism.

H.D. Osiewacz (ed.), Aging of Organisms, 31–53.

Fungi as model systems in aging research

In many cases, fungi are easy to cultivate under laboratory condition. This is common knowledge for baker's yeast, *Saccharomyces cerevisiae*, but holds true for other yeasts and mycelial fungi as well. In fact, mycelial ascomycetes like *Neurospora crassa*, *Podospora anserina*, and *Aspergillus nidulans* represent "classical" experimental systems, which are extensively investigated in different areas of research (e.g., basic genetics, developmental biology). In addition to their easy and inexpensive cultivation, there are a number of other characteristics making them attractive. First, the complexity of their organization is much lower than that of higher organisms. Second, different species are well suited for genetic analysis of different phenomena and pathways. Most important in this respect is that mutants in whom the phenotype of interest is affected can be generated for subsequent analyses. However, in spite of these advantageous characteristics, only a few fungal species are investigated in the field of experimental aging research. This is due to the fact that most fungi, including the vast majority of mycelial fungi, do not show any signs of aging even after prolonged vegetative growth. They may form huge vegetation bodies which, as in a reported case of the fungus *Armillaria bulbosa*, may reach $\frac{1}{2}$ km in diamer [4]. Mycelial fungi may thus represent the biggest and oldest organisms on earth.

However, in sharp contrast, there are a few fungi, which clearly do not propagate indefinitely. This holds true for yeast's like *S. cerevisiae* (baker's yeast) and mycelial fungi like *P. anserina* and some Neurospora strains. In the case of *S. cerevisae*, individuals originating from a single ascospore, the product of a sexual propagation, are only able to a strain-specific number of vegetative (mitotic) propagation processes, termed budding [5, 6]. Today, this fungal system has emerged to one of the most extensively investigated simple aging models. For details and further references about yeast aging the reader is referred to Chapter 1 of this volume.

Whereas yeast aging occurs at the single cell level and thus is evident microscopically, aging of mycelial fungi is easily visible at the macroscopic level. Species in which senescence processes were described in the past are indicated in Table 1.

In this chapter I shall concentrate on only some general aspects of fungal senescence and in particular on aging processes in the filamentous ascomycete *P. anserina*. In this fungus, the basis of aging has been extensively investigated for now exactly half a century. It was the first system for which clear molecular concepts were elaborated that turned out to be valid also in other biological systems. For more details concerning fungal senescence in general the reader is referred to previous publications [29–40].

Discovery and early investigations of aging in *Podospora anserina*

Wild-type cultures of Podospora anserina senesce after a strain-specific period of vegetative propagation

In filamentous fungi, aging research started in the early 1950's when George Rizet described for the first time that cultures of the filamentous ascomycete *P. anserina* do not grow indefinitely but senesce after a strain-specific period of growth [25, 26]. In comparison to many other biological systems this time is conveniently short.

Table 1. Mycelial fungi for which spontaneous ageing processes have been described

Organism	References
Ascobolus stercorarius	7
Aspergillus amstelodami	8–11
Aspergillus glaucus	12
Helminthelsporium victoria	13
Neurospora crassa	14–19
Neurospora intermedia	16, 20–23
Pestalozzia annulata	24
Podospora anserina	25, 26
Podospora curvicolla	27
Podospora setosa	26
Sordaria macrospora	28

For a few examples these processes have been extensively investigated. Only a selected list of references is provided. For further references the reader is referred to the review articles mentioned in the text

Depending on the strain and on the incubation conditions, cultures become senescent after a few weeks to a few months. As one specific example, the wild-type strain s, which has been most extensively analysed, is characterized by a mean life span of 25 days when grown on rich-medium at 27°C. During this period of time, starting from a newly isolated and germinated ascospore, the culture initially a growth with a linear growth rate of 7 mm/day until it reaches the senescent stage (Figure 1). At this stage, the growth rate slows progressively and the phenotype of the culture changes: the formation of hyphae growing into the air, so-called aerial hyphae, decreases, the pigmentation increases, and finally the growth of a senescent culture stops completely and the hyphae die at the growth front. The hyphal tips of senescent cultures are more undulate and slender than those from younger cultures. Moreover, tips from senescent cultures were found to burst frequently [29, 41].

Importantly, the "biological clock" counting the age of a culture can be reset during sexual reproduction. Crosses of two strains of opposite mating type lead to the generation of ascospores from which juvenile mycelia arise. This holds true even if none of the two parental strains is juvenile itself indicating some kind of "filter" mechanism operating during sexual reproduction. Such a mechanism has been demonstrated to lead to a restoration of wild-type mtDNA eliminating specific mutations [42, 43]. The underlying details of that mechanism, although bearing important clues, are not understood yet.

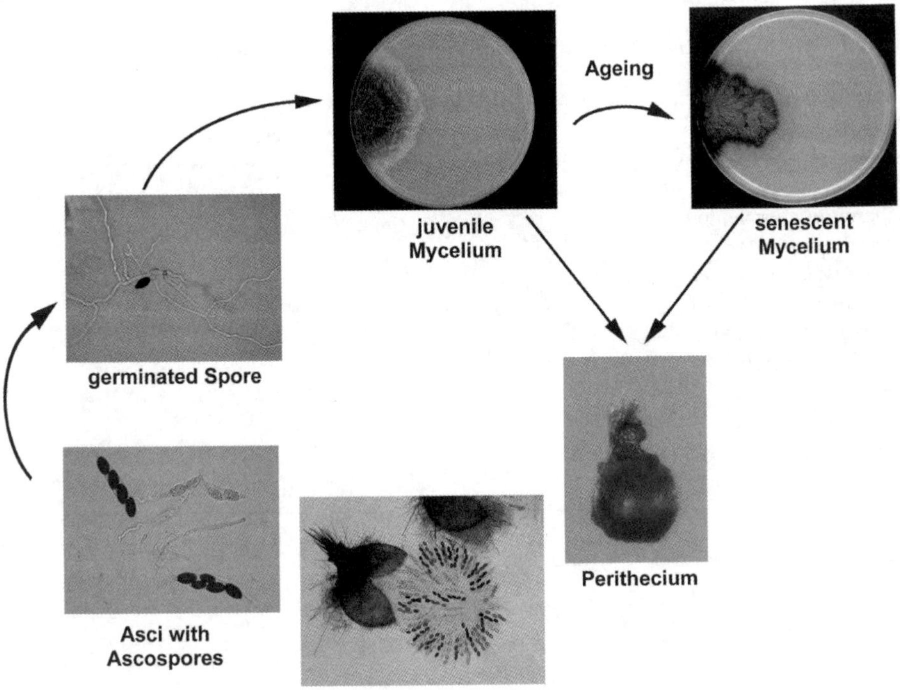

Figure 1. Stages in the life cycle of Podospora anserina. Germination of ascospores leads to the formation of a juvenile culture. Growth of the culture proceeds via hyphal tip elongation and branching. Under appropriate conditions, controlled by the mating-types plus and minus, sexual reproduction takes place leading to the formation of fruiting bodies, the perithecia. Mature ascospores are normally actively discharged from a perithecium and, after passing the intestine of an herbivore, germinate on dung. In the laboratory, ascospores can easily be isolated from perithecia and germinated on acetate containing growth medium. Senescent cultures develop after a strain-specific growth period. They are characterized by a specific senescent phenotype (e.g., darker pigmentation, reduced aerial hyphae, growth stop at the hyphal tips).

Genetic traits involved in life span control

Early genetic investigations of the processes leading to the development of senescent cultures suggested that a "determinant of senescence" accumulates during aging of cultures [7]. This conclusion arose from experiments in which pieces of mycelia from the growth front were cut from a culture at different times after germination of a spore and subsequently inoculated on fresh agar plates. Pieces isolated before reaching a critical point of age ("point of no return") were found to be able to grow again for the strain-specific time (the typical life span of that strain). In sharp contrast, mycelial pieces cut from the growth front that has passed the "point of no return" were only able to grow for a time corresponding to the difference between the strain specific life span and the time elapsed at the point the mycelial piece was collected. Mycelia taken from the edge of a senescent culture do not give rise to a new culture on fresh agar plates. The hyhal tips of these cultures are obviously dead. From these data it can be concluded that the postulated "determinant of senescence" appears to be absent from young cultures and accumulates after the culture passes through the "point of no return."

In another series of experiments, Marcou [7, 44] demonstrated that under conditions which do not allow the migration of nuclei, the fusion of hyphae from a senescent culture leads to a transfer of the accumulated "determinant of senescence" to a young culture. This culture immediately becomes senescence and dies at the growth front. The "determinant of senescence" thus appears to be infectious and cytoplasmic.

This idea of a cytoplasmic genetic factor controlling the onset of senescence was further supported by reciprocal crosses (Figure 2), which can easily be performed in *P. anserina* to genetically discriminate whether a specific phenotype is controlled by nuclear or extranuclear (cytoplasmic) genetic traits. This type of analysis is possible because the sexual "organs" participating in a sexual reaction are of different size. Whereas they provide equal numbers of nuclei, the amount of cytoplasm coming from the two partners of a cross differs significantly. The so-called spermatia, which represent the male gametes, contain only little cytoplasm and a single nucleus. On the other hand, the female gametangia (protoperithecia) which fuse with a spermatium contain most of the cytoplasm going into the dikaryotic phase that arises prior to karyogamy of nuclei, meiosis and ascospore formation. It needs to be mentioned that, although producing both, male gametes as well as female gametangia fertilization does not occur in monokaryotic cultures expressing just one of the two mating-types plus or minus. Only if spermatia of one mating-type come into contact with gametangia of the opposite mating type fertilization takes place and the products of such a process, numerous ascospores, are produced in fruiting bodies. These well-defined characteristics of the sexual reproduction system of *P. anserina* are the basis for a genetic analysis. Experimentally it is easy to collect spermatia from defined cultures just by washing the surface of a culture with sterile water. Pouring aliquots of this suspension from a strain of one mating-type on a culture of the opposite mating-type leads to fertilization of the protoperithecia on that plate. When analysing the inheritance of a particular phenotype (e.g., "juvenile culture" vs. "senescent culture") two different crosses can be performed (Figure 2). First, spermatia from a juvenile culture of one mating-type can be used to fertilize protoperithecia of a senescent culture of the opposite mating-type. Second, in the reciprocal cross spermatia of a senescent culture are poured on the mycelium of the juvenile culture. The ascospores from the corresponding crosses were analysed. It turned out that reciprocal differences occurred. None of the progeny of a cross in which spermatia of a "senescent" strain were used to fertilize protoperithecia of a "juvenile" culture was "senescent." In contrast, up to 90% of the progeny obtained from a cross between "juvenile" spermatia and "senescent" protoperithecia were senescent [45]. This type of inheritance is not consistent with the inheritance of nuclear factors. However, crosses of certain mutants with wild-type strains also indicated that nuclear factors

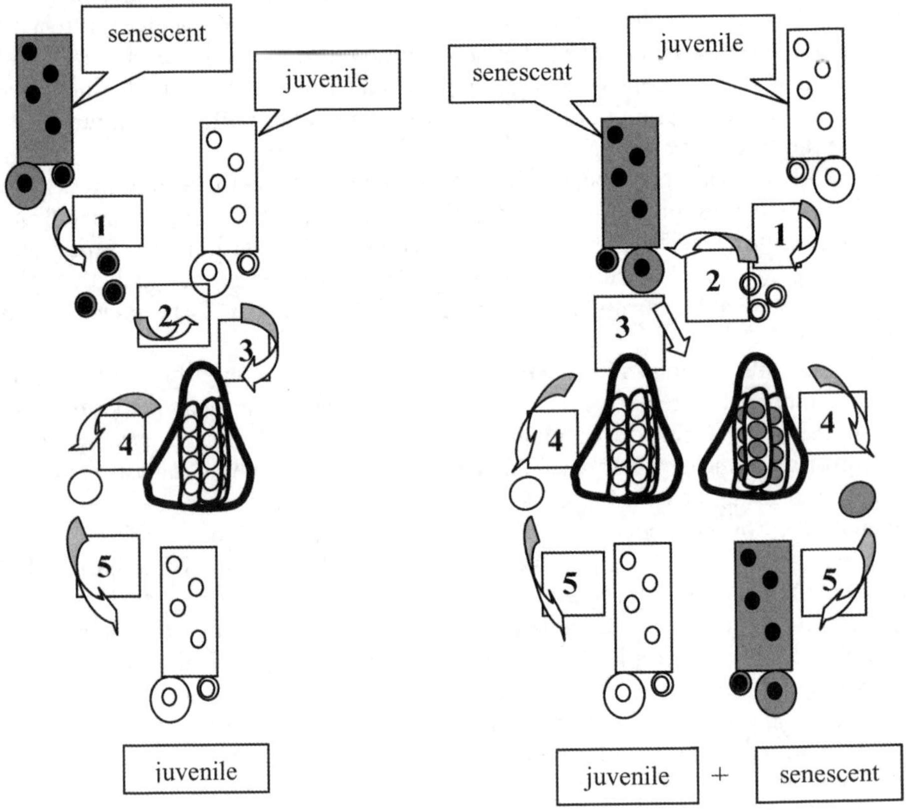

***Figure 2**. Reciprocal crosses indicate that the onset of senescence of P. anserina is controlled by extranuclear (cytoplasmic) genetic traits. Juvenile (white) and senescent (grey) cultures (mycelia) are represented by the rectangles. Each mycelium is able to produce female gametangia (protoperithecia) indicated by the bigger circles and male gametes (spermatia) indicated by smaller circles. Due to the control of the sexual reaction by the two mating-types plus and minus (black and white circles inside the mycelia and in the protoperithecia and spermatia, respectively), fertilization does only occur when spermatia meet protoperithecia from a culture of the opposite mating-type. Note that spermatia do contain very little cytoplasm whereas protoperithecia contain most cytoplasm carried to the next generation. These characteristics of the mating system allow to experimentally performing two types of crosses. On the left of the scheme, spermatia from a senescent culture are isolated (1) and used to fertilise (2) a juvenile culture of the opposite mating-type. After karyogamy and meiosis fruiting bodies (perithecia) are formed (3) containing the haploid products, termed ascospores, of the corresponding sexual reaction. Single ascospores (ellipsoids) can easily be isolated (4). After germination (5) they give rise to a new mycelium (progeny). Fertilization of juvenile cultures with spermatia from senescent cultures (left) leads to juvenile cultures. In the reciprocal cross (right) in which female gametangia from a senescent culture are fertilized by spermatia of a juvenile culture both perithecia containing ascospores, which exclusively give, rise to juvenile or senescent progeny only. The frequency of the two types of perithecia depends on the age of the cultures used in that cross [7, 45].*

are involved in life span control as well. This is clear from crosses between different mutants characterized by clearly different life spans. In reciprocal crosses of such strains no reciprocal differences are observed. In these cases life span of the progeny segregates according to the mendelian rules. Pathways affecting the life span of *P. anserina* are thus encoded by extranuclear and nuclear factors.

Metabolic inhibitors affecting the life span suggested a role of mitochondrial functions

Apart from the demonstration of a genetic control of life span early experiment revealed a clear impact of environmental factors. Importantly, this type of experiments clearly pointed to a crucial role of mitochondria. This became obvious in growth experiments in which different metabolic inhibitors added to the medium significantly affected the life span of *P. anserina* cultures [46]. Inhibitors of mitochondrial ribosomes like kanamycin, neomycin, streptomycin, puromycin, and tiamulin led to an increased life span [47, 48]. The same was demonstrated for compounds like ethidium bromide, acridine, and acriflavine, substances that preferentially intercalate into mtDNA. In addition, growth of senescent cultures on solid medium containing ethidium bromide resulted in the rejuvenation of the culture [49]. Finally, inhibitors of the mitochondrial respiratory chain like mucidin and potassium cyanide also had an life span increasing effect [48]. Importantly, such effects were observed under conditions when the additive did not or only slightly affect the growth rate of the analysed cultures.

P. anserina: a paradigm of a mitochondrial etiology of aging

Age-related rearrangements of the mtDNA

As mentioned above, the onset of senescence in *P. anserina* is controlled by nuclear genes and by extrachromosomal genetic traits [7, 46, 50–52]. The search to elucidate the physical nature of these factors revealed a covalently closed circular DNA species, termed plDNA or α-senDNA, that accumulates in mitochondria of senescent cultures of wild-type strains [53, 54]. This genetic element is derived from the first intron (the pl-intron) of the gene (*CoxI*) coding for the largest subunit of the cytochrome oxidase (COX). In mitochondria of juvenile cultures no or only very low amounts of free plDNA are found (Figure 3). During aging, the pl-intron becomes systematically liberated and amplified and large parts of the mtDNA are deleted [55–57]. For a long time it was believed that senescent in *P. anserina* is strictly dependent on the accumulation of plDNA. However, more recent data demonstrated that *P. anserina* mutants may senesce independent of the amplification of plDNA (see below). It appears that this unusual genetic element is an accelerator of aging.

The underlying molecular mechanism leading to the wild-type specific mtDNA rearrangements has been elucidated in some detail. The basis is the mobility of the pl-intron, which leads to the generation of duplicated pl-intron sequences. Between these sequences homologous recombination processes occur efficiently and systematically. In addition, pl-intron-independent recombination processes were found to occur between short dispersed direct repeats in the mtDNA. The latter take place

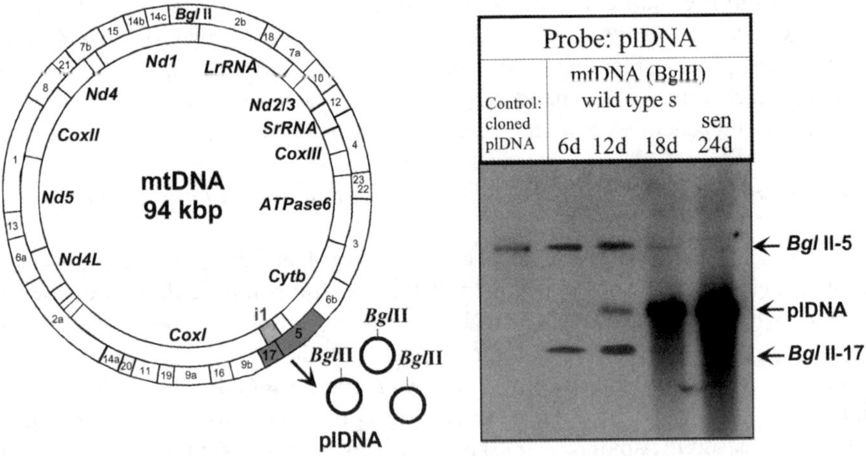

Figure 3. *During aging of wild-type strains of P. anserina the mitochondrial DNA (mtDNA) becomes grossly rearranged. A Southern blot analysis reveals systematic rearrangements occurring in the region containing the first intron of the Cox1. A circular DNA molecule, termed plDNA, is liberated and amplified. At the same time the BglI-5 and BglII-17 mtDNA fragment containing the first CoxI intron disappears. Rearrangements occur also in other regions of the mtDNA which encode genes of the mitochondrial protein synthetic apparatus (e.g., SrRNA, LrRNA coding the RNA of mitochondrial ribosomes) and proteins of the mitochondrial respiratory chain (e.g., cytochrome oxidase: CoxI, II, III; NADH dehydrogenase: Nd1,2/3,4,4L,5; ATPsynthase: ATPase6) [58].*

with a much lower frequency than those leading to the plDNA-dependent processes and are not systematically found in every individual aged culture of wild type strains. However, since the age-related reorganization of the mtDNA is almost quantitative, the vast majority of mtDNA molecules in senescent cultures are extensively rearranged [55, 56, 59–61].

The first step in the pl-intron-dependent pathway, intron transposition, occurs either to a position directly downstream of the first *CoxI* exon ("homing-like" transposition) or to other acceptor sites in the mtDNA ("ectopic" transposition). This process appears to take place frequently and leads to the generation of mtDNA molecules with two or more copies of the 2.5 kbp intron sequence. Depending on the integration site duplications are found either in tandem or dispersed in the same mtDNA molecule [62, 63]. Subsequent homologous recombination between these duplicated sequences seems to give rise to the formation of circular plDNA molecules or to other mtDNA subcircles of different size. If circles are only occasionally generated and do not contain an "origin of replication" they become easily lost during subsequent growth. Only circles generated frequently or those, which replicate autonomously are retained in senescent cultures.

The occurrence of the amplified plDNA appears to be a good marker of transposition processes and of subsequent homologous recombination greatly contributing to the characteristic age-related mtDNA reorganizations observed during senescence of *P. anserina*. These processes depend on different factors. Intron transposition seems to be proceeded by a reverse transcriptase step depending on the activity of a protein encoded by an open reading frame on the pl-intron [57, 59, 64, 65].

Various lines of evidence established a crucial role of mtDNA reorganizations and senescence in *P. anserina*. As one example the rejuvenation of senescent cultures, which was demonstrated to posses grossly rearranged mtDNA, by growing them on ethidium bromide was demonstrated to correlate with the recovery of intact wild-type mtDNA after the incubation of the senescent culture on the drug containing medium [49]. In addition, the analysis of a variety of different long-lived mutants or transgenic strains provided further correlation between processes giving rise to a stabilization of the mtDNA and life span increase. The stabilization may be due for different reasons and may result from changes in mitochondria or may be caused by nuclear mutations. One example is a deletion of the pl-intron or parts of it from the mtDNA [66–68]. As a consequence, the free plDNA giving rise to the characteristic wild-type age-specific mtDNA rearrangements introduced above, is not generated leading to a stabilization of the mtDNA and an increased life span. In another long-lived extrachromosomal mutant, AL2, the mtDNA rearrangements leading to the amplification of the plDNA were found to be delayed [69]. A linear plasmid, pAL2-1, encoding an RNA and a DNA polymerase was demonstrated to interfere with this process [69–74]. Apart from these extrachromosomal mutants, different nuclear long-lived mutants were found in which a mutation in a nuclear gene increase the stability of the mtDNA. One example is the grisea mutant [52, 75]. In this mutant, the pathway leading to the duplication of pl-intron sequences is not affected. However, the last step, homologous recombination between repeated sequences, is impaired [62, 63]. Growing the mutant in media containing increased copper concentrations reverts the molecular phenotype indicating that homologous recombination between the duplicated sequences directly or indirectly is dependent on the availability of copper [76].

Nuclear mutations affecting mitochondrial functions

The isolation and characterization of a number of different longevity mutants of *P. anserina* clearly revealed that apart from extrachromosomal also nuclear genes are involved in determining the life span of cultures. Both, mutations decreasing and increasing life span were determined [29]. Many of the genes specified by the corresponding mutations are still not cloned and their molecular role remains obscure. However, most interestingly, some of the genes that were characterized turned out to be linked to mitochondrial functions.

The selection of specific mutants of *P. anserina* which do age although senescent mitochondria of these strains do not contain amplified plDNA molecules clearly revealed that the amplification of this genetic element is not the prerequisite for aging [77, 78]. This view is supported by a mutant strain in which the pl-intron is exactly deleted from the mtDNA. In contrast to other pl-intron deletion mutants in which also parts of the *Cox1* exon sequences are deleted, this mutant shows a wild-type specific cytochrome c oxidase dependent respiration. Although the pl-intron is absent, the corresponding mutant is not immortal but characterized by an increased life span of about 50% clearly verifying that the pl-intron is not the cause of aging *per se* but rather an accelerator of aging [68]. Consequently, at least one other basic mechanism is responsible for the aging of *P. anserina* cultures. The characterization of independent long-lived mutants provided important clues towards the identification of the relevant factors.

In different long-lived mutants and in one transgenic strain, the cyanide sensitive respiration chain utilizing complex VI with cytochrome oxidase (COX) as the terminal electron transferase is affected. In the grisea mutant, [52, 75] the mutant phenotype is the result of a loss-of-function mutation in the *Grisea* gene [77, 79]. This is a nuclear gene encoding the copper-modulated transcription factor GRISEA. The transcription factor is involved in a tight control of cellular copper levels. At low levels, GRISEA activates the transcription of different target genes including *PaSod2*, *PaGrg1* and *PaCtr3* [80–83]. The latter was found to encode a high affinity copper transporter able to transport copper efficiently across the plasma membrane even if copper concentrations in the environment are very low. In the grisea mutant, due to the mutation in *Grisea*, high affinity copper-uptake is impaired leading to cellular copper-deficiency. Since copper is a cofactor of tyrosinase, an enzyme involved in melanin biosynthesis, the mutant displays a hypopigmentation phenotype: both, mycelia as well as the ascospores are grey instead of dark green to black. In addition, the delivery of copper to mitochondria and the assembly of complex IV requiring copper, are affected. Like in yeast and other systems, this process depends on copper chaperones binding copper immediately after it enters the cell. In yeast, yCOX17, ySCOI and most likely yCOX11 are involved in this process [84–86]. Until now, in *P. anserina* only one gene of this pathway coding for a putative COX17 homologue has been identified. PaCOX17 delivers copper to the mitochondrial intermembrane space (unpublished). Subsequently, yet unidentified proteins like PaSCOI and PaCOX11 may pass copper from PaCOX17 to the two copper binding subunits, COXI and COXII, of complex IV. In principle, since *P. anserina* is an obligate aerobe and thus depending on mitochondrial ATP generation, a defect in the delivery of copper to mitochondria should be lethal. However, in this critical situation a specific response is induced. This response resembles the "retrograde response" first described in *S. cerevisiae* [87–89], a mechanism able to compensate primary defects like impairments of the respiratory chain. In the case of the grisea mutant, which is unable to assemble the cyanide sensitive complex IV, compensation results from the assembly of a cyanide resistant alternative oxidase (AOX) into the inner mitochondrial membrane. In contrast to COX, this enzyme requires iron instead of copper. The AOX pathway branches at the ubiquinone pool. The final step is the transfer of electrons from AOX to oxygen giving rise to the formation of water. Since the AOX is located upstream of complex III, the formation of the electron motive force is completely restricted to complex I. Consequently, the production of ATP is reduced. However, since in the grisea mutant copper-deficiency is not complete and low amounts of copper enter the cell via a low affinity uptake system, the mitochondrial respiratory chain of the grisea mutant seems to respire via both a copper-dependent COX (complex IV) and the iron-dependent AOX respiration. The life span of this mutant is increased by about 60%. Interestingly, mutant ex1, which exclusively respires via the AOX alternative pathway, appears to be immortal.

The induction of a retrograde response in *P. anserina* in different long-lived mutants is intriguing. However, what are the specific reasons for the observed life span extension and why are individual mutants in whom this response is induced characterized by different life spans? It appears that the characteristics of the two types of respiratory pathways are a clue to answer this question. In one study, PaAOX was found to be induced in a transgenic COX deficiency strain. In this strain the generation of ROS was significantly lower than in the wild-type strain respiring via the standard COX-dependent oxidase [90]. These data are in agreement with data from higher plants demonstrating a reduced generation of ROS via an AOX-dependent respiratory chain [91] and link aging of *P. anserina* to the free radical theory of aging [92–98]. The current version of this theory suggests that mitochondria are the main source of cellular ROS generation. ROS are generated mainly at complex I and III of the respiratory chain. They lead to damage and malfunction of all types of biomolecules including nucleic acids, lipids, and proteins. In mitochondria of young organisms, only very low amounts of ROS are produced. However, as mitochondria and in particular the individual components of the respiratory chain become damaged, higher amounts of ROS are generated. Finally, this kind of a "vicious cycle" leads to dysfunctional mitochondria. Lowering mitochondrial ROS generation, as in the long-lived *Cox5* inactivation strain mentioned above, appears to be the reason for life span extension in the two mutants grisea and ex. The differences in the life span of these mutants, 39 days vs. greater than 10 years, seem to result from some basic differences. Both the exclusive respiration via the alternative oxidase and differences in the ROS scavenging system may be responsible. In the immortal ex mutant the *Grisea* gene is functional and thus copper levels are regulated utilizing transcription factor GRISEA. The same holds true for target genes of GRISEA not directly related to copper-uptake, transport, or storing. One example is *PaSod2* coding for the mitochondrial MnSOD. Moreover, copper is available at normal levels for the activation of the apoprotein of the cytoplasmic Cu/ZnSOD. Significantly, this enzyme appears to have also a scavenging function in mitochondria and not exclusively in the cytoplasm [99]. Thus, in the ex mutant, the system directed against oxidative stress appears to be of a greater efficiency than in the mortal long-lived grisea mutant [82, 83].

Given that mitochondrial oxidative stress and the ability of cells to compensate mitochondrial dysfunction are crucial components of the network involved in life span control in *P. anserina*, the significance of the age-related mtDNA rearrangements demonstrated to play an important role during wild-type strain aging needs to be integrated into the network of molecular pathways affecting longevity. Here it needs to be recalled that senescence occurs at the periphery of the mycelium, the hyphal tips. These are the areas in which most energy is required since the hyphal tips are elongating and branching. Consequently, there is a clear demand for functional mitochondria. These organelles are provided by fission of existing organelles. However, after some time of growth, fully functional mitochondria are only available if an effective remodeling system is able to remove damaged components. Since both, the mtDNA and the nuclear DNA encode the proteins of the respiratory chain, remodeling is an integrated action of the nucleus, the cytoplasm and the mitochondrion. The process depends on the integrity of the mtDNA and on various factors encoded in the nucleus and which synthesized in the cytoplasm. However, as mentioned above, in wild-type strains of *P. anserina*, the mtDNA becomes extensively reorganized in a rather short period of time and consequently the synthesis of mtDNA-encoded proteins of the respiratory chain is time-limited. In senescent wild-type strains, synthesis of mitochondrial-encoded proteins thus appears to be a major problem.

Mutant grisea is only one example in which the mutation of a nuclear gene affects mitochondrial functions having an impact on longevity. Since the vast majority of genes coding for different components of functional mitochondria are encoded by the nucleus, many other genes are controlling the biogenesis and function of mitochondria including components of the respiratory chain, the whole set of enzymes of the citric acid cycle, all components of the protein import machinery, those of the biogenesis of Fe/S clusters, the enzymes involved in mtDNA replication and the expression of mitochondrial genes and of other essential mitochondrial functions. The transport of the various gene products and of cofactors like copper and iron into the different compartments of the organelle and the correct assembly of supramolecular complexes appears to be of prime significance for the remodeling of existing mitochondria and for division of mitochondria in actively growing parts of a mycelium. At this time, only very limited data are available concerning age-related aspects of these basic processes. As mentioned above, the delivery of copper to the respiratory chain is one specific example but in addition to the few components (e.g., PaCTR3, PaCOX17) identified in *P. anserina* others remain to be demonstrated and the role of the involved molecular pathways remains to be elucidated in more detail. The mitochondrial protein import machinery plays another important role. This is suggested by the analysis of a long-lived mutant in which the mutated gene was demonstrated to code for PaTOM70, a component of this machinery [100]. Finally, another type of genes, which on first glance appear not to be related to mitochondrial functions but which are components of the cytoplasmic translation apparatus, may indirectly also affect mitochondrial functions. Mutations in different genes of this apparatus are known to affect the life span of the corresponding mutants. Interestingly, the hypothesis has been put forward that mutations in these genes may have effects other than translation fidelity and may lead to a modified spectrum of proteins in mitochondria and consequently a modified mitochondrial metabolism. As a result the mtDNA metabolism and life span are changed [101].

Non-mitochondrial pathways involved in life span control

There is no doubt that also pathways not directly related to mitochondrial functions are affecting the life span of *P. anserina* cultures. Among these, pathways involved in maintenance functions other than those effective in mitochondria are certainly of relevance. Part of such systems are the recently identified metallothionein and, most likely, a putative *o*-methyltransferase of *P. anserina* [102, 103]. Whereas the role of the metallothionein in protecting against copper-related oxidative stress via binding of and shielding free copper is rather clear, the function of the *o*-methyltransferase is still speculative. However, from its biochemical characteristics the protein may catalyse methylation reactions of hydroxyl groups which, if not modified, may be converted to a free radical and participate in radical chain reactions leading to increased levels of ROS.

Certainly in *P. anserina* a machinery identified in other biological systems is the DNA maintenance system specifically dealing with repair of damaged nuclear DNA. Yet experimental data characterizing this system and demonstrating its relevance for life span control are awaited.

Relevance of the *P. anserina* system in the field of aging research

Soon after the first data about the changes occurring during aging *P. anserina* cultures in mitochondria emerged investigations aimed to test such a role of mitochondria also in other systems were initiated. And indeed, also in other filamentous fungi, rearrangements were demonstrated to correlate with senescence processes. However, in no case these processes were found to be related to the activity of a mobile intron, as it is typical for wild-type strains of *P. anserina*. Instead, in several cases homologous recombination between dispersed short direct repeats were found to be involved, rearrangements as they also occur in *P. anserina* during senescence. Since variable pairs of repetition sequences can be involved in such recombination processes, the resulting molecules are different in size. This holds true for senDNA's identified in senescent *Podospora curvicolla* strains [27] and in the "ragged" mutants of *Aspergillus amstelodami*. Also in different laboratory mutants of *N. crassa* displaying a senescence like degenerative phenotype intramolecular homologous recombination between pairs of short direct repeats were reported to lead to gross mtDNA reorganizations [19, 104, 105]. In addition to this type of common mechanism, mtDNA rearrangements were found to occur utilizing different GC-rich palindromes able to form stem-loop secondary structures. These mtDNA rearrangements appear to be the result of single- or double-strand breaks and subsequent ligation processes [15].

Apart from such processes in which sequences of the standard mtDNA are involved, another type of senescent-related mtDNA rearrangement was found in a number of natural isolates in the genus Neurospora. Here mtDNA reorganizations are generated by autonomous linear mitochondrial plasmids. In several Neurospora isolates from different parts of the globe including Hawaii, India, or China such elements have been isolated and characterized. One example is the Kalilo DNA, pKAL, a plasmid of 8.3 kbp containing long terminal inverted repeats and coding for a DNA and an RNA polymerase. Senescence of strains carrying this plasmid was reported to correlate with the integration of the plasmid into the mtDNA and the generation of giant repeats at the integration sites. Several integration sites were demonstrated in essential mtDNA sequences. However, integration of the element into the mtDNA occurs infrequently. Therefore it is surprising that mtDNA molecules containing the integrated plasmid sequence that interrupts essential functions accumulates during senescence. The reason is unsolved.

Finally, in specific Neurospora strains, also circular mitochondrial plasmids were found to integrate into the standard mtDNA and subsequently lead to deletions and insertions. Interestingly, integration of the plasmids most likely proceeds via an RNA intermediate and via reverse transcription. Also in this system, dysfunctional mtDNA's accumulate and consequently lead to senescence [17].

It is very striking that in different filamentous fungi different types of genetic elements affect the stability of the mitochondrial genome leading to the development of degenerative phenotypes. Interestingly, the activity of comparable genetic traits may have different outcomes in different systems. Examples are the senescent inducing linear plasmids in Neurospora and the linear plasmid pAL2-1 of *P. anserina*. Whereas pAL2-1 appears to stabilize the mtDNA leading to longevity of the corresponding Podospora strain, the kalilo plasmid, pKAL, from *N. intermedia* and other linear plasmids lead to increased mtDNA reorganizations and senescence. Moreover, there are a number of species known in which linear plasmids do not give rise to a particular phenotype [106, 107].

The search for age-related DNA rearrangements in mammalian systems was first rather disappointing. However, first data were obtained from a heteroduplex analysis using mtDNA from mice of different age. In these experiments the number of single stranded loops in denaturated and reannealed mtDNA preparations increased when mtDNA of older mice was used. These data suggested an age-related increase in deletions/additions of short mtDNA sequences during aging [108]. Utilizing PCR approaches this became demonstrated repeatedly both in tissues of healthy subjects and of those suffering from a range of different types of diseases [109–116].

Rather recently, a crucial role of mitochondria in life span control emerged in aging models in which mitochondria were first not known to play a major role. Although *S. cereviseae* differs from strictly aerobic organisms like mycelial fungi and mammals because yeast can survive under anaerobic conditions, it is an important finding that the retrograde response that can compensate affected mitochondrial functions was first reported in this organism [88, 89, 117] and was later demonstrated to affect life span [118]. Moreover, very recent investigations demonstrated that also in yeast the accumulation of dysfunctional mitochondria as the result of a breakdown of a postulated "filter" mechanism providing functional mitochondria to daughter cells of older mothers are causatively involved in life span control [119] (see also Chapter 1, this volume). In another aging model, the nematode *Caenorhabditis elegans*, mtDNA reorganizations were found to occur between short direct repeats by using long-extention PCR strategy. The rate by which the deletions occur were found to be significantly slower in the *age-1* long-lived strain as compared to wild-type animals [120]. Recently the relevance of mitochondria in the aging of *C. elegans* were further illuminated by the results of an iRNA analysis of 5690 genes of the nematode demonstrating that about 15% of the analysed genes leading to an extended life span code for mitochondrial functions [121]. It thus appears that where

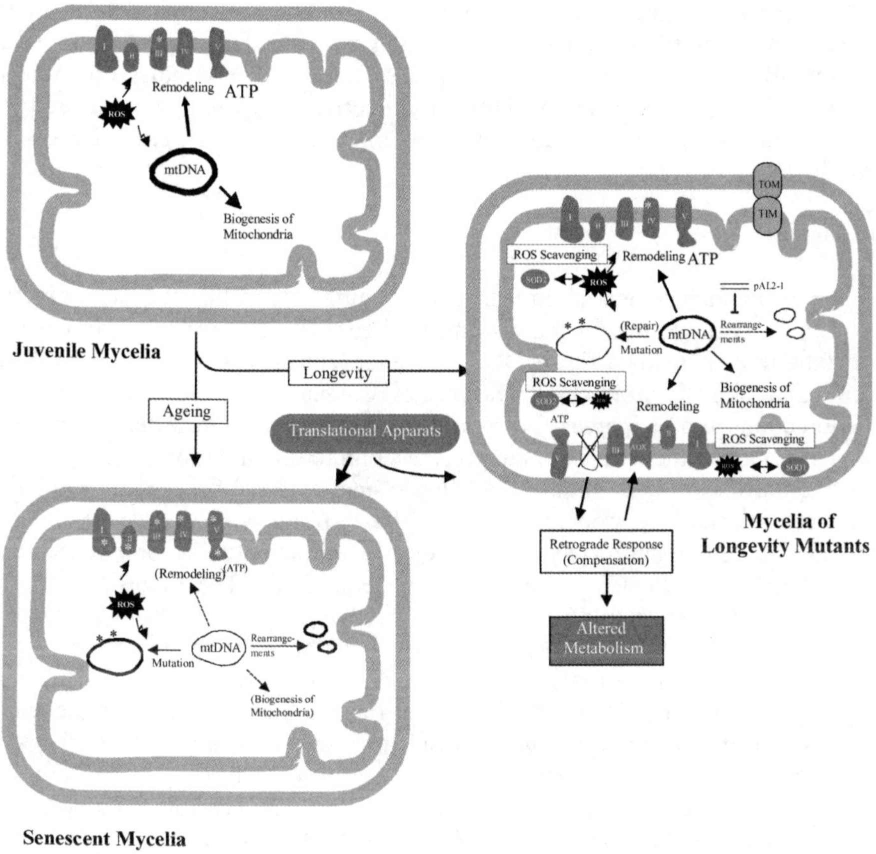

Figure 4. *Mitochondrial functions affecting longevity of Podospora anserina. In juvenile mycelia respiration proceeds via the COX-dependent respiratory chain. Only very low amounts of ROS are produced which during aging lead to damaged macromolecules including proteins and DNA (damaged molecules are indicated by asterisks). Since mitochondrial DNA is present in non-mutated form remodeling of damaged components of the respiratory chain is possible. Note that this remodeling capacity is greatly dependent on the integrity and expression of the nuclear genome (not indicated in the scheme). The remodeling capacity is also crucial for the biogenesis of new mitochondria (by division of existing ones) during growth of a culture. During aging the mtDNA becomes greatly rearranged and mutated. Consequently remodeling of affected respiratory proteins and mitochondria biogenesis is impaired. On the right various processes and modifications which individually or together were found to increase life span in P. anserina are indicated. This are stabilization of the mtDNA (e.g., via the activity of a linear plasmid, pAL2-1), the increase of the ROS scavenging capacity (e.g., SOD1, SOD2, the induction of a retrograde response leading to the compensation of COX deficiencies, or the increased fidelity of the translational apparatus. In addition the mitochondrial protein import machinery (TOM, TIM) band most likely a mitochondrial DNA repair system are effective. The different modifications may lead to an altered metabolism that also contributes to life span increase (for further details, see text)*

ever an appropriate selection procedure or methodology is applied (e.g., heteroduplex analysis and PCR strategies to demonstrate age-related mtDNA deletions and point mutations *C. elegans* and in mammals, systematic iRNA analysis of large sets of genes in *C. elegans*, a screen to identify asymmetric segregation of mitochondria in yeast) an important if not causal role of mitochondria in aging processes is unravelled.

Conclusions

The data and conclusions derived from fungal aging models like *P. anserina* indicate that many aspects of the biogenesis and the remodeling of mitochondria play an important role in aging (Figure 4). In juvenile cultures, low levels of ROS are generated at the respiratory chain. Damage of proteins in the respiratory chain leads to malfunctions and an increased formation of ROS production. Affected mitochondria, depending on the severity of damage, respond to this situation in different ways. Both, the ROS scavenger and a remodeling system as part of a general "caretaker" system are affective. Remodeling of damaged respiratory chains depends on the availability of functional mtDNA. This kind of a mitochondrial repair system has a positive effect on the capacity to remodel damaged proteins. The opposite is true for a mitochondrial DNA recombination system. A reduced efficiency of this system leads to a higher mtDNA stability and consequently to a higher remodeling capacity. Since most mitochondrial proteins are encoded by nuclear genes, the import of cytoplasmic proteins are important. Obviously, mitochondrial remodeling is strongly dependent on well co-ordinated nuclear-mitochondrial interactions. If mitochondrial respiratory chains are severely damaged (e.g., as the result of deficiency in cofactors like copper and iron), rescue of mitochondrial function is only possible via a compensation system able to substitute affected components. In copper-deficient *P. anserina* strains, the induction of the iron-dependent alternative respiratory pathway is such an example.

In aging research, the comparison of the data elucidated in different biological systems allows to deduce some more general pathways that appear to be part of molecular mechanisms involved in the aging. However, also specialities of the individual systems become apparent. These specialities appear to be related to the different organization of the system and the different conditions the different systems are facing. *P. anserina* is a sessile organism naturally growing on herbivorous dung. Since this habitat dries out rather fast, conditions for life of cultures are restricted to a short period time. Certainly, these conditions led to the evolution of a special strategy that allows *P. anserina* to survive even under these unfavourable conditions. A short life cycle is the consequence allowing an efficient reproduction. In the laboratory ascospores are produced in about 10–12 days. Numerous spores are generated and actively discharged from the perithecium. After release, they stick to the herbage in the neighbourhood of the culture from which they originate. Subsequent digestion of an herbivore and passage through the intestine provide a new substrate for progression through the next life cycle. Since the natural conditions are life limiting, it makes no sense to put much of the available energy into running an efficient cellular maintenance system. Instead, according to the concept of the disposable soma theory of aging [122, 123], the energy appears to better used to ensure an efficient formation of reproduction units. This strategy seems to be one, but not the exclusive way to deal with the constraints of the ecological niche to which this fungus has adapted. This is clear because otherwise aging processes similar to those in Podospora would be found in all or most coprophilus fungi. However, although aging of *P. anserina* appears not to be exclusively dependent on the activity of a specific mobile intron the question is still open whether in nature it is not the only mechanism by which *P. anserina* life span is determined. If so, for *P. anserina* the activity of plDNA would thus indeed be the cause of aging as it has been suggested earlier. The mechanisms bringing life to an end in the protected conditions of the laboratory, which are independent of the activity of plDNA, would not be of relevance in the natural habitat.

Apart from species-specific mechanisms, which necessarily will be dealt with, when working with any aging model also conserved pathways can be unravelled when the obtained data are compared with those of other systems. In addition specific strategies can be designed to demonstrate whether or not the pathways demonstrated in one system are valid also in other systems. In fact, this appears to be true for a role of mitochondria in the different systems investigated in aging research. Given that the role of these organelles in bioenergetics is essential, this outcome is no surprise. In fact, pathways related to mitochondrial functions seem to be part of aging mechanisms referred to as "conserved" or "public mechanisms" of aging [124]. However, although important in a wide range of organisms, these mechanisms are not the only ones that are relevant but are linked to various other mechanisms (e.g., scavenging and repair systems, epigenetic processes).

Acknowledgments

The experimental work of the author was supported by a grant of the Deutsche Forschungsgemeinschaft (Bonn, Germany).

References

1. Bartnicki-Garcia S (2002). Hyphal tip growth: outstanding questions. In: Osiewacz HD, ed. *Molecular Biology of Fungal Development*. New York, Basel: Marcel Dekker, pp. 29–58.
2. Glass NL, Saupe SJ (2002). Vegetative incompatibility in filamentous ascomycetes. In: Osiewacz HD, ed. *Molecular Biology of Fungal Development*. New York, Basel: Marcel Dekker, pp. 109–32.
3. Mösch H-U (2002). Pseudohyphal growth in yeast. In: Osiewacz HD, ed. *Molecular Biology of Fungal Development*. New York, Basel: Marcel Dekker, pp. 1–27.
4. Smith ML, Bruhn JN and Anderson JB (1992). The fungus *Armillaria bulbosa* is amongst the largest and oldest living organisms. *Nature* 356: 428–31.
5. Barton AA (1950). Some aspects of cell division in *Saccharomyces cerevisiae*. *J Gen Microbiol.* 4: 84–7.

6. Mortimer RK, Johnston JR (1959). Life span of individual yeast cells. *Nature* 183: 1751–2.
7. Marcou D (1961). Notion de longevite et nature cytoplasmatique du determinant de senescence chez quelques champignons. *Ann Sci Nat Bot*. 653–764.
8. Caten CE, Handley L (1978). Vegetative death syndrome in *Aspergillus glaucus*. *Bull Br Mycol Soc*. 12: 114.
9. Handley L, Caten CE (1973). Vegetative death: a mitochondrial mutation in *Aspergillus amstelodami*. *Heredity* 31: 136.
10. Lazarus CM, Earl AJ, Turner G, Küntzel H (1980). Amplification of a mitochondrial DNA sequence in the cytoplasmically inherited "ragged" mutant of *Aspergillus amstelodami*. *Eur J Biochem*. 106: 633–41.
11. Lazarus CM, Küntzel H (1981). Anatomy of amplified mitochondrial DNA in "ragged" mutants of *Aspergillus amstelodami*: excision points within genes and a common 215 bp segment containing a possible origin of amplification. *Curr Genet*. 4: 99–107.
12. Jinks JL (1956). Naturally occurring cytoplasmic changes changes in fungi. *CR Trav Lab Carlsberg Ser Pysiol*. 26: 183–203.
13. Lindberg GD (1959). A transmissible disease of *Helminthosporium victoriae*. *Phytopathology* 49: 29–52.
14. Bertrand H, Collins RA, Stohl LL, Goewert RR, Lambowitz AM (1980). Deletion mutants of *Neurospora crassa* mitochondrial DNA and their relationship to the "stop-start" growth phenotype. *Proc Natl Acad Sci USA* 77: 6032–6.
15. de Vries H, de Jonge JC, van't Sant S, Agsteribbe E, Arnberg A (1981). A "stopper" mutant of *Neurospora crassa* containing two populations of aberant mitochondrial DNA. *Curr Genet*. 3: 205–11.
16. Bertrand H, Chan BS, Griffiths AJ (1985). Insertion of a foreign nucleotide sequence into mitochondrial DNA causes senescence in *Neurospora intermedia*. *Cell*, 41: 877–84.
17. Akins RA, Kelley RL, Lambowitz AM (1986). Mitochondrial plasmids of Neurospora: integration into mitochondrial DNA and evidence for reverse transcription in mitochondria. *Cell* 47: 505–16.
18. de Vries H, Alzner-DeWeerd B, Breitenberger CA, Chang DD, de Jonge JC, RajBhandary UL (1986). The E35 stopper mutant of *Neurospora crassa*: precise localization of deletion endpoints in mitochondrial DNA and evidence that the deleted DNA codes for a subunit of NADH dehydrogenase. *EMBO J*. 5: 779–85.
19. Bertrand H, Wu Q, Seidel-Rogol BL (1993). Hyperactive recombination in the mitochondrial DNA of the natural death nuclear mutant of *Neurospora crassa*. *Mol Cell Biol*. 13: 6778–88.
20. Rieck A, Griffiths AJ, Bertrand H (1982). Mitochondrial variants of *Neurospora intermedia* from nature. *Can J Genet Cytol*. 24: 741–59.
21. Griffiths AJF, Bertrand H (1984). Unstable cytoplasm in Hawaiian strains of *Neurospora intermedia*. *Curr Genet*. 8: 387–98.
22. Griffiths AJ, Yang X (1995). Recombination between heterologous linear and circular mitochondrial plasmids in the fungus Neurospora. *Mol Gen Genet*. 249: 25–36.
23. Bertrand H, Griffiths AJ, Court DA, Cheng CK (1986). An extrachromosomal plasmid is the etiological precursor of kalDNA insertion sequences in the mitochondrial chromosome of senescent Neurospora. *Cell* 47: 829–37.
24. Chevaugeon J, Digbeu S (1960). Un second facteur cytoplasmique infectant chez le *Pestalozzia annulata*. *CR Acad Sci*. 251: 3043–61.
25. Rizet G (1953). Sur l'impossibilite d'obtenir la multiplication vegetative initerrompuet illimite de l'ascomycete *Podospora anserina*. *CR Acad Sci Paris* 237: 838–55.

26. Rizet G (1953). Sur la longevite des phenomen des souches de *Podospora anserina*. *CR Acad Sci Paris* 237: 1106–9.
27. Böckelmann B, Esser K (1986). Plasmids of mitochondrial origin in senescent mycelia of *Podospora curvicolla*. *Curr Genet*. 10: 803–10.
28. Gagny B, Rossignol M, Silar P (1997). Cloning, sequencing, and transgenic expression of *Podospora curvicolla* and *Sordaria macrospora* eEF1A genes: relationship between cytosolic translation and longevity in filamentous fungi. *Fungal Genet Biol*. 22: 191–8.
29. Esser K, Tudzynski P (1980). Senescence in fungi. In: Thimann KV, ed. *Senescence in Plants*. Boca Raton: CRC Press, pp. 67–83.
30. Esser K, Kück U, Lang-Hinrichs C, *et al.* (1986). *Plasmids of Eukaryotes. Fundamentals and Applications*. Heidelberg, New York, Tokyo: Springer
31. Osiewacz HD (1990). Molecular analysis of aging processes in fungi. *Mutat Res*. 237: 1–8.
32. Griffiths AJ (1992). Fungal senescence. *Annu Rev Genet*. 26: 351–72.
33. Osiewacz HD (1997). Genetic regulation of aging. *J Mol Med*. 75: 715–27.
34. Osiewacz HD, Kimpel E (1999). Mitochondrial-nuclear interactions and lifespan control in fungi. *Exp Gerontol*. 34: 901–9.
35. Bertrand H (2000). Role of mitochondrial DNA in the senescence and hypovirulence of fungi and potential for plant disease control. *Annu Rev Phytopathol*. 38: 397–422.
36. Osiewacz HD, Borghouts C (2000). Mitochondrial oxidative stress and aging in the filamentous fungus *Podospora anserina*. *Ann NY Acad Sci*. 908: 31–9.
37. Osiewacz HD (2002). Mitochondrial functions and aging. *Gene* 286: 65–71.
38. Osiewacz HD (2002). Aging in fungi: role of mitochondria in *Podospora anserina*. *Mech Ageing Dev*. 123: 755–64.
39. Osiewacz HD (2002). Genes, mitochondria and aging in filamentous fungi. *Ageing Res Rev*. 28: 1–18.
40. Osiewacz HD, Scheckhuber CQ (2002). Senescence in *Podospora anserina*. In: Osiewacz HD, ed. *Molecular Biology of Fungal Development*. New York, Basel: Marcel Dekker, pp. 87–108.
41. Delay C (1963). Observations inframicroscopiques sur le mycelium “senescent” du *Podospora anserina*. *CR Acad Sci Paris* 256: 4721–4.
42. Silliker ME, Liotta MR, Cummings DJ (1996). Elimination of mitochondrial mutations by sexual reproduction: two *Podospora anserina* mitochondrial mutants yield only wild-type progeny when mated. *Curr Genet*. 30: 318–24.
43. Silliker ME, Monroe JA, Jorden MA (1997). Evaluation of the efficiency of sexual reproduction in restoring *Podospora anserina* mitochondrial DNA to wild-type. *Curr Genet*. 32: 281–6.
44. Marcou D, Schecroun J (1959). La senescence chez *Podospora* pourrait etre due a des particles cytoplasmatiques infectantes. *CR Acad Sci*. 248: 280–3.
45. Rizet G (1957). Les modifications qui conduisent a la senescence chez *Podospora* sont-elles de nature cytoplasmique. *CR Acad Sci*. 244: 663–5.
46. Tudzynski P, Esser K (1979). Chromosomal and extrachromosomal control of senescence in the ascomycete *Podospora anserina*. *Mol Gen Genet*. 173: 71–84.
47. Esser K, Tudzynski P (1977). Prevention of senescence in the ascomycete *Podospora anserina* by the antibiotic tiamulin. *Nature* 265: 454–6.
48. Tudzynski P, Esser K (1977). Inhibitors of mitochondrial function prevent senescence in the ascomycete *Podospora anserina*. *Mol Gen Genet*. 153: 111–13.

49. Koll F, Begel O, Keller AM, Vierny C, Belcour L (1984). Ethidium bromide rejuvenation of senescent cultures of *Podospora anserina*: loss of senescence-specific DNA and recovery of normal mitochondrial DNA. *Curr Genet.* 8: 127–34.
50. Smith JL, Rubenstein I (1973). Cytoplasmic inheritance of the timing of "senescence" in *Podospora anserina. J Gen Microbiol.* 76: 297–304.
51. Smith JR, Rubenstein I (1973). The development of 'senescence' in *Podospora anserina. J Gen Microbiol.* 76: 283–96.
52. Esser K, Keller W (1976). Genes inhibiting senescence in the ascomycete *Podospora anserina. Mol Gen Genet.* 144: 107–10.
53. Stahl U, Lemke PA, Tudzynski P, Kück U, Esser K (1978). Evidence for plasmid like DNA in a filamentous fungus, the ascomycete *Podospora anserina. Mol Gen Genet.* 162: 341–3.
54. Cummings DJ, Belcour L, Grandchamp C (1979). Mitochondrial DNA from *Podospora anserina*. II. Properties of mutant DNA and multimeric circular DNA from senescent cultures. *Mol Gen Genet.* 171: 239–50.
55. Belcour L, Begel O, Mosse MO, Vierny-Jamet C (1981). Mitochondrial DNA amplification in senescent cultures of *Podospora anserina*: variability between the retained, amplified sequences. *Curr Genet.* 3: 13–21.
56. Kück U, Stahl U, Esser K (1981). Plasmid-like DNA is part of mitochondrial DNA in *Podospora anserina. Curr Genet.* 3: 151–6.
57. Osiewacz HD, Esser K (1984). The mitochondrial plasmid of *Podospora anserina*: A mobile intron of a mitochondrial gene. *Curr Genet.* 8: 299–305.
58. Cummings DJ, McNally KL, Domenico JM, Matsuura ET (1990). The complete DNA sequence of the mitochondrial genome of *Podospora anserina. Curr Genet.* 17: 375–402.
59. Kück U, Osiewacz HD, Schmidt U, *et al.* (1985). The onset of senescence is affected by DNA rearrangements of a discontinuous mitochondrial gene in *Podospora anserina. Curr Genet.* 9: 373–82.
60. Jamet-Vierny C, Boulay J, Begel O, Silar P (1997). Contribution of various classes of defective mitochondrial DNA molecules to senescence in *Podospora anserina. Curr Genet.* 31: 171–8.
61. Jamet-Vierny C, Boulay J, Briand JF (1997). Intramolecular cross-overs generate deleted mitochondrial DNA molecules in *Podospora anserina. Curr Genet.* 31: 162–70.
62. Sellem CH, Lecellier G, Belcour L (1993). Transposition of a group II intron. *Nature* 366: 176–8.
63. Borghouts C, Kerschner S, Osiewacz HD (2000). Copper-dependence of mitochondrial DNA rearrangements in *Podospora anserina. Curr Genet.* 37: 268–75.
64. Michel F, Lang BF (1985). Mitochondrial class II introns encode proteins related to the reverse transcriptases of retroviruses. *Nature* 316: 641–3.
65. Fassbender S, Bruhl KH, Ciriacy M, Kück U (1994). Reverse transcriptase activity of an intron encoded polypeptide. *EMBO J.* 13: 2075–83.
66. Vierny-Jamet C, Keller AM, Begel O, Belcour L (1982). A sequence of mitochondrial DNA is associated with the onset of senescence in a fungus. *Nature* 297: 157–9.
67. Schulte E, Kück U, Esser K (1988). Extrachromosomal mutants from *Podospora anserina*: permanent vegetative growth in spite of multiple recombination events in the mitochondrial genome. *Mol Gen Genet.* 211: 342–9.
68. Begel O, Boulay J, Albert B, Dufour E, Sainsard-Chanet A (1999). Mitochondrial group II introns, cytochrome c oxidase, and senescence in *Podospora anserina. Mol Cell Biol.* 19: 4093–100.

69. Osiewacz HD, Hermanns J, Marcou D, Triffi M, Esser K (1989). Mitochondrial DNA rearrangements are correlated with a delayed amplification of the mobile intron (plDNA) in a long-lived mutant of *Podospora anserina*. *Mutat Res*. 219: 9–15.
70. Hermanns J, Osiewacz HD (1992). The linear mitochondrial plasmid pAL2-1 of a long-lived *Podospora anserina* mutant is an invertron encoding a DNA and RNA polymerase. *Curr Genet*. 22: 491–500.
71. Hermanns J, Asseburg A, Osiewacz HD (1994). Evidence for a life span-prolonging effect of a linear plasmid in a longevity mutant of *Podospora anserina*. *Mol Gen Genet*. 243: 297–307.
72. Hermanns J, Osiewacz HD (1994). Three mitochondrial unassigned open reading frames of *Podospora anserina* represent remnants of a viral-type RNA polymerase gene. *Curr Genet*. 25: 150–7.
73. Hermanns J, Debets F, Hoekstra R, Osiewacz HD (1995). A novel family of linear plasmids with homology to plasmid pAL2-1 of *Podospora anserina*. *Mol Gen Genet*. 246: 638–47.
74. Hermanns J, Osiewacz HD (1996). Induction of longevity by cytoplasmic transfer of a linear plasmid in *Podospora anserina*. *Curr Genet*. 29: 250–6.
75. Prillinger H, Esser K (1977). The phenoloxidases of the ascomycete *Podospora anserina*. XIII. Action and interaction of genes controlling the formation of laccase. *Mol Gen Genet*. 156: 333–45.
76. Marbach K, Fernandez-Larrea J, Stahl U (1994). Reversion of a long-living, undifferentiated mutant of *Podospora anserina* by copper. *Curr Genet*. 26: 184–6.
77. Borghouts C, Kimpel E, Osiewacz HD (1997). Mitochondrial DNA rearrangements of *Podospora anserina* are under the control of the nuclear gene grisea. *Proc Natl Acad Sci USA* 94: 10768–73.
78. Silar P, Koll F, Rossignol M (1997). Cytosolic ribosomal mutations that abolish accumulation of circular intron in the mitochondria without preventing senescence of *Podospora anserina*. *Genetics* 145: 697–705.
79. Osiewacz HD, Nuber U (1996). GRISEA, a putative copper-activated transcription factor from *Podospora anserina* involved in differentiation and senescence. *Mol Gen Genet*. 252: 115–24.
80. Borghouts C, Osiewacz HD (1998). GRISEA, a copper-modulated transcription factor from *Podospora anserina* involved in senescence and morphogenesis, is an ortholog of MAC1 in *Saccharomyces cerevisiae*. *Mol Gen Genet*. 260: 492–502.
81. Kimpel E, Osiewacz HD (1999) *PaGrg1*, a glucose-repressible gene of *Podospora anserina* that is differentially expressed during lifespan. *Curr Genet*. 35: 557–63.
82. Borghouts C, Scheckhuber CQ, Werner A, Osiewacz HD (2002). Respiration, copper availability and SOD activity in *P. anserina* strains with different lifespan. *Biogerontology* 3: 143–53.
83. Borghouts C, Scheckhuber CQ, Stephan O, Osiewacz HD (2002). Copper homeostasis and aging in the fungal model system *Podospora anserina*: differential expression of *PaCtr3* encoding a copper transporter. *Int J Biochem Cell Biol*. 34: 1355–71.
84. Glerum DM, Shtanko A, Tzagoloff A, Gorman N, Sinclair PR (1996). Cloning and identification of HEM14, the yeast gene for mitochondrial protoporphyrinogen oxidase. *Yeast* 12: 1421–5.
85. Hiser L, Di Valentin M, Hamer AG, Hosler JP (2000). Cox11p is required for stable formation of the Cu(B) and magnesium centers of cytochrome c oxidase. *J Biol Chem*. 275: 619–23.

86. Lode A, Kuschel M, Paret C, Rodel G (2000). Mitochondrial copper metabolism in yeast: interaction between Sco1p and Cox2p. *FEBS Lett*. 485: 19–24.
87. Liao X-B, Clare J, Farabaugh P (1987). The upstream activation site of a Ty2 element of yeast is necessary but not sufficient to promote maximal transcription of the element. *Proc Natl Acad Sci USA* 84: 8520–4.
88. Liao X, Butow RA (1993). RTG1 and RTG2: two yeast genes required for a novel path of communication from mitochondria to the nucleus. *Cell* 72: 61–71.
89. Sekito T, Thornton J, Butow RA (2000). Mitochondria-to-nuclear signaling is regulated by the subcellular localization of the transcription factors Rtg1p and Rtg3p. *Mol Biol Cell*. 11: 2103–15.
90. Dufour E, Boulay J, Rincheval V, Sainsard-Chanet A (2000). A causal link between respiration and senescence in *Podospora anserina*. *Proc Natl Acad Sci USA* 97: 4138–43.
91. Wagner AM, Moore AL (1997). Structure and function of the plant alternative oxidase: its putative role in the oxygen defence mechanism. *Biosci Rep*. 17: 319–33.
92. Harman D (1956). A theory based on free radical and radiation chemistry. *J Gerontol*. 11: 298–300.
93. Harman D (1988). Free radicals in aging. *Mol Cell Biochem*. 84: 155–61.
94. Harman D (1992). Free radical theory of aging. *Mutat Res*. 275: 257–66.
95. Miquel J, de Juan E, Sevila I (1992). Oxygen-induced mitochondrial damage and aging. *EXS* 62: 47–57.
96. Beckman KB, Ames BN (1998). The free radical theory of aging matures. *Physiol Rev*. 78: 547–81.
97. Harman D (1998). Aging and oxidative stress. *J Int Fed Clin Chem*. 10: 24–7.
98. Harman D (2001). Aging: overview. *Ann NY Acad Sci*. 928: 1–21.
99. Sturtz LA, Diekert K, Jensen LT, Lill R, Culotta VC (2001). A fraction of yeast Cu,Zn-superoxide dismutase and its metallochaperone, CCS, localize to the intermembrane space of mitochondria. A physiological role for SOD1 in guarding against mitochondrial oxidative damage. *J Biol Chem*. 276: 38084–9.
100. Jamet-Vierny C, Contamine V, Boulay J, Zickler D, Picard M (1997). Mutations in genes encoding the mitochondrial outer membrane proteins Tom70 and Mdm10 of *Podospora anserina* modify the spectrum of mitochondrial DNA rearrangements associated with cellular death. *Mol Cell Biol*. 17: 6359–66.
101. Silar P, Lalucque H, Vierny C (2001). Cell degeneration in the model system *Podospora anserina*. *Biogerontology* 2: 1 17.
102. Averbeck NB, Jensen ON, Mann M, Schägger H, Osiewacz HD (2000). Identification and characterization of PaMTH1, a putative *o*-methyltransferase accumulating during senescence of *Podospora anserina* cultures. *Curr Genet*. 37: 200-208.
103. Averbeck NB, Borghouts C, Hamann A, Specke V, Osiewacz HD (2001). Molecular control of copper homeostasis in filamentous fungi: increased expression of a metallothionein gene during aging of *Podospora anserina*. *Mol Gen Genet*. 264: 604–12.
104. Gross SR, Hsieh TS, Levine PH (1984). Intramolecular recombination as a source of mitochondrial chromosome heteromorphism in Neurospora. *Cell* 38: 233–9.
105. Almasan A, Mishra NC (1988). Molecular characterization of the mitochondrial DNA of a stopper mutant ER-3 of *Neurospora crassa*. *Genetics* 120: 935–45.
106. Kempken F, Hermanns J, Osiewacz HD (1992). Evolution of linear plasmids. *J Mol Evol*. 35: 502–13.
107. Meinhardt F, Kempken F, Kämper J, Esser K (1990). Linear plasmids among eukaryotes: fundamentals and application. *Curr Genet*. 17: 89–95.

108. Bulpitt KJ, Piko L (1984). Variation in the frequency of complex forms of mitochondrial DNA in different brain regions of senescent mice. *Brain Res.* 300: 41–8.
109. Linnane AW, Marzuki S, Ozawa T, Tanaka M (1989). Mitochondrial DNA mutations as an important contributor to ageing and degenerative diseases. *Lancet* 1: 642–5.
110. Wallace DC (1989). Mitochondrial DNA mutations and neuromuscular disease. *Trends Genet.* 5: 9–13.
111. Linnane AW, Baumer A, Maxwell RJ, Preston H, Zhang CF, Marzuki S (1990). Mitochondrial gene mutation: the ageing process and degenerative diseases. *Biochem Int.* 22: 1067–76.
112. Kadenbach B, Müller-Hocker J (1990). Mutations of mitochondrial DNA and human death. Naturwissenschaften 77: 221–5.
113. Osiewacz HD, Hermanns J (1992). The role of mitochondrial DNA rearrangements in aging and human diseases. *Aging* 4: 273–86.
114. Wallace DC (1993). Mitochondrial diseases: genotype versus phenotype. *Trends Genet.* 9: 128–33.
115. Wallace DC (1999). Mitochondrial diseases in man and mouse. *Science* 283: 1482–8.
116. Wallace DC (2001). A mitochondrial paradigm for degenerative diseases and ageing. *Novartis Found Symp.* 235: 247–63.
117. Liao XS, Small WC, Srere PA, Butow RA (1991). Intramitochondrial functions regulate nonmitochondrial citrate synthase (CIT2) expression in *Saccharomyces cerevisiae. Mol Cell Biol.* 11: 38–46.
118. Kirchman PA, Kim S, Lai CY, Jazwinski SM (1999). Interorganelle signaling is a determinant of longevity in *Saccharomyces cerevisiae. Genetics* 152: 179–90.
119. Lai CY, Jaruga E, Borghouts C, Jazwinski SM (2002). A mutation in the ATP2 gene abrogates the age asymmetry between mother and daughter cells of the yeast *Saccharomyces cerevisiae. Genetics* 162: 73–87.
120. Melov S, Lithgow GJ, Fischer DR, Tedesco PM, Johnson TE (1995). Increased frequency of deletions in the mitochondrial genome with age of *Caenorhabditis elegans. Nucleic Acids Res.* 23: 1419–25.
121. Lee SS, Lee RY, Fraser AG, Kamath RS, Ahringer J, Ruvkun G (2002). A systematic RNAi screen identifies a critical role for mitochondria in *C. elegans* longevity. *Nat Genet.* 33: 40–8.
122. Kirkwood TB, Holliday R (1979). The evolution of ageing and longevity. *Proc R Soc Lond B Biol Sci.* 205: 531–46.
123. Kirkwood TB (2002). Evolution of ageing. *Mech Ageing Dev.* 123: 737–45.
124. Martin GM, Austad SN, Johnson TE (1996). Genetic analysis of ageing: role of oxidative damage and environmental stresses. *Nat Genet.* 13: 25–34.

Genetic, Metabolic and Environmental Factors Associated with Aging in Plants

Karin Krupinska[1], Jon Falk[1] and Klaus Humbeck[2]
[1] *Institute of Botany, University of Kiel, Olshausenstrasse 40, D-24098 Kiel, Germany;* [2] *Institute of Plant and Cell Physiology, Martin Luther University Halle-Wittenberg, Weinbergweg 10, D-06099 Halle, Germany*

Introduction

In general biological aging either refers in a broad sense to the chronological progression of an organism's life, or in a more confined way to the "old age," the last phase preceding death, which is characterized by a decline in physiological processes and vigour [1, 2]. In analogy to the theory of aging as defined for animals by Harman [3, 4], in this article aging of plants is used in the broader sense as the process of slow, progressive and sequential alterations a plant or a part of a plant undergoes during its development until death. In contrast to animals, plants have extremely different life spans ranging from a few days up to several thousands of years in case of some trees. Moreover, life spans of different organs of a plant can substantially differ from the whole plant [5].

The article deals with aging at different levels, whole plant aging, aging of leaf organs and briefly also with cellular aging. We will focus the attention thereby on the shoot and photo-synthetically active tissue of higher plants.

Plant development and aging
The shoot system of higher plants after post-embryogenic development passes through several different phases, a juvenile vegetative phase, an adult vegetative phase and a reproductive phase [6]. As the shoot develops, its apical meristem increases in age, changes its position in relation to previously formed shoots, and its offspring cells produce an increasing number of structures, such as leaves and internodes [6]. In parallel to the discrete transitions through the developmental phases, the shoot also undergoes a gradual aging process which is apparent, e.g., by the reduction in growth rate, loss of apical dominance and reduced flower produc-

H.D. Osiewacz (ed.), Aging of Organisms, 55–78.

tion. Genetic and molecular analyses of plant development by mutants have revealed that the lengths of the different phases are regulated by different transition factors. While some of these factors are specific for a certain phase of development, others seem to influence several phase transitions [7, 8].

With regard to reproduction and the possible causes of death, two types of plants are distinguished, monocarpic plants with a single reproduction phase followed by death and polycarpic plants with repeated reproduction phases [9–11]. Annuals and biennials are monocarpic, whereas with a few exceptions perennials are polycarpic. While death of monocarpic plants is tightly coupled with reproduction, death of polycarpic plants may be much more controlled by environmental factors [11].

The last developmental phase before death of a monocarpic plant or a plant organ usually is termed senescence [9, 10, 12]. In annuals such as *Arabidopsis thaliana*, which is the model of plant genetics and molecular biology, this phase may parallel or follow reproduction [13, 14].

In polycarpic plants also three phases of development are distinguished: juvenility (pre-reproductive plants), maturity (reproductive plants) and old age [2]. A characteristic of old age as of monocarpic senescence is the decline in vegetative growth which eventually may lead to death [11]. Although the parts of woody plants can age, senesce and die, the whole individual does not senesce the way that monocarpic plants do [2, 11]. Nevertheless, the term polycarpic senescence is commonly used for the terminal decline in vegetative growth and function of polycarpic plants [11]. In polycarpic plants this decline is an attrition over a long time, while in monocarpic plants the decline occurs much faster in tight coupling with reproduction. This suggests that the causes of the decline in vigour leading to death differ between monocarpic and polycarpic plants.

In many monocarpic plants the decline in vigour is tightly linked to reproduction, e.g., in soybean where the removal of pods delays leaf senescence [11, 15]. In trees, however, aging and reproduction seem not to be coupled. Old trees and shrubs are quite different in appearance and growth compared with young and mature trees [2]. Typically, the old trees have thicker leaves and do not form primary branches. These old growth characters emerge as the trees approach maximal height. These age related changes in structure and function of trees may continue long after sexual maturity. A polycarpic plant will obviously not die as long as it keeps growing and forming new leaves. A reduction in organ production and cessation of vegetative growth during development is reflecting a decrease in activity of meristems [11], a possible cause of which could be the accumulation of somatic mutations [16].

An intriguing question related to aging is how time is measured. It has been proposed that in plants developmental time is either measured as real time or as size respective organ number [7.] The *psd* (*paused*) mutant of *Arabidopsis thaliana* is unable to form the first true leaf, but continues with the formation of the following leaves in the same schedule as the wild type. This suggests that vegetative development is regulated by a clock independent of leaf initiation [7]. As also shown with mutants of *Arabidopsis thaliana* the shift from vegetative growth to flower formation is under control of the biological clock, too [17].

The relationship between aging and senescence

In plant sciences the term senescence is not used synonymously with the term aging as in animal sciences, but to characterize the last phase in development preceding death [10, 16]. Plant senescence is a genetically programmed process which occurs at any of several levels, from tissues and organs to whole organisms. It is a unique feature of multicellular plants that this phase is not only characterized by degenerative processes but also by active processes required for remobilization of valuable compounds. Re-mobilized compounds are exported from senescing parts of a plant and may be reused for building up new parts, such as young leaves, seeds and fruits. Molisch [9] was among the first botanists recognizing that the degeneration of plant organisms and their parts is under endogenous control. This endogenous program in its later stage resembles a cell death program [14, 18]. The senescence program can be divided into several phases [14]. An initiation phase is characterized by the decline in vegetative growth of whole plants and photosynthesis of leaves which may be a consequence of aging [13]. The later degeneration phase of senescence is characterized by the degradation of the photosynthetic apparatus as evident by chlorophyll decline and remobilization of valuable compounds. In its terminal stage senescence involves cell death [18, 19]. Since recycling processes during senescence are important for plants, the degeneration leading finally to death occurs rather late and in a more abrupt way than in animals [11].

In contrast to senescence, which is a programmed phase in plant development, the term aging in plants has been used mainly to refer to a wide array of passive and non-regulated degenerative processes driven primarily by exogenous factors such as stress or starvation [2, 10, 12, 16].

The relationship between senescence and aging is far from being clear. There are examples for aging and death without senescence and examples for senescence without death [16]. Since leaf senescence at least in its initial phase is a reversible process [16, 20–22] it does not inevitably lead to death. In fruits it has been shown that induction of ripening by ethylene requires a certain age of the fruit [18]. Similarly, flower induction depends on a certain age of the plant [23] and the rate of senescence associated chlorophyll degradation in wheat leaves depends on their physiological age [24]. In all these cases aging obviously establishes the competence for developmental programs such as ripening, reproduction and senescence.

During development of leaves the decline in photosynthesis on one hand seems to be the result of aging whereas chlorophyll degradation on the other hand is a feature of senescence. Both events are not coupled, as evident by the observation that depodding of soybean delayed chlorophyll breakdown, while the photosynthetic decline was not affected [15]. Even, the age-related decline in photosynthesis was proposed to be the trigger of senescence including chlorophyll breakdown of the rosette leaves of *Arabidopsis thaliana* [13] (see also Figure 1 and section on "Reactive oxygen species as a by-product of photosynthesis").

In this review we use the term aging to refer to the alterations plants undergo during time dependent development. We want to stress that aging in plants as in animals has a genetic basis as evident by the different life spans of plants and the existence of mutants with altered life spans. Moreover the aging process in plants is influenced by metabolic activities and environmental factors.

Factors associated with whole plant aging

In the plant kingdom there is a much broader range of different life spans than in the animal kingdom. Some trees can even live up to several thousands of years. In plants like animals maximal life span is a characteristic of the species. In dioecious plants, female plants usually live longer than male plants [9, 25] (Table 1). Besides genetic factors and sex, metabolism and environmental conditions regulate longevity. In plants of temperate climate zones, death and production of seeds can be seen as a part of a preparation for an adverse season. Given the evidence that longevity is a characteristics of the species and is genetically determined, it may be inferred that some sort of time-measuring mechanism limits life span. It seems that the time-keeper mechanism is connected to the developmental program, since clonal species which propagate by continued vegetative growth and renewal of assimilatory organs may persist almost indeterminately [11]. There are however some indications for a time-measuring mechanism even in clonal species. For instance the asexual off-springs of old duckweed clones have less vigour than the off-springs of young clones [11].

***Table 1**. Maximal life spans of higher plants*

Plant	Maximal life span
Perennials:	
Yellow wood anemone (*Anemone ranunculoides*)	7 years
Scandinavian thyme (*Thymus chamaedrys*)	14 years
English ivy (*Hedera helix*)	200 years
European white birch (*Betula verucosa*)	120 years
Apple (*Pyrus malus*)	200 years
Scots pine (*Pinus silvestris*)	500 years
Olive (*Olea europaea*)	700 years
Giant sequoia (*Sequoia gigantea*)	3200 years
Bristlecone pine (*Pinus aristata*)	4600 years
Annuals:	
Cannabis sativa, male	Around 4 months
Cannabis sativa, female	About 2–4 weeks longer
Arabidopsis thaliana	8–10 weeks
Amaryllis lucida	10 days

Plants are categorized with regard to their life cycle patterns and life spans [9]. Some plants have a very short life span, e.g., desert annuals, weeds and bulbous plants, whereas woody perennials accumulate substantial reserves over many years before reproduction and death [8]. Examples for perennial and annual plants with different life spans are provided in Table 1.

Sources for the maximal life spans of perennials are listed in Noodén [11]. Data on approximate life spans of annuals are from Molisch [9] and from Hensel *et al.* [13].

Genetic determinants of life span

In the last two decades *Arabidopsis thaliana* became an invaluable tool for analysis of the genetic basis of developmental processes such as embryogenesis [26], flowering [23] and senescence [27]. As a typical representative of short-living ruderal plants, its life strategy is characterized by a high mortality risk with early reproductive development and high fecundity [13]. Meanwhile a considerable number of mutants with defined alterations in developmental processes are available. In the following we will present examples of Arabidopsis mutants with altered life spans (Table 2).

Table 2. *Mutants of Arabidopsis thaliana with altered life spans*

Mutant name		References
Long-living mutants:		
lfy	*leafy*	28
gi	*gigantea*	29, 30
clv2-4, clv3-	*clavata*	31
vam1 enh (rev-1)	*revoluta*	32,33
Short-living mutants:		
tfl1	*terminal flower1*	8
apg 9-1	*autophagy gene 9-1*	34
efs	*early flowering in short days*	35

Life span of annual plants such as *Arabidopsis thaliana* is intimately linked to the timing of flowering and seems to depend here solely on the time of transition from the vegetative to the reproductive phase. Consequently, late-flowering mutants can be considered as extended longevity mutants and *vice versa* [13, 23]. Extreme examples are the *gi* (*gigantea*) [30] and the *lfy* (*leafy*) mutants [28], which continue to produce leaves and lateral shoots when the wild-type begins to produce flowers. *Lfy* is a gene important for the transition from the vegetative to the reproductive phase [28]. The *gigantea* gene is controlled by the circadian clock and encodes a nuclear protein involved in phytochrome signalling [36]. The nucleoplasmic localization of the protein suggests a possible role in regulation of transcription. Indeed circadian controlled genes have altered expression patterns in the *gi* mutant. The *efs* mutant belongs to the group of early flowering mutants [35] indicating that EFS is an inhibitor of flowering. A mutation in this gene reduces the duration of the adult vegetative phase.

Of special interest with regard to aging is the *tfl1* mutant, where in contrast to the *lfy* and *efs* mutants, transition through all the different phases of the life-cycle is retarded [8]. To investigate whether variations in the levels of TFL1 are also influencing the life span of other plants, genes orthologous to the *tfl1* gene of *Arabidopsis thaliana* have to be identified and modified in their expression in diverse species with different life strategies. So far, the impact of the *tfl1* gene on life span has been also investigated in transgenic plants overexpressing this gene [8]. As expected, in contrast to the mutant these transgenic lines have an extended life span. Using a collection of Arabidopsis mutants with altered shoot development, Noodén and Penney [37] investigated whether life span of the whole plant is linked to leaf longevity (see section on "Genetic determinants of leaf longevity"). The findings of this study indicate that growth mutations such as *clv2-4, clv3-2, det3, vam1 enh* and *dark green* have prolonged life span through the production of new leaves and stems with inflorescences rather than by extension of leaf longevity [37].

Among the mutants with altered life spans to our knowledge only *apg*-9 has a dramatic impact on leaf longevity as well [34] (see also section on "Genetic determinants of leaf longevity"). Less dramatic influences on leaf longevity were observed in case of the mutants *dark-green* and *vam1 enh* [37].

The existence of genes that when mutated can alter life span suggests that one or few of genetically controlled processes may be critical in aging [5]. So far, mutants with altered life spans are not available from polycarpic plants. However, there are at least two findings indicating a genetic control of aging in trees and shrubs, too. On one hand changes in the physiology and structure of leaves are linked to changes in gene expression during maturation [reviewed in ref. 38], on the other hand, mature scions grafted onto juvenile root stocks stay mature [2, 38].

Metabolism in relation to aging

Genetic studies of longevity mutants of animals as divergent as nematodes, Drosophila and mammals have suggested that in addition to accumulating DNA damage and telomere shortening alterations in energy metabolism with subsequent oxidative damage play a major role in determining life span [39]. Indeed, in many cases a positive correlation has been found between life span and antioxidant capacity, e.g., in yeast, Podospora, Caenorhabditis, Drosophila and mammals [40–44] (see also Chapter 2, this volume). Hence the genetically determined capacity to cope with oxidative stress may be among the genetic factors of aging.

Since oxidative damage is a by-product of aerobic metabolism, nutritional deprivations may extend life span. This relationship has been originally proposed by Pearl [44] as rate of living hypothesis and indicates that organisms are metabolic integrators. It is possible that the timekeeper mechanism measures the amount of a product of metabolism. It has been proposed that a similar mechanism could underlie the aging of polycarpic plants, but its significance for death of monocarpic plants has been questioned [11]. Molisch [9] proposed that plants age due to the accumulation of minerals, silicilic acid and presumably toxic compounds as a consequence of metabolism. From the current knowledge such toxic byproducts of metabolism may be reactive oxygen species accumulating as a consequence of aerobic metabolism.

While in animal cells aerobic metabolism and most generation of reactive oxygen species occurs in mitochondria, in plant cells chloroplasts are the major source of reactive oxygen species [46, 47]. The concentration of oxygen in chloroplasts is even one to three orders of magnitude higher than in animal cells [47]. Not only in monocarpic plants such as *Arabidopsis thaliana*, but also in most of trees and shrubs photosynthetic production declines with age [2, 11]. The reasons for this decline are far from being clear (see also section on "Reactive oxygen species as a by-product of photosynthesis").

To investigate whether in plants as in animals long life span correlates with a higher resistance towards oxidative stress, Kurepa *et al.* [29] tested the response of a group of late-flowering mutants to oxidative stress induced by paraquat which is known to raise the level of superoxide anions. Among the mutants tested, the *gi* mutant (Table 2) showed a very high tolerance towards both paraquat and hydrogen peroxide. The existence of such long-lived and oxidative stress tolerant mutants in *Arabidopsis thaliana* points at a similar link between longevity and oxidative stress tolerance as in animals [29]. Moreover, the results obtained with the mutant *gi* suggest that in plants genes involved in the defence mechanisms against reactive oxygen species might have a role in the control of longevity.

In the last years many reports on transgenic plants with enhanced anti-oxidative capacity due to over-expression of SOD or enzymes of glutathione metabolism have been published. Most of these studies have been performed with tobacco and other monocarpic plants [46]. The results of these studies are rather inconsistent. Though an increase in oxidative stress tolerance could be observed in certain cases [48], no effects on longevity of the plants have been reported. Since compartmentation is critical for the function of antioxidants, the transgenic plants with one of the anti-oxidative systems altered are not expected to show changes in overall stress resistance [48] and in longevity.

The only polycarpic plant used for similar studies is poplar. Analyses of transgenic poplar lines over-expressing SOD and enzymes involved in glutathione biosynthesis [49] were primarily devoted to stress tolerance towards herbicides [50, 51]. The results from such analyses indicate that the transgenic trees have a higher antioxidant capacity of their leaves which seems to improve their capacity to withstand oxidative stress [50]. A systematic analysis of such transgenic poplar plants with regard to life span and leaf longevity has however not been reported so far.

Environmental factors influencing life span

The aging process in plants can be dramatically influenced by changing environmental conditions. On one hand, external biotic and abiotic factors may affect certain developmental phases during ontogeny of a plant [for reviews see refs 6, 52] and by this alter the basic timing of aging. On the other hand, environmental factors may directly contribute to the accumulation of reactive oxygen which is proposed to underlie aging in plants.

It has been shown that manipulation in mutants of the timing of flowering during development affects life span of annual plants (see section on "Genetic determinants of life span"). In general induction of flowering is influenced by light and temperature [53, 54]. Analysis of flowering mutants revealed that light via the phytochrome system in connection with a time measuring process regulates central genes responsible for initiation of flowering [53].

Various stress situations are known to negatively influence life span by inducing a premature leaf senescence. Among the stress factors are drought, chilling, ultraviolet radiation, pathogen attack, high light intensity, salt, heavy metals, ozone and sulphur dioxide [55]. They all have in common an impairment of the photosynthetic electron reactions with a concomitant production of reactive oxygen species [56]. Via this surplus generation of reactive oxygen species as a consequence of disturbances in the photosynthetic apparatus these stressors may negatively influence life span of leaves and consequently of whole plants. It is however difficult to discriminate whether the stress-induced increase in reactive oxygen species results in a general speeding up of the processes underlying development of plants, e.g., the transition from the juvenile to the adult phase [6], or is just inducing lethal damages of essential plant functions which then lead to premature death of the whole plant.

It has been suggested that plants exhibit a decreasing tolerance against oxidative stress with increasing age [57]. Taking this into account it is understandable that the sensitivity towards stress increases with age [11, 55]. In case of trees it has been shown that older and larger trees respond differently to ozone than seedlings which are however mainly used in investigations of stress effects [58].

Though most stress conditions may accelerate aging by giving rise to the production of reactive oxygen species, in some instances it has been observed that exposure to a specific stress situation such as salt [59] or UV-light [60] prolongs the life of an organism. A possible explanation is that a specific stress situation contributes to an elevation of the anti-oxidative defence systems which in turn enhance tolerance to multiple causes of oxidative stress [48].

Aging, senescence and life span of leaves

As already mentioned, in plants the life span of organs such as leaves can differ distinctly from the life span of the whole plant. In case of annual plants aging of the whole plant and aging of leaves are overlapping in time. Age related alterations in leaves after maturation are however almost impossible to distinguish from changes occurring as a part of the program of leaf senescence (see section on "The relationship between aging and senescence").

As longevity of the whole plant, longevity of leaves is genetically determined and differs between species. A comprehensive overview on leaf longevity, ranging from several months in case of annual plants over a few years in case of conifers up to a maximal leaf life span of about 100 years in case of the leaf of *Welwitschia mirabilis*, has been already given by Molisch [9]. Shortest leaf life spans have been observed in the group of monocots. Leaves of the bulbous plant *Amaryllis lucida* last less than 10 days which actually is the life span of the whole plant (Table 1). Often, different types of leaves from one plant may have different life spans. Usually, cotyledones have a shorter life span than the other leaves.

At a certain time of their life, leaves undergo senescence as most obvious by chlorophyll decline. Usually the different leaves of an annual plant undergo a progressive pattern of senescence paralleling their chronological age, e.g., in *Arabidopsis thaliana* and in cereals. This means that older leaves undergo senescence earlier than younger leaves. In many annual plants, this progressive temporal development correlates with a different spatial situation of the leaves, whereby old leaves are located at the base of the shoot and the youngest leaf at the tip [12]. Apart from chronological age, many other factors may determine the timing of senescence [14, 27]. In soybean under the influence of reproduction all leaves, though being of different age, are senescing at once [11], and in some pea varieties senescence even occurs earlier on young leaves [61]. These findings demonstrate the existence of senescence promoting factors acting superimposed on age, e.g., hormones and sugars. In the frame of this review only the age dependent factors of leaf senescence are of interest; actions of hormones and other compounds on senescence are not treated.

Since analysis of aging and longevity at the whole plant level is complicated with reproduction, the leaf organ of *Arabidopsis thaliana* has been considered to be a convenient model to study aging. In contrast to the leaves of many other monocarpic plants such as soybean the developmental program of the rosette leaves of *Arabidopsis thaliana* is rather independent of the timing of reproduction [13, 62]. This has been convincingly demonstrated by analysis of life spans of whole plants and rosette leaves of a broad collection of Arabidopsis mutants with altered growth and timing of reproduction [37].

In principle, the categories made for whole plant development may be also applied to leaf development [2.] Immature leaves are juvenile, and fully expanded leaves are mature. In case of *Arabidopsis thaliana* juvenile leaves can be easily distinguished from adult leaves by the lack of abaxial trichomes [8, 63]. When the plant enters the reproductive phase these trichomes are lost [63] and photosynthesis declines [13], whereby degradative and remobilization processes involved in senescence are induced during this stage of development. In *Arabidopsis thaliana* age-related and senescence specific processes are hardly to discriminate since aging and senescence merge into one another.

Genetic determinants of leaf longevity

Mutants with prolonged leaf longevity which are called *stay green* mutants are known from various different cultivated plant species, such as different cereals, temperate grasses, soybean, bean and tobacco [for reviews see refs 64, 65]. Although the stay-green phenotype is seemingly similar in the different species, the genetic reasons and physiological consequences are diverse. In one type of the *stay-green* mutants simply chlorophyll catabolism is impaired. In the so-called functional *stay-green* mutants, the timing of senescence initiation or the regulation of its progression is altered [64, 65]. So far, the genes affected in most of these physiologically and bio-chemically well characterized functional *stay-green* mutants have not been identified.

In the last years, mutants of *Arabidopsis thaliana* with altered timing of leaf senescence and thereby altered life spans of leaves became a powerful tool to analyze the molecular mechanisms underlying the control of leaf longevity. The first mutant of *Arabidopsis thaliana* described to be a functional *stay-green* mutant is *etr1* which has a reduced sensitivity towards ethylene [66] (Table 3). Later on, Oh *et al.* [67] identified three genetic loci (*ore1, 3, 9*) controlling leaf senescence and leaf longevity in *Arabidopsis thaliana.* The three *ore* mutants isolated by Oh *et al.* [67] and a further *ore* mutant isolated later on by Woo *et al.* [68] are all characterized by delayed leaf senescence (Table 3). Since in all four mutants both chlorophyll breakdown and photosynthesis are delayed they belong to the group of functional *stay-green* mutants. The *ore3* mutant was identified as being allelic with *ein2-1*, an ethylene insensitive mutation. The characteristics of both mutants, *etr-1* and *ein2-1,* indicate that leaf senescence is controlled by ethylene. The *ore9* gene was identified to encode an F-box protein, suggesting that ORE9 functions to limit leaf longevity by removing, through ubiquitin-dependent proteolysis target proteins, that are required to delay the leaf senescence program [68]. The *ore*1, 3, 9 mutants showed delayed age-dependent senescence as well as delayed senescence when they were subjected to treatments with various senescence promoting factors, including darkness, abscisic acid, methyljas-monate and ethylene [67, 68]. Recently, another mutant with a phenotype delayed both in age-mediated and dark-induced senescence has been described [69] (Table 3). This *dls1* mutant has a T-DNA insertion in the gene encoding arginine-tRNA:protein ariginyltransferase for proteolysis by the N-end rule pathway, one of the several pathways of the ubiquitin system [69]. Both mutants, *ore9* and *dls1* point at the ubiquitin-proteasome system playing a functional role in leaf senescence. Since the three *ore* genes and the *dls1* gene function not only in age-dependent senescence but also in senescence induced by different treatments in younger leaves, they most likely are not involved in the process of aging.

In contrast to the previously described *ore* mutants and the *dls* mutant, the *ore* 4-1 mutation does exclusively affect age-mediated senescence and has no effect on dark-induced senescence [62]. Molecular analysis revealed that the *ore*4-1 mutant has a T-DNA insertion in the promoter of the nuclear gene r*ps17* leading to a reduced expression of the plastid ribosomal protein RPS17. Mutant chloroplasts have an impaired translational apparatus and are therefore disturbed in chloroplast functions including photosynthesis. As a consequence, in this mutant leaf growth is decreased. The timing of maturation is however similar to that of the wild-type. Analogously, reduction of photosynthesis in *rbcS* antisense plants coincides with retarded leaf senescence [71]. Indeed an inverse relationship between photosynthesis and longevity has been observed with a broad range of plants from different ecosystems [72–74] being in accordance with the rate of living hypothesis (see section "Metabolism in relation to aging").

In the *apg* mutant of *Arabidopsis thaliana* not only leaf senescence as shown by chlorophyll decline is accelerated, but also bolting, making this mutant also an early

Table 3. Mutants with altered leaf longevity

Mutant		References
Delayed leaf senescence:		
etr-1	*ethylene response 1*	66
ore1,3,9	*oresara**	67, 68
ore4-1	*oresara* 4-1*	62
dls1	*delayed leaf senescence*	69
Accelerated leaf senescence:		
apg 9-1	*autophagy gene 9-1*	34
old 1,2,3	*onset of leaf death 1,2,3*	70

**oresara* = "long-living" in Korean

flowering type mutant respective a mutant with reduced life span [34] (Tables 2 and 3). The authors hypothesize that autophagy may help to maintain viability during senescence and that without it, plant cells may degrade themselves by other mechanisms than senescence and die earlier. Further research on the cytological basis of autophagy in plants is required in order to elucidate its role for aging.

In addition to these mutants, several mutants with accelerated cell death have been reported, e.g., the mutants *adc1* [75] and *mod1* (mosaic death) [76]. These mutants form spontaneous lesions which influence life span of the leaf. Since accelerated cell death in these mutants may occur both in juvenile and mature tissues, the corresponding genes rather play a role in the plant's defence system and are most likely not involved in aging.

With the exception of the *apg* mutant all mutations reported in *Arabidopsis thaliana* either affect life span of the whole plant or longevity of rosette leaves. It will be interesting to see whether besides leaves in the *apg* mutant also longevity of other parts is likely affected. So far, it seems that the *apg* gene is the only described gene which is involved in a general aging mechanism applying to both the organ and the whole-plant level. Among the other mutants described to affect leaf senescence only the *ore4-1* mutant seems to be of interest with regard to leaf aging. To investigate whether the effects of the *apg* and *ore4-1* mutations are unique to *Arabidopsis thaliana* and other rapidly growing plants, the impact of orthologous genes on leaf life span in other species has to be analyzed, either in similar mutants or in transgenic plants with altered expression of the corresponding genes.

Similarly as the *apg* mutant, transgenic Arabidopsis plants over-expressing the WRKY53 gene which encodes a transcription factor involved in the regulation of senescence [77] show both accelerated leaf development and accelerated flowering (Zentgraf, personal communication). To our knowledge this is the first transgenic line with an accelerated development both at the whole plant level and at the level of leaves. The finding suggests that WRKY transcription factors may be involved in regulation of genes involved in overall development and aging.

Mutants and transgenic lines with alterations in the concentrations of hormones or sugars are not considered here. The modulating influence of hormones and sugars on development and senescence is well-known [11, 27]. The respective mutants rather give insight in senescence regulation by these signalling compounds than in the basic mechanisms of aging.

Reactive oxygen species as a by-product of photosynthesis

The reasons underlying the decline in photosynthesis with age of whole plants and leaves are far from being clear. As suggested by Molisch in 1929 changes during aging of plants and leaves are due to a kind of attrition "Abnutzung." Today we know that during photosynthesis parts of the complex apparatus are continuously destroyed by attack of reactive oxygen species and have to be repaired and replaced. The protein with the highest turnover is the D1 protein in photosytem II [78, 79]. Theoretically, the potential for photodamage in the photosynthetic apparatus is supposed to largely increase in the initial phase of leaf senescence when photosynthesis declines while chlorophylls are excited by light. The decline in photosynthesis is paralleled by degradation of reaction centre complexes while the light-harvesting complex is rather stable and declines parallel with the bulk chlorophylls [14, 80, 81]. Hensel *et al.* [13] proposed that the decline in photosynthesis initiates the senescence program. It has been further suggested that the mechanism underlying the induction of the senescence program in response to declining rates of photosynthesis may involve sugars or other photosynthetic metabolites as signalling compounds [27, 82, 83]. Alternatively or in addition, reactive oxygen species as by-products of aerobic metabolism or accumulating at certain environmental stress situations may be involved in triggering senescence. These reactive oxygen species are proposed to contribute to the overall pool of reactive oxygen species overcoming a certain threshold level (Figure 1A). Indeed, reactive oxygen species such as hydrogen peroxide [56, 84] may act as signalling molecules in initiation of senescence. Recently, it has been reported that hydrogen peroxide indeed accumulates transiently before the onset of leaf senescence in *Arabidopsis thaliana* (Zentgraf, personal communication, see also page 71 of this chapter). This hydrogen peroxide burst may be a consequence of decreasing photosynthesis (Figure 1C). In accordance with this model several senescence associated genes are induced by reactive oxygen species such as ozone [85] or hydrogen peroxide [86]. Whether the burst in hydrogen peroxide before senescence onset indeed occurs in chloroplasts and is related to the decrease in photosynthesis as proposed by our model (Figure 1C) has still to be demonstrated.

The inverse relationship between leaf longevity and photosynthesis (see section on "Genetic determinants of leaf longevity") suggests that turnover of the photosynthetic machinery may act as a time measuring device. It is likely that damages to the photosynthetic apparatus accumulate with time while many anti-oxidative systems decline with age [57]. This situation may contribute to the increase in oxidative stress in old leaves. Among the anti-oxidative systems declining with age is vitamin C [87].

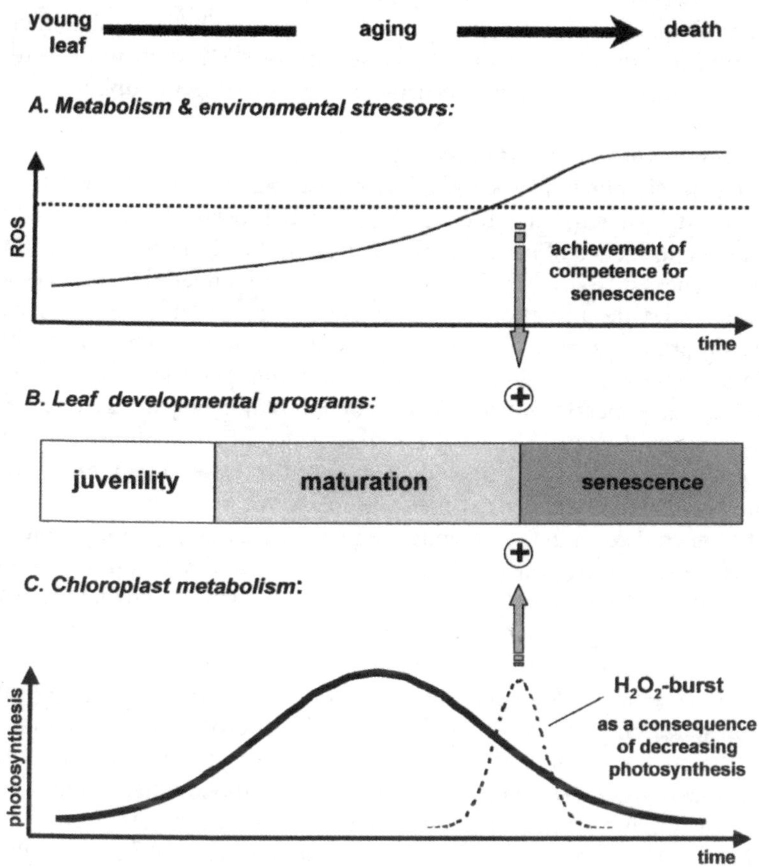

Figure 1. *Model of the age-dependent decline in photosynthesis of leaves in relation to initiation of senescence. (A) Reactive oxygen species (ROS) accumulate with time as a result of aerobic metabolism and environmental stress situations experienced by the leaf. (B) Development of the leaf can be divided into three stages. (C) Photosynthesis declines with age before senescence onset and causes an oxidative burst triggering the onset of senescence by overcoming the threshold level of reactive oxygen species depicted in part A. A stimulating effect on senescence is indicated by a (+).*

In contrast the vitamin E content of leaves increases during aging [88]. In barley flag leaves the tocopherol content increases earlier than the decline in chlorophylls suggesting that this change is associated with aging rather than with senescence [89]. Whether this increase in vitamin E is a response towards increasing oxidative stress has yet to be investigated.

So far, over-expression of genes encoding components of the anti-oxidative defence system, e.g., SOD and enzymes involved in glutathione metabolism did neither have an influence on life span of whole plants (see section on "Metabolism in relation to aging") nor on aging of leaves, in monocarpic plants as well as in poplar.

Environmental modulation of leaf longevity

As many other developmental processes, leaf senescence and thereby leaf longevity can be influenced by environmental factors. This has been already described by Molisch [9] as the influence of climate and location on leaf longevity. As outlined in reviews on leaf senescence [14, 82], the regulatory network underlying leaf senescence is quite complex and the initiation of the senescence program depends as well on autonomous (internal) as on environmental (external) signals. The environmental cues include a whole set of different factors such as temperature, light conditions, drought, ozone, heavy metals, nutrient deficiency, wounding or pathogens [27, 82]. In the following we will focus only on a few aspects of environment-dependent regulation of leaf senescence being aware that there are much more articles describing stress related alterations of the leaf senescence program.

Among the various external factors influencing the senescence process light plays a major role. Extreme shading of leaves or darkness, and also day-length are important signals for initiation of senescence processes [14, 27, 82]. It was shown that not only light intensity but also light quality serves as a signal for triggering the senescence program. Rousseaux *et al.* [90] could demonstrate that a reduced red/far-red ratio induces premature senescence indicating the action of the phytochrome system. The hypothesis of senescence initiation by a decline in photosynthesis (Figure 1, see section on "Reactive oxygen species as a by-product of photosynthesis") suggests that photosynthesis and thereby light could act as a metabolic timer in plants.

It is well-known that high light intensities may increase the risk of photo-oxidative stress and thereby induce premature senescence of leaves [57, 91]. Alternatively or in addition, high light intensities may induce premature leaf senescence simply by accelerating leaf development. A change in the light environment hence may influence leaf development at different levels. Actually, it could be shown that changes in light intensity could really change the timing of senescence or even revert the senescence process [92].

Another well-known phenomenon is that drought induces or accelerates the senescence program [e.g., ref. 93]. Pic *et al.* [94] applied a mild water stress and could show that in pea the drought induced degradation processes in leaves of stressed plants follow the same sequence of macroscopic, biochemical and molecular events as in leaves of control plants.

Besides abiotic stressors also biotic factors can induce degradation processes which are similar to those observed during leaf senescence. In accordance with these observations several (but not all) senescence associated genes have been shown to be induced also after pathogen attack or wounding [e.g., refs 95–97]. These results indicate that there is a considerable overlap in the complex regulatory networks of the senescence program and the pathogen defence reactions.

Most different stress situations inducing premature senescence have in common the accumulation of reactive oxygen species as a consequence of disturbances in the photosynthetic apparatus [55]. Since the capacity to withstand oxidative stress declines with age, the sensitivity towards stress increases with age of the leaves. For instance, young leaves of poplar trees are only slightly affected by ozone while mature leaves experience a 75% decrease in photosynthesis [98]. Similarly, the capacity for the synthesis of UV screening compounds declines with age of the leaves [99]. These and many other studies clearly show that the timing of the stress in relation to development determines its impact on leaf senescence [55, 100].

Plant cell aging

Plant cells in culture

While senescence, which is characterized by remobilization of nutrients from senescing cells and translocation of the valuable resources to other cells within one plant, is a typical feature of multi-cellular organisms [10, 11], aging occurs also in single cells [101]. Cultures of plant cells are commonly used to investigate the control of cell division or to analyze biochemical processes such as chlorophyll catabolism [102]. Peters *et al.* [101] used autotrophic cell cultures of *Chenopodium rubrum* to investigate aging and death occurring after the phase of cell division. Their data show that after a certain time of growth, rejuvenilization of the cells by addition of mitogenic compounds such as auxin and cytokinines is no more possible.

The reasons for aging in plant cells are far from being clear. Whereas in mammalian cells telomere shortening is one of the mechanisms responsible for cellular senescence [103], the evidence for a similar mechanism in plant cells is weak [104]. In barley it has been shown that telomeres shorten indeed during differentiation and aging while dedifferentiation in callus culture resulted in an increase in the length of telomere sequences [105]. The authors hypothesized that the lengths of the telomeres might regulate the level of certain telomere binding proteins playing a role in regulation of gene expression. In contrast to the situation in barley, in *Arabidopsis thaliana* the length of the telomeres does however not change significantly during differentiation and senescence [106]. Using a gel mobility shift assay Zentgraf *et al.* [106] detected a protein specifically binding to the telomeres in senescing leaves of *Arabidopsis thaliana*. A systematic study on the changes in the lengths of telomeres and proteins binding to the telomeres in various plants during development has not been performed, however.

Plant cells in leaf tissues

In general, senescence of leaves occurs systemically in the whole organ irrespective of the different age of individual cells. This is most obvious in leaves of monocots such as barley where cell divisions occur only in a basal meristem resulting in a gradient of cells with increasing age from the base to the tip [107, 108]. Hence, these leaves are particularly well suited to investigate the relationship between cell age and senescence. As shown by analysis of chlorophyll content and photosynthetic capacity along the primary foliage leaf, senescence is a systemic feature of the whole leaf blade and is not determined by the age of the cells (Thomsen and Krupinska, unpublished). This co-ordinated decline in the function of cells of different age requires the existence of age-related signals distributed within the entire organ.

An exception to the co-ordinated decline in leaf cells are guard cells which even in yellow falling leaves remain green and operable [109]. Since guard cells are not connected with the mesophyll they may not perceive the signals supposed to be distributed within the whole leaf. Recently, Yang *et al.* [110] identified in tobacco two leucine zipper type transcription factors, TBZF and TBZ17, which could specifically play a role in senescence retardation in guard cells.

The contribution of reactive oxygen species from different sources to aging

In mammalian cells oxidative damage as a consequence of aerobic metabolism seems to be a major contributor to cellular degeneration during aging [3, 41, 111]. Common opinion is that in plants oxidative damages are much less important for aging, since plants possess a much more effective potential for radical detoxification [112] than fungi and animals. Nevertheless, as also discussed in the previous chapters there are several findings clearly indicating that also in plants reactive oxygen species may play a role for the aging process. For instance it has been shown that during aging cell suspension cultures of *Chenopodium rubrum* accumulate lipid oxidation products [113] suggesting that the antioxidant systems decline during cell aging. During leaf senescence biochemical reactions giving rise to the production of reactive oxygen species become more prominent and are proposed to promote the oxidative membrane deterioration leading to cell death [114].

While in animal and fungal cells mitochondria are the most important site of oxygen metabolism and a privileged site of the production of reactive oxygen species [41], in plants most reactive oxygen species are produced in chloroplasts as a consequence of photosynthesis (see section on "Reactive oxygen species as a by-product of photosynthesis"). At the level of the leaf, the age of the whole organism has an impact on photosynthesis as well as on longevity of the organ [115]. Munné-Bosch and Alegre [57] analyzed leaves of the same age before approaching the stage of senescence in 1, 3 and 7 years old Cistus plants. In detail these analyses showed that the level of malondialdehyde (MDA), which is a marker for lipid peroxidation, is enhanced in leaves of 7 years old plants, while photosynthesis and chlorophyll content as well as levels of antioxidants (α-tocopherol, β-carotine) are decreased in leaves from the older plants. These findings suggest that oxidative stress contributes to aging in plants before leaf senescence occurs, and that oxidative stress associated with aging in plants occurs at least partly in chloroplasts [57]. Unfortunately, with this study no data on longevity of the leaves are provided. In earlier studies an increased accumulation of hydrogen peroxide and malondialdehyde in aging leaves was observed to be accompanied by an increase of activities involved in the production of hydrogen peroxide [47].

Besides chloroplasts, plant cells have further sources of reactive oxygen species. Important during aging and senescence seems to be also the production of superoxide anions and hydrogen peroxide in peroxisomes [18]. In conjunction with photosynthetic activity in chloroplasts, photorespiration in peroxisomes is an additional source for hydrogen peroxide. Noctor *et al.* [116] calculated that the photorespiratory hydrogen peroxide production even could account for about 70% of total hydrogen peroxide in leaves of C3 plants. Interestingly, during senescence the number of peroxisomes and mitochondria is rising up to fivefold in pea leaves [117, 118]. In addition, levels of key enzymes of the glyoxylate cycle and purine catabolism increase in peroxisomes of senescing leaves [118]. By the various reactions leading to the production of reactive oxygen species peroxisomes appear to have an important function in leaf senescence. Among the reactive oxygen species especially hydrogen peroxide is rather stable and is considered as a signalling molecule [84]. Hydrogen peroxide leaving peroxisomes may induce alterations in gene expression [118]. Reactive oxygen species actively produced as signalling compounds could be shown to control also programmed cell death, abiotic stress responses and pathogen defence [56]. The observation that leaves before senescence onset exhibit a burst of hydrogen peroxide (Zentgraf, personal communication) is in accordance with a signalling role of this compound during aging.

Peroxisomes are also a site of NO biosynthesis [119]. Nitric oxide (NO), a potentially toxic, free radical gas has gained much interest as an endogenous maturation and senescence regulating factor [120], effecting also seed germination, deetiolation [121], plant defence response [122] and acting as a scavenger of reactive oxygen species [123].

Changes in the peroxisome population and their biochemical equipment in relation to aging have not gained much recognition in animals. It is however likely that reactive oxygen species accumulate as by-products of purine catabolism during aging in animals likewise as in plants.

In fungi and animals where mitochondria are the major source of reactive oxygen species, defects in mitochondrial DNA and their consequences on energy metabolism are involved in aging [42]. Investigations with isolated mitochondria and peroxisomes from pea leaves suggest that during senescence mitochondria are affected by oxidative damage earlier than peroxisomes [124]. A significance of reactive oxygen production in mitochondria could be shown for programmed cell death [125].

Conclusions and future prospects

As evident by the existence of mutants with altered life spans, aging in plants is under genetic control as in animals. As discussed for animals [39] the genetic basis of aging in fact may be the genetically controlled capacity of plants to deal with oxidative stress as a consequence of metabolism and environmental stress situations. This view contrasts with models implying that a genetic program causes aging.

Analyses of mutants of *Arabidopsis thaliana* point at the importance of some genes for aging, e.g., genes involved in autophagy. The significance of such genes for aging of plants in general has to be investigated by identification of corresponding genes in other plants and modification of their expression in transgenic lines.

Several findings suggest that oxidative stress is involved in plant aging, although plants are supposed to have efficient anti-oxidative defence mechanisms enabling them to cope with oxidative stress. However, the capacity to cope with oxidative stress obviously declines at later stages of development. In contrast to animals, in plants photosynthesis in chloroplasts and photorespiration in peroxisomes seem to be the major metabolic reactions producing reactive oxygen species which may be involved in aging. Apart from damaging important cellular constituents, reactive oxygen species seem to have a signalling role for developmental processes such as leaf senescence.

Oxidative damages accumulating due to metabolism may function as a time measuring device. So far, it is not clear which kinds of damages are critical for aging in plants. We hypothesize that changes in the photosynthetic apparatus may be part of the time measuring system and finally may be responsible for the decline in photosynthesis which eventually leads to initiation of senescence by reactive oxygen species. To test this hypothesis, detailed analyses of photo-oxidative damages in the photosynthetic apparatus and the repair capacity during development have to be performed.

So far, transgenic plants with enhanced capacity of different anti-oxidative systems yielded inconsistent results with regard to longevity of the whole plant and the leaf. It is probably not sufficient to increase the capacity of one particular anti-oxidative system in a single cell compartment. Future studies on aging aiming at identifying the key oxidants and their sub-cellular location have to be performed. In addition general factors involved in the regulation of the balance of antioxidants and oxidants in plant cells have to be identified. Based on such knowledge new approaches to test in transgenic plants the significance of reactive oxygen species for plant aging have to be designed.

Acknowledgments

We are very grateful to Dr. Ulrike Zentgraf (ZMBP, University of Tübingen, Germany) for communicating unpublished results and for stimulating discussions.

References

1. Partridge L, Barton NH (1993). Optimality, mutation and the evolution of ageing. *Nature* 362: 305–11.
2. Bond BJ (2000). Age-related changes in photosynthesis of woody plants. *Trends Plant Sci.* 5: 349–53.
3. Harman D (1981). The aging process. *Proc Natl Acad Sci USA* 78: 7124–8.
4. Harman D (1991). The aging process: major risk factor for disease and death. *Proc Natl Acad Sci USA* 88: 5360–3.
5. Guarente L, Ruvkun G, Amasino R (1998). Aging, life span, and senescence. *Proc Natl Acad Sci USA* 95: 11034–6.
6. Poethig RS (1990). Phase change and the regulation of shoot morphogenesis in plants. *Science* 250: 923–30.
7. Lawson EJ, Poethig RS (1995). Shoot development in plants: time for a change. *Trends Genet.* 11: 263–8.

8. Ratcliffe OJ, Amaya I, Vincent CA, *et al.* (1998). A common mechanism controls the life cycle and architecture of plants. *Development* 125: 1609–15.
9. Molisch H, ed. (1929). *Die Lebensdauer der Pflanzen.* Verlag Gustav Fischer, Jena.
10. Noodén LD (1988a). The phenomena of senescence and aging. In: Noodén LD, Leopold AC, eds. *Senescence and Aging in Plants.* San Diego: Academic Press Inc, pp. 1–50.
11. Noodén LD (1988b). Whole plant senescence. In: Noodén LD, Leopold AC, eds. *Senescence and Aging in Plants.* San Diego: Academic Press Inc., pp. 391–439.
12. Leopold AC (1961). Senescence in plant development. *Science* 134: 1727–32.
13. Hensel LL, Grbic V, Baumgarten DA, Bleecker AB (1993). Developmental and age-related processes that influence the longevity and senescence of photosynthetic tissues in *Arabidopsis*. Plant Cell, 5: 553–64.
14. Noodén LD, Guiamét JJ, John I (1997). Senescence mechanisms. *Physiol Plant* 101: 746–53.
15. Wittenbach VA (1982). Effect of pod removal on leaf senescence in soybeans. *Plant Physiol.* 70: 1544–8.
16. Thomas H (2002). Ageing in Plants. *Mech Ageing Dev.* 123: 747–53.
17. Barak S, Tobin EM, Andronis C, Sugano S, Green RM (2000). All in good time: the *Arabidopsis* circadian clock. *Trends Plant Sci.* 5: 517–21.
18. Dangl JI, Dietrich RA, Thomas H (2000). Senescence and programmed cell death. In: Buchanan B, Gruissem W, Jones R, eds. *Biochemistry and Molecular Biology of Plants.* Rockville, MD: American Society of Plant Physiologist, pp. 1044–100.
19. Pennell RI, Lamb C (1997). Programmed cell death in plants. *Plant Cell* 9: 1157–68.
20. Mothes K, Baudisch W (1958). Untersuchungen über die Reversibilität der Ausbleichung grüner Blätter. *Flora* 146: 521–32.
21. Kleber-Janke T, Krupinska K (1997). Isolation of cDNA clones for genes showing enhanced expression in barley leaves during dark-induced senescence as well as during senescence under field conditions. *Planta* 203: 332–40.
22. Thomas H, Donnison I (2000). Back from the brink: plant senescence and its reversibility. In: Bryant JA, Hughes SG, Garland JM, eds. *Programmed Cell Death in Animals and Plants.* Oxford: BIOS Scientific Publishers Ltd, pp. 149–62.
23. Koornneef M, Alonso-Blanco C, Peeters AJ, Soppe W (1998). Genetic control of flowering time in *Arabidopsis. Annu Rev Plant Physiol Plant Mol Biol.* 49: 345–70.
24. Fischer A, Feller U (1994). Senescence and protein degradation in leaf segments of young winter wheat: influence of leaf age. *J Exp Bot.* 45: 103–9.
25. Mothes K, Engelbrecht L (1952). Über geschlechtsverschiedenen Stoffwechsel zweihäusiger einjähriger Pflanzen. *Flora* 139: 1–27.
26. Wolpert, L (ed). (1998). *Principles of Development.* Oxford: Oxford University Press.
27. Nam HG (1997). The molecular genetic analysis of leaf senescence. *Curr Opin Biotechnol.* 8: 200–7.
28. Blasquez MA, Soowal LN, Lee I, Weigel D (1997). LEAFY expression and flower initiation in *Arabidopsis. Development* 124: 3835–44.
29. Kurepa J, Smalle J, VanMontagu M, Inzé D (1998). Oxidative stress tolerance and longevity in *Arabidopsis*: the late-flowering mutant *gigantea* is tolerant to paraquat. *Plant J.* 14: 759–64.
30. Koornneef M, Hanhart CJ, van der Veenj JH (1991). A genetic and physiological analysis of late flowering mutants in *Arabidopsis thaliana. Mol Gen Genet.* 229: 57–66.
31. Clark SE, Jacobsen SE, Levin JZ, Meyerowitz EM (1996). The clavata and shoot meristemless loci competitively regulate meristem activity in *Arabidopsis. Development* 122: 1567–75.

32. Otsuga D, DeGuzman B, Prigge MJ, Drews GN, Clark SE (2001). Revoluta regulates meristem initiation at lateral positions. *Plant J.* 25: 223–36.
33. Pogany JA, Simon EJ, Katzman RB, *et al.* (1998). Identifying novel regulators of shoot meristem development. *J Plant Res.* 111: 307–13.
34. Hanaoka H, Noda T, Shirano Y, *et al.* (2002). Leaf senescence and starvation-induced chlorosis are accelerated by the disruption of an *Arabidopsis* autophagy gene. *Plant Physiol.* 129: 1181–93.
35. Soppe WJ, Bentsink L, Koornneef M (1999). The early-flowering mutant *efs* is involved in the autonomous promotion pathway of *Arabidopsis thaliana*. *Development* 126: 4763–70.
36. Huq E, Tepperman M, Quail PH (2000) *Gigantea* is a nuclear protein involved in phytochrome signaling in *Arabidopsis*. *Proc Natl Acad Sci USA* 97: 9789–94.
37. Noodén LD, Penney JP (2001). Correlative controls of senescence and plant death in *Arabidopsis thaliana* (Brassicaceae). *J Exp Bot.* 52: 2151–9.
38. Greenwood MS (1995). Juvenility, maturation in conifers: current concepts. *Tree Physiol.* 15: 433–8.
39. Guarente L (1997). What makes us tick? *Science* 275: 943–4.
40. Braeckman BP, Houthoofd K, Vanfleteren JR (2001). Insulin-like signaling, metabolism, stress resistance and aging in *Caenorhabditis elegans*. *Mech Ageing Dev.* 122: 673–693.
41. Beckman KB, Ames BN (1998). The free radical theory of aging matures. *Physiol Rev.* 78: 547–81.
42. Osiewacz HD (1997). Genetic regulation of aging. *J Mol Med.* 75: 715–27.
43. Osiewacz HD, Kimpel E (1999). Mitochondrial-nuclear interactions and life span control in fungi. *Exp Gerontol.* 34: 901–9.
44. Osiewacz HD, Stumpferl SW (2001). Metabolism and aging in the filamentous fungus *Podospera anserina*. *Arch Gerontol Geriatr.* 32: 185–97.
45. Pearl R (1928). *The Rate of Living*. London: University of London Press.
46. Allen RD, Webb RP, Shale SA (1997). Use of transgenic plants to study antioxidant defenses. *Free Rad Biol Med.* 23: 473–9.
47. Polle A (1997). Defense against photooxidative damage in plants. In: Scandalios JG, ed. *Oxidative Stress and the Molecular Biology of Antioxidant Defenses.* Cold Spring Harbor Laboratory Press, pp. 623–66.
48. Bray EA, Bailey-Serres J (2000). Responses to abiotic stresses. In: Buchanan BB, Gruissem W, Jones RL, eds. *Biochemistry and Molecular Biology of Plants*. Rockville, MD, USA: American Society of Plant Physiologists, pp. 1158–203.
49. Arisi AM, Noctor G, Foyer CH, Jouanin L (1997). Modification of thiol contents in poplars (*Poplus tremula, Poplus alba*) overexpressing enzymes involved in glutathione synthesis. *Planta* 203: 362–72.
50. Foyer CH, Souriau N, Perret S, *et al.* (1995). Overexpression of glutathione reductase but not glutathione synthetase leads to increases in antioxidant capacity and resistance to photoinhibition in poplar trees. *Plant Physiol.* 109: 1047–57.
51. Gullner G, Kömives T, Rennenberg H (2001). Enhanced tolerance of transgenic poplar plants overexpressing γ-glutamylcysteine synthese towards chloroacetanilide herbicides. *J Exp Bot.* 52: 971–9.
52. Casal JJ (2002). Environmental cues affecting development. *Curr Opin Biotechnol.* 5: 37–42.
53. Pidkowich MS, Klenz JE, Haughn GW (1999). The making of a flower: control of floral meristem identity in *Arabidopsis*. *Trends Plant Sci.* 4: 64–70.

54. Reeves PH, Coupland G (2000). Response of plant development to environment: Control of flowering by day length and temperature. *Curr Opin Plant Biol.* 3: 37–42.
55. Pell EJ, Dann MS (1991). Multiple stress-induced foliar senescence and implications for whole-plant longevity. In: *Responses of Plants to Multiple Stresses*. New York: Academic Press, pp. 189–204
56. Mittler R (2002). Oxidative stress, antioxidants and stress tolerance. *Trends Plant Sci.* 7: 405–10.
57. Munné-Bosch S, Alegre L (2002). Plant aging increases oxidative stress in chloroplasts. *Planta* 214: 608–15.
58. Kolb TE, Fredrickson TS, Steiner KC, Skelly JM (1998). Issues in scaling tree size and age responses to ozone: a review. *Environ Pollution* 98: 195–208.
59. Mittova V, Tal M, Volokita M, Guy M (2002). Salt stress induces up-regulation of an efficient chloroplast antioxidant system in the salt-tolerant wild tomato species *Lycopersicon penellii* but not in the cultivated species. *Physiol Plant* 115: 393–400.
60. Kondo N, Kawashima M (2000). Enhancement of the tolerance to oxidative stress in cucumber (*Cucumis sativus* L.) seedlings by UV-B irradiation: possible involvement of phenolic compounds and antioxidative enzymes. *J Plant Res.* 113: 311–17.
61. Proebstig WM, Davies PJ, Marx GA (1976). Photoperiodic control of apical senescence in a genetic line of peas. *Plant Physiol.* 58: 800–2.
62. Woo HR, Goh CH, Park JH, *et al.* (2002). Extended leaf longevity in the *ore*4-1 mutant of *Arabidopsis* with a reduced expression of a plastid ribosomal protein gene. *Plant J.* 31(3): 331–40.
63. Telfer A, Bollman KM, Poethig RS (1997). Phase change and the regulation of trichrome distribution in *Arabidopsis thaliana. Development* 124: 645–54.
64. Thomas H, Smart CM (1993). Crops that stay green. *Ann Appl Biol.* 123: 193–219.
65. Thomas H, Howarth CJ (2000). Five ways to stay green. *J Exp Bot.* 51: 329–37.
66. Grbic V, Bleecker AB (1995). Ethylene regulates the timing of leaf senescence in *Arabidopsis. Plant J.* 8: 595–602.
67. Oh SA, Park JH, Lee GI, Paek SH, Park SK, Nam HG (1997). Identification of three genetic loci controlling leaf senescence in *Arabidopsis thaliana. Plant J.* 12: 527–35.
68. Woo HR, Chung KM, Park JH, *et al.* (2001). ORE9, an F-box protein that regulates leaf senescence in *Arabidopsis. Plant Cell* 13: 1779–90.
69. Yoshida S, Ito M, Callis J, Nishida I, Watanabe A (2002). A delayed leaf senescence mutant is defective in arginyl-tRNA: protein arginyltransferase, a component of the N-end rule pathway in *Arabidopsis. Plant J.* 32: 129–37.
70. Jing HC, Sturre MJG, Hille J, Dijkwel PP (2002). Arabidopsis *onset of leaf death* mutants identify a regulatory pathway controlling leaf senescence. *Plant J.* 32: 51–63.
71. Miller AC, Schlagnhaufer C, Spalding M, Rodermel S (2000). Carbohydrate regulation of leaf development: prolongation of leaf senescence in Rubisco antisense mutants of tobacco. *Photosyn Res.* 63: 1–8.
72. Freeland RO (1952). Effect of age of leaves upon the rate of photosynthesis in some conifers. *Plant Physiol.* 27: 685–90.
73. Oren RE, Schulze ED, Matyssek R, Zimmermann R (1986). Estimating photosynthetic rate and annual carbon gain in conifers from specific leaf weight and leaf biomass. *Oecologica* 70: 187–93.
74. Reich PB, Walters MB, Ellsworth DS (1997). From tropics to tundra: global convergence in plant functioning. *Proc Natl Acad Sci USA* 94: 13730–4.
75. Greenberg JT, Ausubel FM (1993) *Arabidopsis* mutants compromised for the control of cellular damage during pathogenesis and aging. *Plant J.* 4: 327–41.

76. Mou Z, He Y, Dai Y, Liu X, Li J (2000). Deficiency in fatty acid synthase leads to premature cell death and dramatic alterations in plant morphology. *Plant Cell* 12: 405–17.
77. Hinderhofer K, Zentgraf U (2001). Identification of a transcription factor specifically expressed at the onset of leaf senescence. *Planta* 213: 469–73.
78. Melis A (1990). Dynamics of photosynthetic membrane composition and function. *Biochim Biophys Acta* 1058: 87–106.
79. Andersson B, Aro EM (2001). Photodamage and D1 protein turnover in photosystem II. *Adv Photosyn Respir.* 11: 377–93.
80. Humbeck K, Quast S, Krupinska K (1996). Functional and molecular changes in the photosynthetic apparatus during senescence of flag leaves from field-grown barley plants. *Plant Cell Environ.* 19: 337–44.
81. Krupinska K, Humbeck K (2003). Photosynthesis and chloroplast breakdown. In: Noodén LD, ed. *Cell Death in Plants.* San Diego: Academic Press Inc., in press.
82. Gan S, Amasino RM (1997). Making sense of senescence – molecular genetic regulation and manipulation of leaf senescence. *Plant Physiol.* 113: 313–19.
83. Wingler A, von Schaewen A, Leegood RC, Lea PJ, Quick WP (1998). Regulation of leaf senescence by cytokinin, sugars, and light. Effects on NADH-dependent hydroxypyruvate reductase. *Plant Physiol.* 116: 329–35.
84. Neill S, Desikan R, Hancock J (2002). Hydrogen peroxide signalling. *Curr Opin Plant Biol.* 5: 388–95.
85. Miller JD, Arteca RN, Pell EJ (1999). Senescence-associated gene expression during ozone-induced leaf senescence in Arabidopsis. *Plant Physiol.* 120: 1015–24.
86. Falk J, Krauss N, Dähnhardt D, Krupinska K (2002). A senescence associated gene of barley encoding 4-hydroxyphenylpyruvate dioxygenase is expressed during oxidative stress. *J Plant Physiol.* 159: 1245–53.
87. Kunert KJ, Ederer M (1985). Leaf aging and lipid peroxidation: the role of antioxidants vitamin C and E. *Physiol Plant* 65: 85–8.
88. Lichtenthaler HK (1966). Verbreitung und Konzentration des α-Tocopherols in Chloroplasten. *Ber Dtsch Bot Ges.* 79: 111–17.
89. Chrost B, Falk J, Kernebeck B, Mölleken H, Krupinska K (1999). Tocopherol biosynthesis in senescing chloroplasts – a mechanism to protect envelope membranes against oxidative stress and a prerequisite for lipid remobilization? In: Argyroudi-Akoyunoglou JH, Senger H, eds. *The Chloroplast: From Molecular Biology to Biotechnology.* The Netherlands: Kluwer Academic Publishers, pp. 171–6.
90. Rousseaux MC, Sanchez RA (1996). Rar-red enrichment and photosynthetically active radiation level influence leaf senescence in field-grown sunflower. *Physiol Plant* 96: 217–24.
91. Prochazkova D, Sairam RK, Srivastava GC, Singh DV (2001). Oxidative stress and antioxidant activity as the basis of senescence in maize leaves. *Plant Sci.* 161: 765–71.
92. Humbeck K, Krupinska K (2003). The abundance of minor chlorophyll a/b-binding proteins CP29 and LHCI of barley (*Hordeum vulgare* L.) during leaf senescence is controlled by light. *J Exp Bot.* 54: 375–83.
93. DeSouza PI, Egli DB, Bruening WP (1997). Water stress during seed filling and leaf senescence in soybean. *Agron J.* 89: 807–12.
94. Pic E, Teyssendier de la Serve B, Tardieu F, Turc O (2002). Leaf senescence induced by mild water deficit follows the same sequence of macroscopic, biochemical, and molecular events as monocarpic senescence in pea. *Plant Physiol.* 128: 236–46.

95. Obregón P, Martin R, Sanz A, Castresana C (2001). Activation of defence-related genes during senescence: a correlation between gene expression and cellular damage. *Plant Mol Biol.* 46: 67–77.
96. Ruperti B, Cattivelli L, Pagni S, Ramina A (2002). Ethylene-responsive genes are differentially regulated during abscission, organ senescence and wounding in peach (*Prunus persica*). *J Exp Bot.* 53: 429–37.
97. Robatzek S, Somssich IE (2002). Targets of AtWRKY6 regulation during plant senescence and pathogen defense. *Genes Dev.* 16: 1139–49.
98. Strohm M, Eiblmeier M, Langebartels C, *et al.* (1999). Responses of transgenic poplar (*Populus tremula* × *P. alba*) overexpressing glutathione synthetase or glutathione reductase to acute ozone stress: visible injury and leaf gas exchange. *J Exp Bot.* 50: 365–74.
99. Bilger W, Johnsen T, Schreiber U (2001). UV-excited chlorophyll fluorescence as a tool for the assessment of UV-protection by the epidermis of plants. *J Exp Bot.* 52: 2007–14.
100. Orendi G, Zimmermann P, Baar C, Zentgraf U (2001). Loss of stress-induced expression of catalase3 during leaf senescence in *Arabidopsis thaliana* is restricted to oxidative stress. *Plant Sci.* 161: 301–14.
101. Peters W, Ritter J, Tiller H, *et al.* (2000). Growth, ageing and death of a photoautotrophic plant cell culture. *Planta* 210: 478–87.
102. Schoch S, Vielwerth FX (1983). Chlorophyll degradation in a senescent tobacco cell culture (*Nicotiana tabacum* var. Samsun). *Z Pflanzenphysiol.* 110: 309–17.
103. Greider CW (1998). Telomeres and senescence: the history, the experiment, the future. *Curr Biol.*8: R178–81.
104. Shippen DE, McKnight TD (1998). Telomeres, telomerase and plant development. *Trends Plant Sci.* 3: 126–30.
105. Kilian A, Stiff C, Kleinhofs A (1995). Barley telomeres shorten during differentiation but grow in callus culture. *Proc Natl Acad Sci USA* 92: 9555–9.
106. Zentgraf U, Hinderhofer K, Kolb D (2000). Specific association of a small protein with the telomeric DNA-protein complex during the onset of leaf senescence in *Arabidopsis thaliana. Plant Mol Biol.* 42: 429–38.
107. Leech RM (1984). Chloroplast development in angiosperms: current knowledge and future prospects In: Baker NR, Barber J, eds. *Chloroplast Biogenesis*. Amsterdam: Elsevier Science Publishers, pp. 1–21.
108. Krupinska K, Falk J (1994). Changes in RNA-polymerase activity during biogenesis, maturation and senescence of barley chloroplasts. Comparative analysis of transcripts synthesized either in run-on assays or by transcriptionally active chromosomes. *J Plant Physiol.* 143: 298–305.
109. Zeiger E, Schwartz A (1982). Longevity of guard cell chloroplasts in falling leaves: implication for stomatal function and cellular aging. *Science* 218: 680–2.
110. Yang SH, Berberich T, Sano H, Kusano T (2001). Specific association of transcripts of *tbzF* and *tbz17*, tobacco genes encoding basic region leucine zipper-type transcriptional activators, with guard cells of senecing leaves and/or flowers. *Plant Physiol.* 127: 23–32.
111. Harman D (1956). Aging: a theory based on free radical and radiation chemistry. *J Gerontol.* 11: 298–300.
112. Noctor G, Foyer CH (1998). Ascorbate and Glutathione: keeping active oxygen under control. *Annu Rev Plant Physiol Plant Mol Biol.* 49: 249–79.
113. Meyer W, Spiteller G (1997). Oxidized phytosterols increase by ageing in photoautotrophic cell cultures of *Chenopodium rubrum*. *Phytochem.* 45: 297–302.

114. Thompson JE, Ledge RL (1987). The role of free radicals in senescence and wounding. *New Phytol.* 105: 317–44.
115. Schoettle AW (1994). Influence of tree size on shoot structure and physiology of *Pinus contorta* and *Pinus aristata*. *Tree Physiol.* 14: 1055–68.
116. Noctor G, Veljovic-Jovanovic S, Driscoll S, Novitskaya L, Foyer CH (2002). Drought and oxidative load in the leaves of C3 plants: a predominant role for photorespiration? *Ann Bot.* 89: 841–50.
117. Pastori GM, del Rio LA (1994). An activated-oxygen-mediated role for peroxisomes in the mechanism of senescence of pea leaves. *Planta* 193: 385–91.
118. del Rio LA, Pastori GM, Palma JM, *et al.* (1998). The activated oxygen role of peroxisomes in senescence. *Plant Physiol.* 116: 1195–200.
119. Corpas FJ, Barroso JB, del Rio LA (2001). Peroxisomes as a source of reactive oxygen species and nitric oxide signal molecules in plant cells. *Trends Plant Sci.* 6: 145–50.
120. Leshem YY, Wills RH, Ku VV (1998). Evidence for the function of the free radical gas – nitric oxide ($NO^{\cdot}$) – as an endogenous maturation and senescence regulating factor in higher plants. *Plant Physiol Biochem.* 36: 825–33.
121. Beligni MV, Lamattina L (2001). Nitric oxide in plants: the history is just beginning. *Plant Cell Environ.* 24: 267–78.
122. Klessig DF, Durner J, Noad R, *et al.* (2000). Nitric oxide and salicylic acid signaling in plant defense. *Proc Natl Acad Sci USA* 97: 8849–55.
123. Beligni MV, Lamattina L (1999). Nitric oxide counteracts cytotoxic processes mediated by reactive oxygen species in plant tissues. *Planta* 208: 337–44.
124. Jiménez A, Hernández JA, Pastori G, del Rio LA, Sevilla F (1998). Role of the ascorbate-glutathione cycle of mitochondria and peroxisomes in the senescence of pea leaves. *Plant Physiol.* 118: 1327–35.
125. Lam E, Pontier D, del Pozo O (1999). Die and let live – programmed cell death in plants. *Curr Opin Plant Biol.* 2: 502–7.

Aging in Sponges

Heinz C. Schröder, Matthias Wiens and Werner E.G. Müller
Institut für Physiologische Chemie, Abteilung Angewandte Molekularbiologie, Johannes Gutenberg Universität, Duesbergweg 6, D-55099 Mainz, Germany

Introduction

The sponges (phylum Porifera) are the lowest and phylogenetically oldest, still extant phylum of Metazoa [1, 2]. These animals are considered as the evolutionary earliest living relicts of the successful transition from the (perhaps) unicellular to the multicellular state [2]. Evidence has been presented by molecular phylogenetic analysis that sponges share a common ancestor with the other metazoan phyla, the hypothetical Urmetazoa [2, 3].

Investigations on the expressed sponge genome (for results of an ongoing study, see: <http://spongebase.uni-mainz.de/cgi-bin/blast/blastserver.cgi>) revealed that sponges are provided with molecules that are characteristic for metazoan phyla, e.g., G-protein linked transmembrane receptors, transmembrane adhesion receptors and extracellular adhesion molecules involved in cell-cell and cell-matrix interactions, signal transduction molecules, immune molecules and morphogenetic factors [2, 4, 5]. Most of these discoveries have been made in marine sponges which are subdivided into three classes, the Hexactinellida, the Demospongiae and the Calcarea.

The isolation, expression and functional characterization of genes typical for metazoa from sponges allowed to conclude that these animals are not merely nonorganized cell aggregates but are able to organize themselves according to a body plan. The formation of a distinct body plan in sponges is supported by the presence of skeletal elements, the spicules. In Demospongiae and Hexactinellida the spicules are composed of amorphous, noncrystalline silica. The spicule formation is assumed to be controlled by specific proteins, one of them, silicatein, being present in their axial filament [6]. At least one morphogen has been identified, myotrophin, which is likely involved in axis formation and regulation of collagen production; the collagen is cemented together with the spicules to the functional sponge skeleton [7]. In the class of Calcarea, the spicules are composed of calcium carbonate.

H.D. Osiewacz (ed.), Aging of Organisms, 79–98.

Sponges are considered as long living animals. At least for some sponge species it could be demonstrated that they can reach a life span of more than 1500 years [8].

Growth, differentiation and life span

In metazoan organisms proliferation of cells is mostly associated with differentiation. The latter process usually requires cellular interactions such as cell-cell or cell-matrix adhesion. The theory of terminal differentiation [9] and the theory of genetic instability [10] are two major theories that explain cellular differentiation and aging. More recently, it has been proposed that the shortening of telomeric DNA of chromosomes is involved in the process of aging. Analysis of telomeric DNA in cell systems that allow to study aging *in vitro* (human fibroblasts) revealed that telomere length decreases as a function of serial passages [11]. The hypothesis that the telomeres play a major role during aging is supported by the fact that the loss of telomeric DNA can only be observed in (mortal) somatic cells of higher Metazoa [12] but not in (immortal) reproductive cells and tumor cells which are characterized by an unlimited replicative potential. The potential role of telomeres in sponges can be studied in the primmorph system.

Primmorphs

Primmorphs are a special form of aggregates formed by sponge single cells in which the cells retain the ability to synthesize DNA and to undergo cell division (Figure 1). This cell/tissue culture method has been successfully established primarily with demosponges, including *Suberites domuncula*, *Dysidea avara* and *Geodia cydonium* [13–15]. After incubation of sponge single cells, e.g., from *S. domuncula*, obtained by dissociation of sponge tissue in Ca^{2+}- and Mg^{2+}-free artificial seawater containing EDTA, in Ca^{2+}- and Mg^{2+}-containing seawater-medium for 5 days, round-shaped primmorphs, 1 to 5 mm in size are formed. The cells within the primmorphs are able to proliferate, as demonstrateded by incorporation studies with 5-bromo-2′-deoxy-uridine (BrdU; Figure 1). This is in contrast to sponge cells kept in the single cell state after dissociation in Ca^{2+} and Mg^{2+}-free artificial seawater; these cells lose their capacity to incorporate BrdU and undergo apoptosis [14, 16]. In contrast, the percentage of apoptotic cells in primmorphs is comparably low (Figure 1).

The primmorphs show a characteristic histology. They are surrounded by an almost complete single cell layer of epithelium composed of pinacocytes [14]. The cells inside the primmorphs are primarily spherulous cells with some few other cells, mainly amoebocytes and archaeocytes.

After aggregation of sponge single cells (primmorph formation) and during growth of primmorphs a differential gene expression is observed. This process is most likely involved in cell differentiation. For example, a homeobox containing gene, *Iroquois*, is expressed in the primmorph system; the expression of this gene coincides with the onset of canal formation in primmorphs [17]. Also one gene encoding a growth promoting protein, the myotrophin-like polypeptide, has been identified in *S. domuncula*; this protein stimulates cell proliferation and induces collagen expression in primmorphs [7].

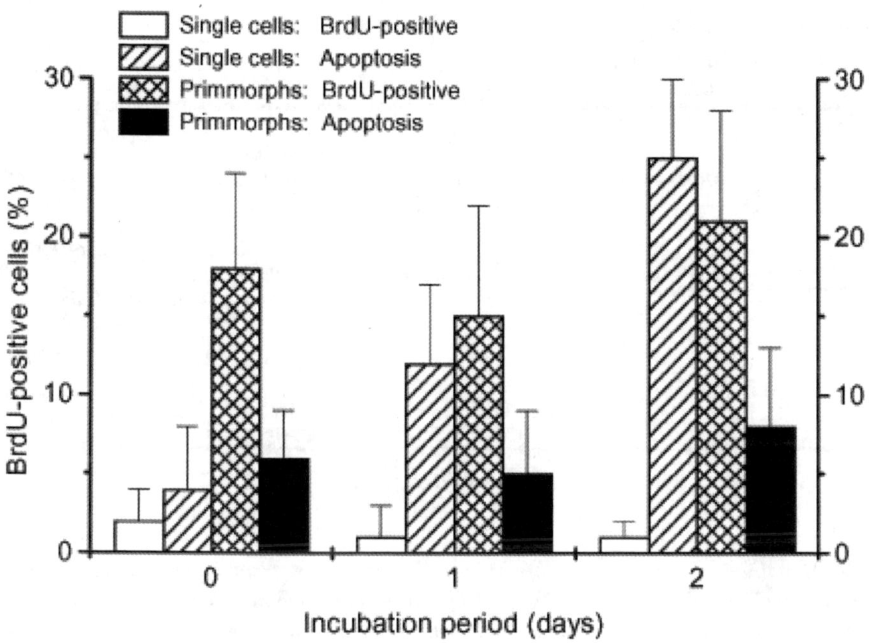

Figure 1. BrdU-positive cells and apoptotic cells in single cell suspensions and in primmorphs of S. domuncula during an incubation period for 0–2 days. The cells remained after dissociation either in Ca^{2+}*- and* Mg^{2+}*-free artificial seawater (single cells) or were transferred to* Ca^{2+}*- and* Mg^{2+}*-containing seawater-medium (formation of primmorphs). Mean values* $\pm$ *SD.*

It should be emphasized that proliferation of sponge cells and increased gene expression, as outlined above, only occur if cell-cell contact is allowed. The effect of cell-cell contact and subsequent events during primmorph formation has also been demonstrated for the expression of other genes, including those for integrins, Ras and calmodulin [18].

Replicative life span

The results mentioned above suggest that the replicative life span of cells from sponges is somewhat different to the senescence of mammalian cells [19] or yeast cells [20]. In the latter systems the number of cell divisions depends on the number of daughter cells that are produced. Sponge cells (within primmorphs or sponge tissue) can proliferate indefinitely, due to their stem cell-like, pluri/totipotent character especially of the archeocytes [21]. This has already been suggested [22] and experimentally supported later [14, 23]. The state of sponge cells ("immortal" cells showing (almost) unlimited proliferation and "mortal" cells which can be triggered to apoptosis) can be changed by intrinsic and extrinsic factors (Figure 2) which can be imitated by defined experimental conditions (see above).

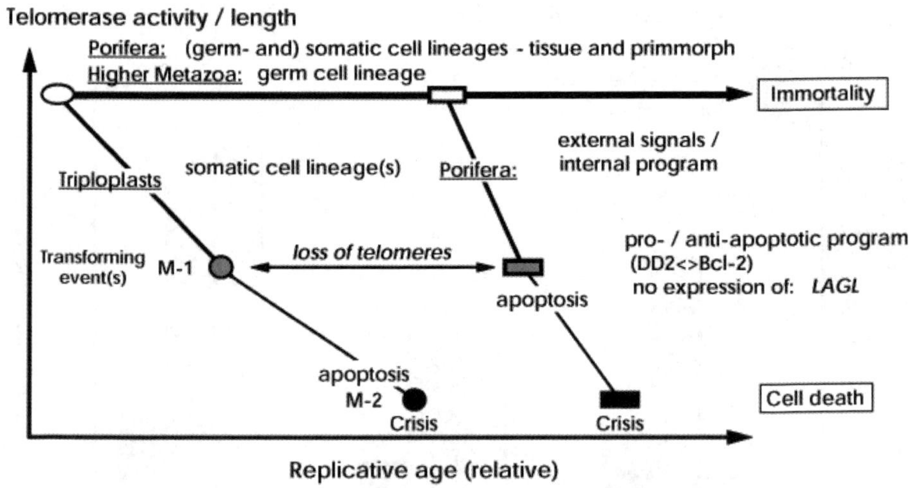

Figure 2. Hypothetical determinants of immortality in species from higher Metazoa and in sponges. The lack of telomerase activity, and in consequence telomere loss, in somatic cells of higher metazoans determines their fate to senescence (circles) via the two phases: the "Mortality Phase 1" (M-1) – cell cycle arrest – and after transformation "Mortality Phase 2" (M-2). Cells of the germ lineage from higher Metazoa remain telomerase-positive and are immortal. In sponges (squares), it is proposed that the switch from immortal "somatic" cells, present in tissue and tissue like assemblies (primmorphs) to mortal cells (in the single cell stage) is triggered both by external and internal programs. The mortal cells are eliminated by the process of apoptosis which is controlled by both pro- and anti-apoptotic programs. During the process of apoptosis in sponge cells, the expression of the gene SDLAGL, encoding the putative longevity assurance-like polypeptide, is downregulated [Adapted from Ref. 23 and 77].

Gerontogenes and proteins

A series of "gerontogenes" [24] has been identified which may affect the rate of aging. Among these candidate genes are those of the p53 gene family, the pRB and INK family, members coding for the Ras/Rad/MEK/MAP kinase pathway, telomerase or mortalin [25] and the longevity gene *LAG* [26]. It can be accepted that with the split of diploid cells in Metazoa in germ cells, introducing the vertical genetic axis, and in soma cells, the proliferative ability of the latter ones had to be limited. Since in sponges the differentiation to germ and soma cells is plastic and is apparently primarily determined by extracellular factors [23], they can be considered as model system for the identification of potential gerontogenes which control the finite life span of their cells. Matrix and receptor molecules could control mortality/immortality of sponge cells.

Telomerase

The telomerase is the enzyme responsible for the maintenance of the telomeres. In protostomians and deuterostomians, only immortal cells, present in the reproductive lineage [27], or cancer cells [28], express high levels of telomerase activity [29]. In contrast to those immortal cells, the somatic, mortal cells of these metazoans lack telomerase activity resulting in a shortening of telomeres with each cell division. In consequence, telomerase-negative cells show a limited number of cell divisions of 50 to 100 times, the Hayflick limit [19]. If the cells reach the Hayflick limit, which is also called mortality phase 1 (M-1) [12], the critical telomere loss on chromosomes initiates a signaling event which results in an irreversible cell cycle arrest [11]. In some instances spontaneous immortalization, due to transforming events or viral oncoproteins, allow metazoan somatic cells to bypass this mortality phase 1 without activating telomerase and to reach mortality phase 2 ("crisis"; M-2) (Figure 2). In M-2 cells often undergo apoptosis or necrosis as a consequence of high genomic instability. Rare clones of cells can activate again telomerase and acquire "secondary" indefinite growth capacity [29]. Extension of cell life span can be achieved after transformation with viral oncoproteins, agents that typically do not fully immortalize the cells [27]. Without transformation by viral oncoproteins or acquisition of other mutational events the transition from the immortal state to the mortal state is very rare.

Surprisingly, a high level of telomerase activity is present in the somatic cells of tissue from demosponges, e.g., *G. cydonium* and *S. domuncula* [23]. Therefore it can concluded that the somatic cells of sponges have an unlimited potency of proliferation and differentiation [23]. This feature is unique in the metazoan kingdom. In higher Metazoa the telomerase can only be detected in cells of the germ line and in tumor cells; in somatic cells only a very low or no activity of this enzyme is found. Therefore, these cells have only a limited potential for proliferation.

As outlined above, telomerase-positive cells in sponge tissue can be triggered to telomerase-negative cells by dissociating the tissue into single cells. Sponge single cells, e.g., from *S. domuncula*, lack the ability to proliferate and undergo apoptosis, in contrast to cells present in primmorphs.

Longevity assurance-like polypeptide

The technique of differential display of mRNA has been used to identify mRNA species which are present in primmorphs from the sponge *S. domuncula* [16], that are composed of proliferating, telomerase-positive cells [14] but absent in non-proliferating, telomerase-negative single cells. Cloning of one of these mRNAs by means of polymerase chain reaction (PCR) revealed a sequence with a high similarity to the longevity assurance genes *LAG* from yeast, mouse, human and a tunicate [16]. The complete cDNA, *SDLAGL*, encoding the putative sponge longevity assurance-like protein LAGL_SUBDO, has a potential open reading frame of 330 amino acids (aa) (Figure 3A). The deduced polypeptide has a calculated M_r of 37 987 and an isoelectric point (pI) of 8.95. In Northern blot analysis one prominent band of approximately 1.4 kb is seen using a *SDLAGL* probe (Figure 3B,C). The Lag1p motif, which is characteristic for the longevity assurance-like polypeptides [26], is present in

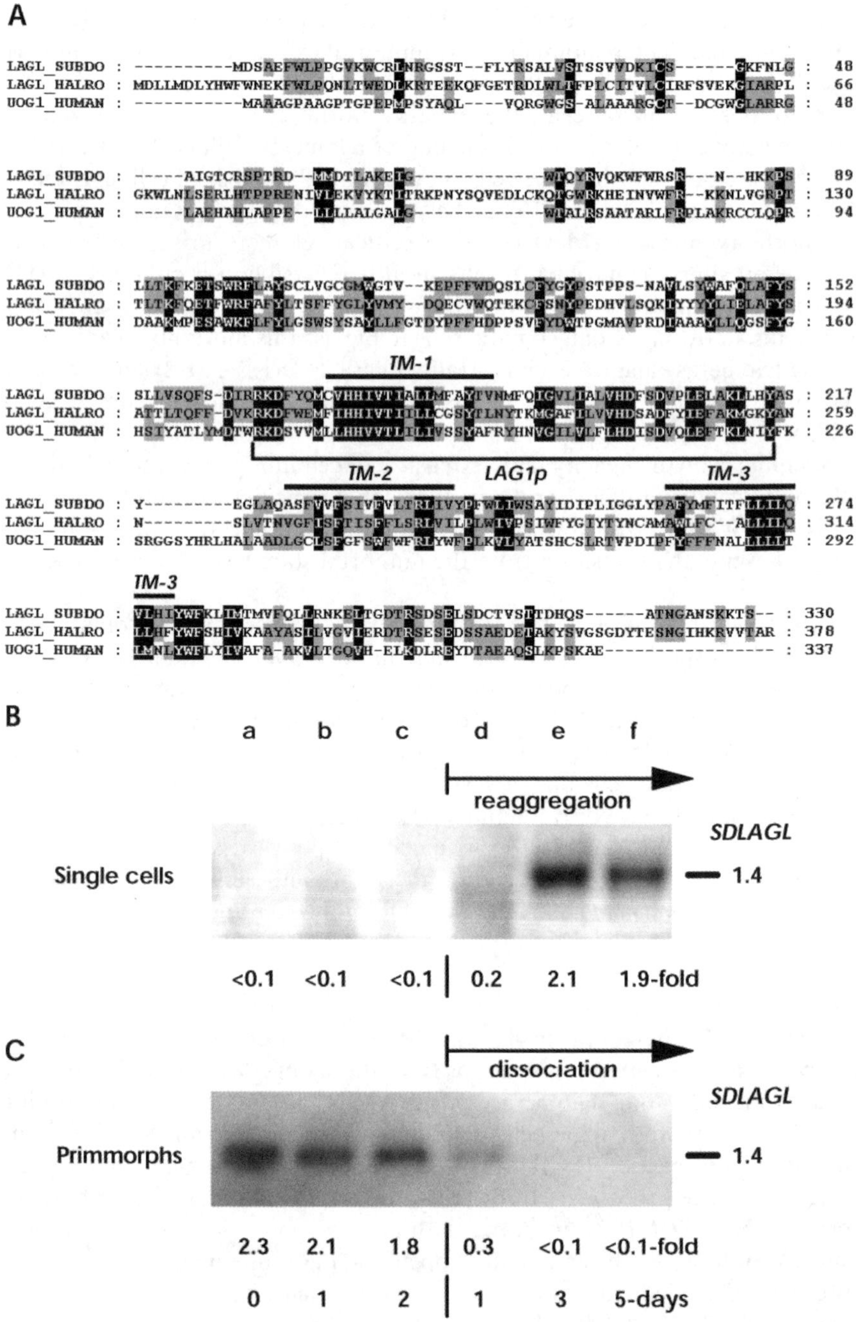
A
LAGL_SUBDO : ----------MDSAEFWLPPGVKWCRLNRGSST--FLYRSALVSTSSVVDKICS------GKFNLG : 48
LAGL_HALRO : MDLLMDLYHWFWNEKFWLPQNLTWEDLKRTEEKQFGETRDLWLTFPLCITVLCIRFSVEKGIARPL : 66
UOG1_HUMAN : -----------MAAAGPAAGPTGPEPMPSYAQL----VQRGWGS-ALAAARGCT--DCGWGLARRG : 48
LAGL_SUBDO : -----AIGTCRSPTRD--MMDTLAKEIG------------WTQYRVQKWFWRSR---N--HKKPS : 89
LAGL_HALRO : GKWLNLSERLHTPPRENIVLEKVYKTITRKPNYSQVEDLCKQTGWRKHEINVWFR--KKNLVGRPT : 130
UOG1_HUMAN : -----LAEHAHLAPPE--LLLLALGALG------------WTALRSAATARLFRPLAKRCCLQPR : 94
LAGL_SUBDO : LLTKFKETSWRFLAYSCLVGCGMWGTV--KEPFFWDQSLCFYGYPSTPPS-NAVLSYWAFQLAFYS : 152
LAGL_HALRO : TLTKFQETFWRFAFYLTSFFYGLYVMY--DQECVWQTEKCFSNYPEDHVLSQKIYYYYLIELAFYS : 194
UOG1_HUMAN : DAAKMPESAWKFLFYLGSWSYSAYLLFGTDYPFFHDPPSVFYDWTPGMAVPRDIAAAYLLQGSFYG : 160
TM-1
LAGL_SUBDO : SLLVSQFS-DIRRKDFYQMCVHHIVTIALLMFAYTVNMFQIGVLIALVHDFSDVPLELAKILHYAS : 217
LAGL_HALRO : ATTLTQFF-DVKRKDFWEMFIHHIVTIILLCGSYTLNYTKMGAFILVVHDSADFYIEFAKMGKYAN : 259
UOG1_HUMAN : HSIYATLYMDTWRKDSVVMLLHHVVTLILIVSSYAFRYHNVGILVLFLHDISDVQLEFTKLNIYFK : 226
LAG1p
TM-2
TM-3
LAGL_SUBDO : Y---------EGLAQASFVVFSIVFVLTRLIVYPFWLIWSAYIDIPLIGGLYPAFYMFITFLLILQ : 274
LAGL_HALRO : N---------SLVTNVGFISFTISFFLSRLVILPLWIVPSIWFYGIYTYNCAMAWLFC--ALLILQ : 314
UOG1_HUMAN : SRGGSYHRLHALAADLGCLSFGFSWFWFRLYWFPLKVLYATSHCSLRTVPDIPFYFFFNALLLLLT : 292
TM-3
LAGL_SUBDO : VLHIYWFKLIMTMVFQLLRNKELTGDTRSDSELSDCTVSTTDHQS------ATNGANSKKTS-- : 330
LAGL_HALRO : LLHFYWFSHIVKAAYASILVGVIERDTRSESEDSSAEDETAKYSVGSGDYTESNGIHKRVVTAR : 378
UOG1_HUMAN : LMNLYWFLYIVAFA-AKVLTGQVH-ELKDLREYDTAEAQSLKPSKAE---------------- : 337
B
a
b
c
d
e
f
reaggregation
SDLAGL
Single cells
1.4
<0.1
<0.1
<0.1
0.2
2.1
1.9-fold
C
dissociation
SDLAGL
Primmorphs
1.4
2.3
2.1
1.8
0.3
<0.1
<0.1-fold
0
1
2
1
3
5-days

the sponge polypeptide. Like the *Saccharomyces cerevisiae* deduced polypeptide [30] the sponge LAGL_SUBDO sequence comprises one cAMP- and cGMP-dependent protein kinase phosphorylation site, five protein kinase C phosphorylation sites, five casein kinase II phosphorylation sites and six potential *N*-myristoylation sites.

A databank search with the deduced *S. domuncula* aa sequence, LAGL_SUBDO, revealed an identity (similarity) to the putative longevity assurance polypeptides; highest are the values to the related sequences from mouse, human, tunicate (*Halocynthia roretzi*) and nematode (*Caenorhabditis elegans*) with 34 to 40% of physico-chemicial similar aa. More distantly related are the sequences from yeast [30] and *Schizosaccharomyces pombe* [31]; the latter one was chosen as an outgroup for the rooted phylogenetic tree to demonstrate the distinction between the metazoan, including sponge, and fungal sequences (Figure 4).

The calculation of the hydropathy plot for the sponge longevity assurance-like polypeptide revealed three potential transmembrane regions [16]. A comparison of these regions with the corresponding sequences present in human, *S. cerevisiae* and *C. elegans* [26, 30] shows a high similarity. All these molecules do not contain a signal peptide. Therefore, it can be suggested that the longevity assurance polypeptides are membrane-associated.

The fungal longevity assurance polypeptide is assumed to play a role in determining yeast longevity. In age-synchronized yeast cells, expression of the *LAG* gene was found to correlate with the age of the cells and is not seen in "senescent" (18-generation-old) cells [30]. A similar function has been proposed for the human protein [26]. Complementation studies revealed that the human gene can replace the yeast gene [26]. The *LAG* genes are expressed early in the life spans of the cells. Their expression is blocked after a higher number of cell divisions. Consequently, *LAG* may set a threshold for the number of possible cell divisions [30].

Figure 3. *(opposite) Longevity assurance-like polypeptide from S. domuncula. (A) Alignment of the deduced aa sequence from the S. domuncula cDNA, LAGL_SUBDO, with the related sequences from Halocynthia roretzi (LAGL_HALRO) and from human (UOG1_HUMAN). Residues conserved (similar or related with respect to their similar physico-chemical properties) in all sequences are shown in white on black and those in at least two sequences in black on gray. The potential transmembrane segments (TM-1 to TM-3) and the motif for longevity assurance-like polypeptides (LAG1p) are marked. (B) and (C) Expression of the sponge SDLAGL gene, encoding the longevity assurance-like polypeptide, determined by Northern blotting. (B) Single cells were obtained by dissociation in Ca^{2+}- and Mg^{2+}-free artificial seawater containing EDTA and then kept for 3 days in Ca^{2+}- and Mg^{2+}-free artificial seawater prior to the two days incubation period where samples were taken (lane a–c). In a separate series of experiments the single cells (after the 3 days pre-incubation period) were transferred in Ca^{2+}- and Mg^{2+}-containing seawater and allowed to re-aggregate for up to five days (lane d–f). (C) Primmorphs were analyzed for 0 to 2 days (lane a–c); separately 5 days old primmorphs were dissociated in Ca^{2+}- and Mg^{2+}-free artificial seawater containing EDTA; the resulting single cells were analyzed after a 1 to 5 days incubation period in Ca^{2+}- and Mg^{2+}-free artificial seawater (lane d–f). The intensities of the transcripts for SDLAGL are correlated with the expression of β-tubulin [Adapted from ref. 16].*

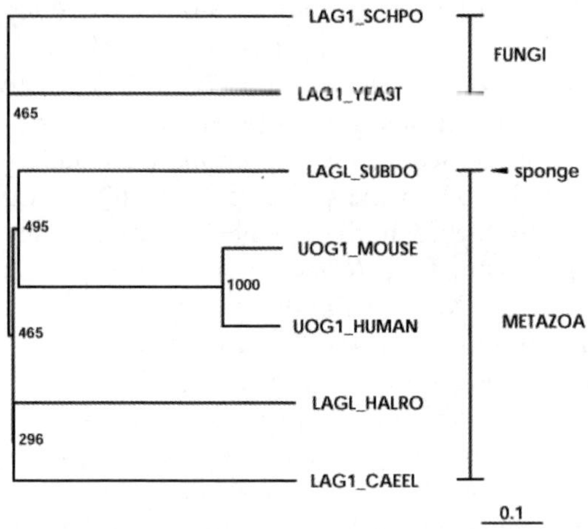

Figure 4*. Rooted phylogenetic tree, constructed after alignment of the sequences shown in Figure 3 together with the related sequences from mouse (UOG1_MOUSE), from C. elegans (LAG1_CAEEL) and from yeast (LAG1_YEAST). The longevity assurance polypeptide from S. pombe (LAG1_SCHPO) was used as outgroup. The numbers at the nodes are an indication of the level of confidence – given in percentage – for the branches as determined by bootstrap analysis (1000 bootstrap replicates). The scale bar indicates an evolutionary distance of 0.1 aa substitutions per position in the sequence.*

Expression studies revealed that the transcription of the *SDLAGL* gene is highly upregulated in proliferating cells present in *S. domuncula* primmorphs [16, 32]. In single cells, which have been dissociated from tissue or from primmorphs (Figure 3B; lane a–c), no transcripts from *SDLAGL* can be identified. However, if the single cells are transferred into Ca^{2+}-containing seawater-medium, which allows re-aggregation, the *SDLAGL* gene is expressed (Figure 3B; lane d–f). The high level of *SDLAGL* transcripts in primmorphs (Figure 3C; lane a–c) drops after dissociation into single cells, by transfer into Ca^{2+}-free and EDTA-containing seawater (Figure 3C; lane d–f).

Sponges can form gemmules which are asexual reproduction bodies composed of cells with a high but transiently dormant proliferation potency [33]. The conclusion that only sponge cells present in the cell-cell interacting state but not sponge single cells express *SDLAGL* is supported by the finding that the dormant cells from sponge gemmules also show high levels of *SDLAGL* [32].

Apoptotic genes and proteins

In the last years it became obvious that apoptosis is not restricted to those Metazoa which have separate germ and somatic cells but evolved already during the transition from the common ancestor of all metazoan phyla to the phylogenetically oldest metazoan taxon, the sponges [3].

All sponge species have a characteristic body plan ("Bauplan" [33]) with a defined final size and shape in spite of the fact that (almost all) cells are telomerase-positive and presumably provided with an unlimited potency for cell proliferation. Consequently, an apoptotic mechanism has to be predicted that allows sponges to eliminate unwanted cells which accumulate during development. The existence of such a mechanism is a prerequisite for the survival of these animals since otherwise, once the development is completed, a controlled maintenance and renewal of the diverse sets of cell types cannot be guaranteed. It has been demonstrated that marine sponges, e.g., *G. cydonium* and *S. domuncula,* undergo apoptosis during formation of gemmules [34] and in response to environmental stress [35].

Like in other metazoans [36] apoptosis in sponges can be induced by intrinsic and extrinsic signals, which act in a positive (pro-apoptotic) or negative (anti-apoptotic) manner. For example, (mortal) sponge cells can be triggered to apoptosis if they (i) lose cell-cell- and/or cell-matrix contact [23], (ii) become exposed to cadmium or bacteria [34], (iii) are treated with estrogens [37], or (iv) are exposed to xenobiotics, e.g., tributyltin and methyl mercury [35]. Besides this exogenously induced apoptosis, it must be postulated that the immortal sponge cells are also provided with the ability to induce apoptosis by an internal program to maintain the specific Bauplan. Genes and proteins involved in apoptosis have been isolated from sponges, including a potential pro-apoptotic molecule, DD2, which comprises two death domains, the potential anti-apoptotic Bcl-2 homologous proteins and one caspase.

Proapoptotic DD2 molecule

Transplantation studies revealed that a tumor necrosis factor (TNF)-like molecule is induced in sponge allo- and xenografts [38]. In vertebrate cells, TNF binds to receptors belonging to the TNF receptor superfamily [39]. These receptors are characterized by the presence of a death domain [40–42], e.g., the TNFR-associated death domain, the Fas-associated death domain (FADD) also termed MORT1, or the receptor-interacting protein [43] (RIP). Until recently neither a TNFR molecule nor one of its associated molecules have been cloned from invertebrates. A screening for a cDNA encoding such a domain in sponges was successful and resulted in the isolation of a potential proapoptotic molecule, called DD2 [44].

The complete cDNA from *G. cydonium,* encoding the potential proapoptotic molecule DD2 was isolated, using a degenerate primer and applying the PCR technique [44]. Interestingly, the sponge DD2 contains not only one but two death domains, termed DD2 domain I and II [44], in contrast to any other known molecules involved in the FasL/TNF death signaling pathway [42]. Figure 5A shows an alignment of the two death domains of the sponge DD2 molecule with the mammalian domains including the human death domain-containing protein FADD/MORT1 [41, 45, 46] and the human Fas/APO-1 [47]. In this alignment an ankyrin segment [45] that displays a similarity to the death domains [48], was included. Ankyrin domains are known to link spectrin-based membrane skeletal proteins to the cytoplasmic domains of several integral membrane proteins. The death domain from Fas/APO-1 and FADD/MORT1 shared with domain I and domain II from the sponge DD2 molecule (Figure 5A).

The phylogenetic analysis revealed that the sponge domain originated from an ankyrin building block from which the death domain of mammalian Fas-FADD/MORT1 evolved [44, 48] (Figure 5B). Both sponge domains show closer relationship to the death domain in Fas/APO-1. The sponge death domain I forms the basis of the branch that includes the sponge death domain II and the death domains known from Fas/APO-1. Our results suggest that the two sponge death domains were created during evolution by gene duplication [48] (Figure 5C). It should be noted that a death domain-containing molecule has not been identified in the *C. elegans* genome.

Sponges are able to discriminate between self and non-self as shown in autograft fusion and allograft rejection experiments [49]. Determinations of the level of expression of sponge DD2 molecule in autografts (showing low percentage of apoptosis-positive cells) and allografts (high percentage of apoptosis-positive cells) from *G. cydonium* revealed that the level of DD2 is strongly upregulated in allografts [44].

Bcl-2-related molecules

The Bcl-2 protein and related cytoplasmic proteins are major molecules known to regulate cell survival [50]. The members of the Bcl-2 family comprise both pro-survival molecules, e.g., Bcl-2 [51] and pro-apoptotic proteins, e.g., Bax [52]. They are able to interact with each other under formation of homo- and heterodimers [53].

Two Bcl-2-related molecules which were termed Bcl-2 homologous proteins (BHP) were identified from sponges [48, 54] (Figure 6). The *G. cydonium* and *S. domuncula* cDNAs are the first ones encoding putative Bcl-2 homologous proteins that had been identified from an animal (sponges) evolutionary older than the triploblastic Metazoa [44, 48, 54].

In vertebrates, the Bcl-2 proteins contain up to four Bcl-2 homology regions (BH) [52, 55]. The sponge molecules are provided with two of the Bcl-2 homology regions, BH1 and BH2 [48]. The analysis of the gene structure of the *G. cydonium* Bcl-2 homologous protein BHP1 revealed one single intron which is located within the BH2 domain at the same position as in Bcl-xL from rat [56] and Ced-9 from *C. elegans* [57] (Figure 7). In addition, a *C*-terminal transmembrane segment is present in the sponge molecule [48].

The alignment of the BH1 domain of the *G. cydonium* and *S. domuncula* BHP1 molecules and the BH1 domain of representative members of the Bcl-2 superfamily showed that the sponge molecules share almost all aa that are characteristic for BH1 domains [58] with the members of the superfamily, as well as with mouse Mtd, frog NR-13, and human A1-BFL-1, MCL-1 and Bak (Figure 8A). In contrast, the similarity of the sponge BH1 domains to the BH1 domain of the mouse Bad, human Hrk, and Ced-9 from *C. elegans* is less. The unrooted phylogenetic tree revealed that the sponge sequences are grouped among the mammalian BH1 domains but are only distantly related to the *C. elegans* BH1 (Figure 8B). Similar results were obtained for the BH2 domains [48]. The Bcl-2-related molecules, as well as the proapoptotic molecule DD2, have no obvious homologues in the *S. cerevisiae* genome, even though apoptotic mechanisms exist in these multicellular organisms [59].

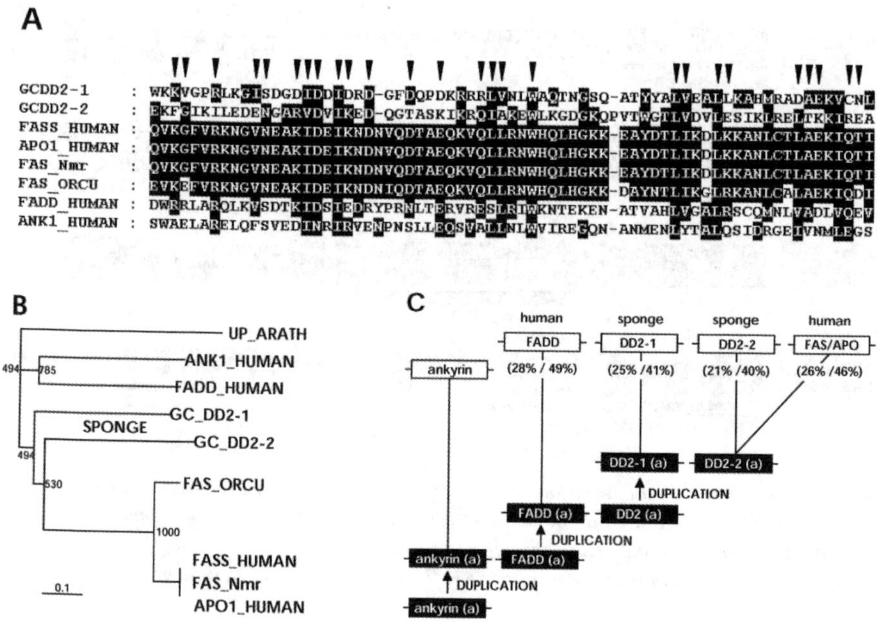

Figure 5. *Deduced polypeptide sequence containing two death domains from G. cydonium (GC DD2 domains I and II). (A) Alignment of the two potential sponge death domains (GC DD2-1 and GC DD2-2) with the following mammalian death domains: the death domain-like sequence, present in human ankyrin (ANK1 HUMAN), human death domain-containing protein FADD/MORT1 (FADD HUMAN), Fas antigen from Oryctolagus cuniculus (FAS ORCU), human Fas soluble protein (FASS HUMAN), human Fas death domain, NMR analysis (FAS Nmr), and human apoptosis antigen 1 (APO1 HUMAN). Those aa that share similarity with the Fas/APO sequences are shown in white letters on black. Those aa that are characteristic for the death domains are marked (V). (B) Rooted phylogenetic tree, constructed after alignment of the death domains. The unknown protein from Arabidopsis thaliana (UP ARATH) was used as out-group. (C) Proposed scheme for the evolution of the sponge and mammalian death domains by gene duplication. The putative ancestral domains are labeled (a). The numbers in parentheses refer to the percentages of aa that are identical or similar, respectively, by comparison with the (next) related domains listed.*

Functional studies revealed that the level of expression of the sponge (*G. cydonium*) Bcl-2-related gene (*BHP2*) drastically increased following exposure to the stressors, tributyltin (TBT) and heat shock, while the level of septin (used as a reference gene to determine cell viability and proliferation) remained unchanged [54]. The sponge *BHP2* gene is even able to confer resistance against stress to human cells. It was found that human HEK-293 cells, which had been stably transfected with the sponge *BHP2*, are highly resistant against stress by serum starvation and TBT [54].

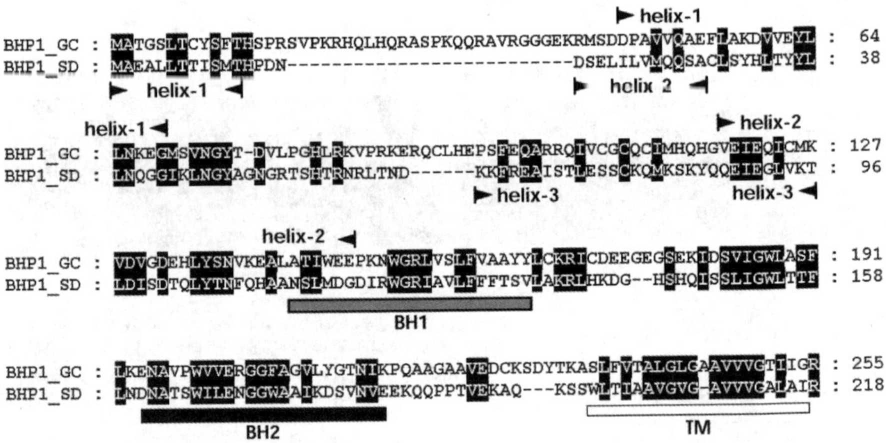

Figure 6. *Aligment of the deduced aa sequences of the sponge Bcl-2 homology proteins from G. cydonium (BHP1 GC) and S. domuncula (BHP1 SD). Similar aa residues are shown in white letters on a black background. The characteristic BH1 and BH2 domains, the putative transmembrane segments (TM) and regions of putative helical conformation are indicated.*

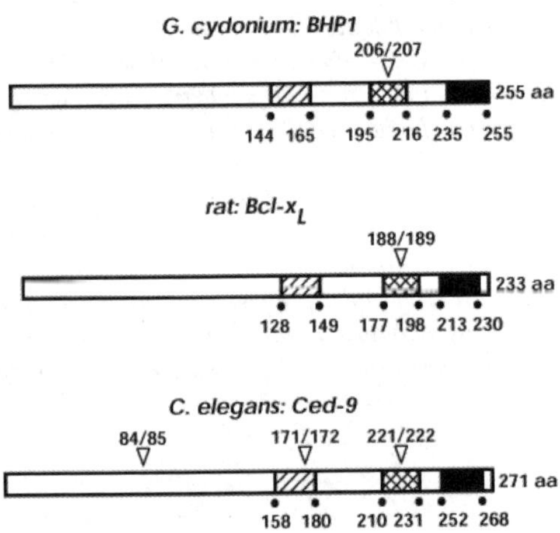

Figure 7. *Gene structure in the sponge Bcl-2 BHP1 from G. cydonium, in rat Bcl-xL, and in the apoptosis regulator Ced-9 from C. elegans. The locations of the introns (shaded) and the positions of BH1, BH2, and the transmembrane regions (black) are marked.*

Caspases

The caspases are involved in the execution of the death signals [60]. In vertebrates [61] this process is more complex than in *C. elegans*, where the apoptotic process is executed by only one enzyme, the CED-3 [57].

The caspases can be divided into two classes [62]. Class I caspases possess a long *N*-terminal prodomain and associate via protein:protein interactions with their prodomains to the proteins/receptors that comprise the death domain [63], while class II caspases have only a short or no prodomain and are involved in downstream functions of the death cascade and in cleavage of the target proteins [61].

Two cDNAs encoding two forms of caspase-3-like proteins were isolated from *G. cydonium* by PCR cloning (Wiens *et al.* submitted). These cDNAs (a long form and a short form) are identical at the open reading frame but differ in the *N*-terminal part of the deduced polypeptides. They are distinguished by the presence of a caspase recruitment domain (CARD). The CARD domains are known to interact with the death effector domain [64]. Such a domain, which is characteristic of class I caspases [62], is found only in the long form but not in the short form of the sponge enzyme. In consequence, the long form of sponge caspase-3 can be grouped to the class I caspases, and the short form to the class II caspases (Wiens *et al.* submitted).

Other proteins

A further family of proteins which are involved in apoptosis, but also in other cellular processes, are the 14-3-3 proteins [65]. The 14-3-3 protein family comprises seven isoforms [65]. A *14-3-3* cDNA was cloned from a *G. cydonium* cDNA library [66]. The deduced sponge polypeptide is closely related to the η and γ isoforms of 14-3-3 protein. Another gene/cDNA which is activated during apoptosis, the *MA-3* gene, has been cloned from *S. domuncula* [34].

Mediators

In mammalian cell systems, TNF causes an activation of arachidonic acid metabolism, resulting in the production of prostaglandins, thromboxanes and leukotrienes [67]. Leukotriene B_4 (LTB_4) is a chemotactic and proinflammatory factor [68, 69] which stimulates the production of reactive oxygen species [70]. This factor could be identified in sponge tissue [44]. The level of LTB_4 drastically increases in sponge allografts, paralleling the increase in expression of the proapoptotic molecule DD2; this increase was not found in autografts [44].

LTB_4 can be converted to less active metabolites, either by cytochrome P-450 LTB_4 ω-oxidation to 20-hydroxy/carboxy-LTB_4 or by the LTB_4 12-hydroxy-dehydrogenase to 12-oxo-LTB_4. The latter enzyme was cloned from *G. cydonium* [44]. Besides the increase in LTB_4 level, an increase in the level of expression of LTB_4 12-hydroxy-dehydrogenase was found in sponge allografts, suggesting that LTB_4 is one metabolite which is involved in the initiation of apoptosis in sponges [44].

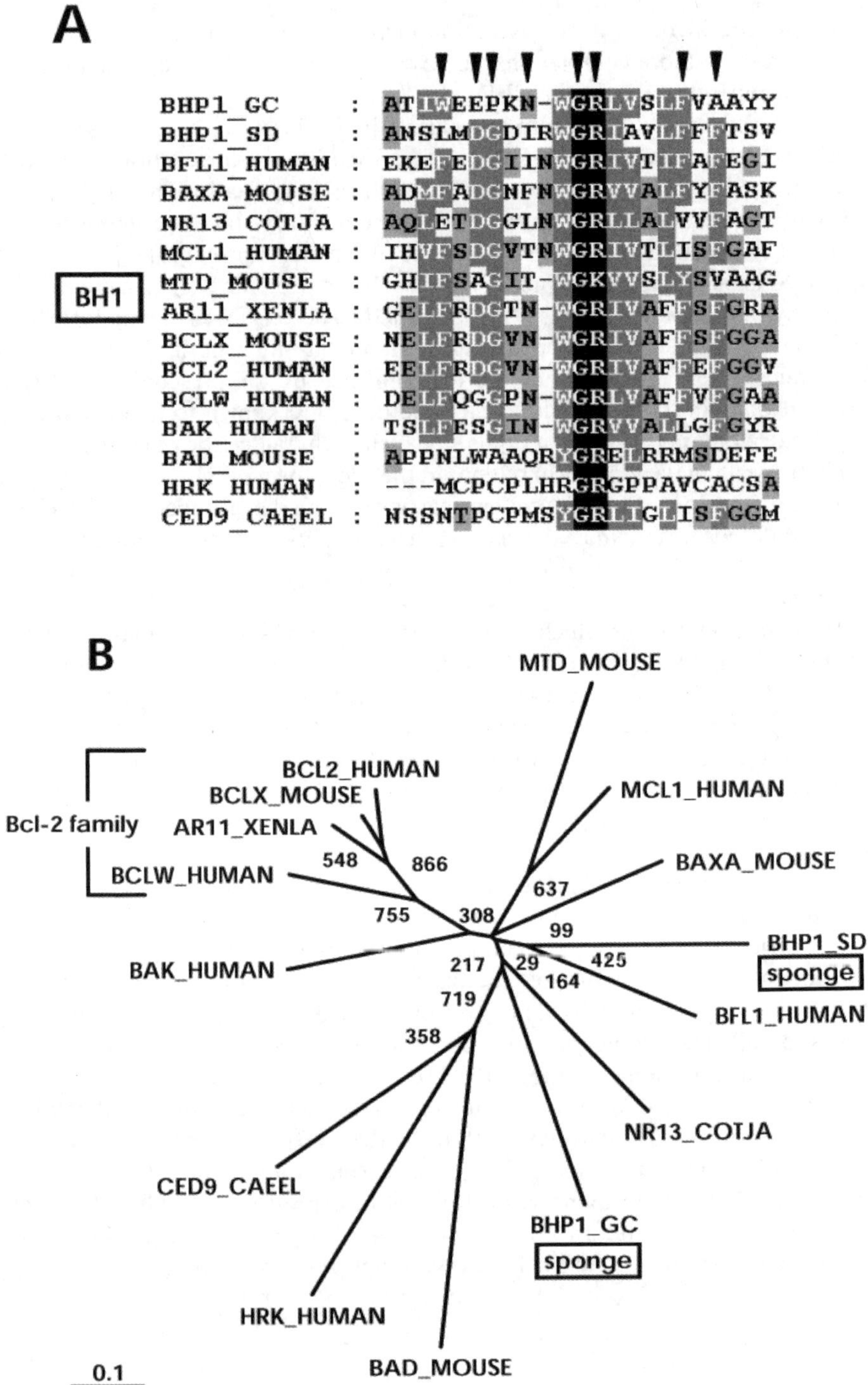
A
BH1
BHP1_GC : ATIWEEPKN-WGRLVSLFVAAYY
BHP1_SD : ANSLMDGDIRWGRIAVLFFFTSV
BFL1_HUMAN : EKEFEDGIINWGRIVTIFAFEGI
BAXA_MOUSE : ADMFADGNFNWGRVVALFYFASK
NR13_COTJA : AQLETDGGLNWGRLLALVVFAGT
MCL1_HUMAN : IHVFSDGVTNWGRIVTLISFGAF
MTD_MOUSE : GHIFSAGIT-WGKVVSLYSVAAG
AR11_XENLA : GELFRDGTN-WGRIVAFFSFGRA
BCLX_MOUSE : NELFRDGVN-WGRIVAFFSFGGA
BCL2_HUMAN : EELFRDGVN-WGRIVAFFEFGGV
BCLW_HUMAN : DELFQGGPN-WGRLVAFFVFGAA
BAK_HUMAN : TSLFESGIN-WGRVVALLGFGYR
BAD_MOUSE : APPNLWAAQRYGRELRRMSDEFE
HRK_HUMAN : ---MCPCPLHRGRGPPAVCACSA
CED9_CAEEL : NSSNTPCPMSYGRLIGLISFGGM
B
MTD_MOUSE
MCL1_HUMAN
BAXA_MOUSE
BHP1_SD
sponge
BFL1_HUMAN
NR13_COTJA
BHP1_GC
sponge
BAD_MOUSE
HRK_HUMAN
CED9_CAEEL
BAK_HUMAN
BCLW_HUMAN
AR11_XENLA
BCLX_MOUSE
BCL2_HUMAN
Bcl-2 family
548
866
755
308
637
99
425
29
164
217
719
358
0.1

Elimination of apoptotic cells
In higher Metazoa apoptotic cells can be eliminated by phagocytosis, especially by macrophages. In sponges no macrophage-related cells have been detected. On the other hand, the marine environment is rich in bacteria and predators, which could settle on apoptotic tissue and could cause fouling.

Recent results revealed that during apoptosis a cytotoxic compound is synthesized in *S. domuncula* tissue, allowing the controlled removal of apoptotic tissue areas by a mollusk [71]. Extracts from apoptotic tissue from this sponge display a strong neurotoxicity [71]. Since this effect could be neutralized by MK-801, a non-competitive antagonist of the NMDA receptor, it was concluded that an agonist of the NMDA receptor is present in the extract from apoptotic sponge tissue. It was identified as quinolinic acid [71]. This excitotoxin exerts its toxicity via activation of the ionotropic NMDA receptor [72]. Also the key enzyme of the quinolinic acid pathway, the 3-hydroxyanthranilate 3,4-dioxygenase, which causes the conversion of 3-hydroxyanthranilic acid to quinolinic acid, was identified in sponges [71]. It was found that the expression of the gene encoding this enzyme is upregulated in apoptotic tissue from *S. domuncula* [71].

These results show that sponges possess a mechanism that allows the selective and controlled removal of apoptotic tissue but does not influence the non-affected, non-apoptotic tissue. Quinolinic acid acts differentially on eukaryotic epibionts; as an attractant to the mollusk which is obviously not influenced by quinolinic acid and as a neurotoxin on other epibionts.

Conclusion

Sponges express a longevity assurance gene which has been proposed to be involved in the shift of the immortal telomerase-positive cells to the telomerase-negative mortal sponge cells. These lowest multicellular animals are already provided with apoptotic pathways, similar to those present in deuterostomes and apparently absent in protostomes. This was shown by the identification of the potential proapoptotic molecule DD2 and the cell survival molecule Bcl-2. The Bcl-2-related gene from *G. cydonium* was even found to confer resistance to environmental stress to human cells.

***Figure 8**. BH1 domains, present in BHP from G. cydonium (BHP1 GC) and S. domuncula (BHP1 SD). (A) Alignment with the corresponding BH1 domains from higher metazoan phyla: mouse Mtd (MTD MOUSE), human MCL-1 (MCL1 HUMAN), frog R11 (AR11 XENLA), the three Bcl-2 sequences Bcl-xL (BCLX MOUSE), Bcl-W (BCLW HUMAN), and Bcl-2 (BCL2 HUMAN), human A1-BFL-1 (BFL1 HUMAN), mouse Bax (BAXA MOUSE), NR-13 from quail (NR13 COTJA), human bak (BAK-HUMAN), mouse Bad protein (BAD MOUSE), human activator of apoptosis Hrk (HRK HUMAN), and Ced-9 from C. elegans (CED9 CAEEL). Residues of aa similar (with respect to physicochemical properties) among all sequences are shown in white letters on black, those similar in at least 50% are in white letters on gray, and those similar in at least 25% are shaded. The highly conserved aa are marked (arrowheads). (B) Unrooted phylogenetic tree constructed from the sequences shown in A.*

Furthermore, we could demonstrate that in sponges caspases exist, which comprise the features of the related enzymes from higher metazoa. Studies can now be performed to determine their role during apoptosis in the sponge system. Moreover, the establishment of the primmorph culture will make it possible to study the mechanisms of aging in sponges in an *in vitro* system.

Analyses of the deduced polypeptide sequences from (marine) sponges revealed a closer similarity/homology to the corresponding genes from human than to those from *C. elegans* or from *Drosophila melanogaster* [73, 74]. Moreover, some proteins found in sponges and human do not exist in *C. elegans*, e.g., 2-5A synthetase [75] or lens βγ-crystallin [76]. The lack of similarity between sponges and protostomians is especially observed for those molecules which are involved in the apoptotic pathway [3]. It can therefore be expected that the results achieved using the sponge model system may also have a (comparably) high relevance for human.

Acknowledgments

This work was supported by grants from the Bundesministerium für Bildung und Forschung (project "Cell Culture of Sponges" and Center of Excellence BIOTEC-marin) and the European Commission (EVK3-CT-1999-00005).

References

1. Müller WEG (1995). Molecular phylogeny of Metazoa (animals): monophyletic origin. *Naturwiss* 82: 321–9.
2. Müller WEG (1998). Origin of Metazoa: sponges as living fossils. *Naturwiss* 85: 11–25.
3. Müller WEG (2001). How was metazoan threshold crossed: the hypothetical Urmetazoa. *Comp Biochem Physiol A* 129: 433–60.
4. Müller WEG (1997). Origin of metazoan adhesion molecules and adhesion receptors as deduced from their cDNA analyses from the marine sponge *Geodia cydonium. Cell Tissue Res.* 289: 383–95.
5. Müller WEG, Blumbach B, Müller IM (1999). Evolution of the innate and adaptive immune systems: relationships between potential immune molecules in the lowest metazoan phylum [Porifera] and those in vertebrates. *Transplantation* 68: 1215–27.
6. Krasko A, Lorenz B, Batel R, Schröder HC, Müller IM, Müller WEG (2000). Expression of silicatein and collagen genes in the marine sponge *Suberites domuncula* is controlled by silicate and myotrophin. *Eur J Biochem.* 267: 4878–87.
7. Schröder HC, Krasko A, Batel R, *et al.* (2000). Stimulation of protein (collagen) synthesis in sponge cells by a cardiac myotrophin-related molecule from *Suberites domuncula. FASEB J.* 14: 2022–31.
8. Lehnert H, Reitner J (1997). Lebensdauer und Regeneration bei *Ceratoporella nicholsoni* (Hickson, 1911) und *Spirastrella* (Acanthochaetetes) *wellsi* (Hartman & Goreau, 1975). *Geol Bl NO-Bayern* 47: 265–72.
9. Goldstein S (1990). Replicative senescence: the human fibroblast comes of age. *Science* 249: 1129–33.
10. Strehler BL (1986). Genetic instability as the primary cause of human aging. *Exp Gerontol.* 21: 283–319.

11. Harley CB, Futcher AB, Greider CW (1990). Telomeres shorten during ageing of human fibroblasts. *Nature* 345: 458–60.
12. Harley CB (1995). Telomeres in aging. In: Blackburn EH, Greider CW, eds. *Telomeres.* Cold Spring Harbor: Cold Spring Harbor Laboratory Press, pp. 247–63.
13. Custodio MR, Prokic I, Steffen R, *et al.* (1998). Primmorphs generated from dissociated cells of the sponge *Suberites domuncula*: a model system for studies of cell proliferation and cell death. *Mech Ageing Dev.* 105: 45–59.
14. Müller WEG, Wiens M, Batel R, *et al.* (1999). Establishment of a primary cell culture from a sponge: primorphs from *Suberites domuncula. Mar Ecol Prog Ser.* 178: 205–19.
15. Müller WEG, Böhm M, Batel R, *et al.* (2000). Application of cell culture for the production of bioactive compounds from sponges: synthesis of avarol by primmorphs from *Dysidea avara. J Nat Prod.* 63: 1077–81.
16. Kruse M, Batel R, Steffen R, Schröder HC, Müller IM, Müller WEG (2000). Sponge homologue of human and yeast gene encoding the longevity assurance polypeptide: differential expression in telomerase-positive and telomerase-negative cells of *Suberites domuncula. Mech Ageing Dev.* 118: 115–27.
17. Müller WEG, Müller IM (2003). Analysis of the sponge (Porifera) gene repertoire: implications for the evolution of the metazoan body plan. In: Müller WEG, ed. *Marine Molecular Biotechnology: Sponges (Porifera)*. Berlin: Springer Press, pp. 1–33.
18. Wimmer W, Perovic S, Kruse M, *et al.* (1999). Origin of the integrin-mediated signal transduction – functional studies with cell cultures from the sponge *Suberites domuncula. Eur J Biochem.* 260: 156–65.
19. Hayflick L, Moorhead PS (1961). The serial cultivation of human diploid cell strains. *Exp Cell Res.* 25: 585–621.
20. Mortimer RK, Johnston JR (1959). Life span of individual yeast cells. *Nature* 183: 407–11.
21. Diaz JP (1979). Variations, differenciations et fonctions des categories cellulaires de la demosponge d'eaux saumatres, *Suberites massa*, Nardo, au cours du cycle biologique annuel et dans des conditions experimentales. Université des Sciences et Techniques du Languedoc (Montpellier), p. 332.
22. Finch CE (1990). *Longevity, Senescence, and the Genome.* Chicago: The University of Chicago Press.
23. Koziol C, Borojevic R, Steffen R, Müller WEG (1998). Sponges (Porifera) model systems to study the shift from immortal to senescent somatic cells: the telomerase activity in somatic cells. *Mech Ageing Dev.* 100: 107–20.
24. Johnson TE, Shook D, Murakami S, Cypser J (1999). Increased resistance to stress is a marker for gerontogenes leading to increased health and longevity in nematodes. In: Bohr VA, Clark BFC, Strevnsner T, eds. *Molecular Biology of Aging*. Copenhagen: Munksgaard, pp. 25–34.
25. Wadhawa R, Kaul SC, Mitsui Y (1999). Cellular mortality and immortalization: a complex interplay of multiple gene functions. *Prog Molec Subcell Biol.* 24: 191–204.
26. Jiang JC, Kirchman PA, Zagulski M, Hunt J, Jazwinski SM (1998). Homologs of the yeast longevity gene LAG1 in *Caenorhabditis elegans* and human. *Genome Res.* 8: 1259–72.
27. Counter CM, Avilion AA, LeFeuvre CE, *et al.* (1992). Telomere shortening associated with chromosome instability is arrested in immortal cells which express telomerase activity. *EMBO J.* 11: 1921–9.
28. Hastie ND, Dempster M, Dunlop AG, Thompson AM, Green DK, Allshire RC (1990). Telomere reduction in human colorectal carcinoma and with ageing. *Nature* 346: 866–8.

29. Harley CB (1999). Telomerase immortalize normal human cells. In: Bohr VA, Clark BCF, Stevnsner T, eds. *The Molecular Biology of Aging*. Copenhagen: Munksgaard.
30. D'Mello NP, Childress AM, Franklin DS, Kale SP, Pinswasdi C, Jazwinski SM (1994). Cloning and characterization of LAG1, a longevity-assurance gene in yeast. *J Biol Chem.* 269: 15451–9.
31. Chanda ER, Lingner C, Ko Z, Young PG. Direct Submission to the EMBL data bank (NOV-1996).
32. Schröder HC, Kruse M, Batel R, Müller IM, Müller WEG (2000). Cloning and expression of the sponge longevity gene *SDLAGL*. *Mech Dev.* 95: 219–20.
33. Simpson TL (1984). *The Cell Biology of Sponges*. New York: Springer-Verlag.
34. Wagner C, Steffen R, Koziol C, *et al.* (1998). Apoptosis in marine sponges: a biomarker for environmental stress (cadmium and bacteria). *Mar Biol.* 131: 411–21.
35. Batel R, Bihari N, Rinkevich B, *et al.* (1993). Modulation of organotin-induced apoptosis by the water pollutant methyl mercury in a human lymphoblastoid tumor cell line and a marine sponge. *Mar Ecol Prog Ser.* 93: 245–51.
36. Thompson CB (1995). Apoptosis in the pathogenesis and treatment of disease. *Science* 267: 1456–62.
37. Wiens M, Seack J, Koziol C, *et al.* (1999). 17β-Estradiol-dependent regulation of chaperone expression and telomerase activity in the marine sponge *Geodia cydonium*. *Mar Biol.* 133: 1–10.
38. Pfeifer K, Schröder HC, Rinkevich B, *et al.* (1992). Immunological and biological identification of tumor necrosis factor in sponges: role of this factor in the formation of necrosis in xenografts. *Cytokine* 4: 161–9.
39. Smith CA, Farrah T, Goodwin RG (1994). The TNF receptor superfamily of cellular and viral proteins: activation, costimulation, and death. *Cell* 76: 959–62.
40. Itoh N, Nagata S (1993). A novel protein domain required for apoptosis. Mutational analysis of human Fas antigen. *J Biol Chem.* 268: 10932–7.
41. Tartaglia LA, Ayres TM, Wong GHW, Goeddel DV (1993). A novel domain within the 55 kd TNF receptor signals cell death. *Cell* 74: 845–53.
42. Cleveland JL, Ihle JN (1995). Contenders in FasL/TNF death signaling. *Cell* 81: 479–82.
43. Ashkenazi A, Dixit VM (1998). Death receptors: signaling and modulation. *Science* 281: 1305–8.
44. Wiens M, Krasko A, Müller IM, Müller, WEG (2000). Increased expression of the potential proapoptofic molecule DD2 and increased synthesis of leukotriene B4 during allograft rejection in a marine sponge. *Cell Death Differ.* 7: 461–9.
45. Boldin MP, Varfolomeev EE, Pancer Z, Mett IL, Camonis JH, Wallach D (1995). A novel protein that interacts with the death domain of Fas/APO1 contains a sequence motif related to the death domain. *J Biol Chem.* 270: 7795–8.
46. Chinnaiyan AM, O'Rourke K, Tewari M, Dixit VM (1995). FADD, a novel death domain-containing protein, interacts with the death domain of Fas and initiates apoptosis. *Cell* 81: 505–12.
47. Mason DM, Morris PJ (1986). Effector mechanisms in allograft rejection. *Annu Rev Immunol.* 4: 119–45.
48. Wiens M, Krasko A, Müller CI, Müller WEG (2000). Molecular evolution of apoptotic pathways: cloning of key domains from sponges (Bcl-2 homology domains and death domains) and their phylogenetic relationships. *J Mol Evol.* 20: 520–31.
49. Blumbach B, Diehl-Seifert B, Seack J, Steffen R, Müller IM, Müller WEG (1999). Cloning and expression of novel molecules belonging to the immunoglobulin superfamily from the

marine sponge *Geodia cydonium:* putative cell recognition molecules. *Immunogenetics* 49: 751–63.
50. Adams JM, Cory S (1998). The Bcl-2 protein family: arbiters of cell survival. *Science* 281: 1322–6.
51. Cleary ML, Sklar J (1985). Nucleotide sequence of a t(14;18) chromosomal breakpoint in follicular lymphoma and demonstration of a breakpoint-cluster region near a transcriptionally active locus on chromosome 18. *Proc Natl Acad Sci USA* 82: 7439–43.
52. Oltvai ZN, Milliman CL, Korsmeyer SJ (1993). Bcl-2 heterodimerizes *in vivo* with a conserved homolog, Bax, that accelerates programmed cell death. *Cell* 74: 609–19.
53. Sato T, Irie S, Krajewski S, Reed JC (1994). Cloning and sequencing of a cDNA encoding the rat Bcl-2 protein. *Gene* 140: 291–2.
54. Wiens M, Diehl-Seifert B, Müller WEG (2001). Sponge Bcl-2 homologous protein (BHP2-GC) confers distinct stress resistance to human HEK-293 cells. *Cell Death Diff.* 8: 887–98.
55. Williams GT, Smith CA (1993). Molecular regulation of apoptosis: genetic controls on cell death. *Cell* 74: 777–9.
56. Shiraiwa N, Inohara N, Okada S, Yuzaki M, Shoji SI, Ohta S (1996). An additional form of rat Bcl-x, Bcl-xb, generated by an unspliced RNA, promotes apoptosis in promyeloid cells. *J Biol Chem.* 271: 13258–65.
57. Hengartner MO, Horvitz HR (1994) *C. elegans* cell survival gene ced-9 encodes a functional homolog of the mammalian proto-oncogene bcl-2. *Cell* 76: 665–76.
58. Yang E, Korsmeyer SJ (1996). Molecular thanatopsis: a discourse on the BCL2 family and cell death. *Blood* 88: 386–401.
59. Madeo F, Fröhlich E, Ligr M, *et al.* (1999). Oxygen stress: a regulator of apoptosis in yeast. *J Cell Biol.* 145: 757–67.
60. Earnshaw WC, LM Martins, SH Kaufmann (1999). Mammalian caspases: structure, activation, substrates, and functions during apoptosis. *Ann Rev Biochem.* 68: 383–424.
61. Cory S, Adams JM (1998). Matters of life and death: programmed cell death at Cold Spring Harbor. *Biochim Biophys Acta* 1377: R25–44.
62. Kumar S, Colussi PS (1999). Prodomains – adaptors – oligomerization: the pursuit of caspase activation in apoptosis. *Trends Biochem Sci.* 24: 1–4.
63. Villa P, Kaufmann SH, Earnshaw WC (1997). Caspases and caspase inhibitors. *Trends Biochem Sci.* 22: 388–93.
64. Hofmann K, Tschopp J, Bucher P (1997). The CARD domain: a new apoptotic signalling motif. *Trends Biochem Sci.* 22: 155–6.
65. Aitken A, Jones D, Soneji Y, Howell S (1995). 14-3-3 proteins: biological function and domain structure. *Biochem Soc Trans.* 23: 605–11.
66. Wiens M, Koziol C, Hassanein HMA, Batel R, Schröder HC, Müller WEG (1998) Induction of gene expression of the chaperones 14-3-3 and HSP70 by PCB118 (2,3′,4,4′,5-pentachlorobiphenyl) in the marine sponge *Geodia cydonium*: novel biomarkers for polychlorinated biphenyls. *Mar Ecol Prog Ser.* 165: 247–57.
67. Coffey RG. ed. (1992). *Granulocyte Response to Cytokines.* New York: Marcel Dekker, New York, pp. 47–76.
68. Needleman P, Turk J, Jakschik BA, Mordson AR, Lekowith JB (1986). Arachidonic acid metabolism. *Annu Rev Biochem.* 55: 69–102.
69. Gallin JI, Goldstein IM, Snyderman R, eds. (1992). *Inflammation: Basic Principles and Clinical Correlates*, 2nd edn. New York: Raven Press.
70. Chabannes B, Perraut C, El-Habib R, Moliere P, Pacheco Y, Lagarde M (1997). Correlation between arachidonic acid oxygenation and luminol-induced chemilumines-

cence in neutrophils: inhibition by diethyldithiocarbamate. *Biochem Pharmacol.* 53: 927–35.

71. Schröder HC, Sudek S, de Caro S, *et al.* (2002). Synthesis of the neurotoxin quinolinic acid in apoptotic tissue from *Suberites domuncula*: cell biological, molecular biological and chemical analysis. *Mar Biotechnol.* 4: 546–58.
72. Krogsgaard-Larsen P, Madsen U, Ebert B, Hansen JJ (1992). Excitatory amino acid receptors: multiplicity and ligand specificity of receptor subtypes. In: Krogsgaard-Larsen P, Hansen JJ, eds. *Excitatory Amino Acid Receptors.* New York: Ellis Horwood Press, pp. 34–55.
73. Gamulin V, Peden JF, Müller IM, Müller WEG (2001). Codon usage in the siliceous sponge *Geodia cydonium*: highly expressed genes in the simplest multicellular animal prefer C- and G-ending codons. *J Zool Evol Res.* 39: 1–6.
74. Müller WEG, Schröder HC, Skorokhod A, Bünz C, Müller IM, Grebenjuk VA (2001). Contribution of sponge genes to unravel the genome of the hypothetical ancestor of Metazoa (Urmetazoa). *Gene* 276: 161–73.
75. Wiens M, Kuusksalu A, Kelve M, Müller WEG (1999). Origin of the interferon-inducible (2′-5′)oligoadenylate synthetases: cloning of the (2′-5′)oligoadenylate synthetase from the marine sponge *Geodia cydonium. FEBS Lett.* 462: 12–18.
76. Krasko A, Müller IM, Müller WEG (1997). Evolutionary relationships of the metazoan βγ-crystallins, including the one from the marine sponge *Geodia cydonium. Proc Royal Soc Lond B* 264: 1077–84.
77. Harley CB (1991). Telomere loss: mitotic clock or genetic time bomb? *Mutat Res.* 256: 271–82.

Aging and Environmental Conditions in Insects

Klaus-Günter Collatz
Institut für Biologie I (Zoologie) der Universität, Haupstr. 1, D-79104 Freiburg, Germany

Introduction

Insects as tools for comparative aging studies

When we speak of insects as model systems for aging studies we first have to recognize that the number of insect species taken as model system up to now is remarkably low. In spite of the existence of about 1.2 million insect species not more than about 50 species were listed for their mean life-span in the classical report of Rockstein on insect aging in the seventies [1]. The situation does not have fundamentally changed since that time. A summary of the range of life-spans in adult insects for the major orders of *Insecta* together with arachnids was given as graph in a recent review on insect biodemography [2]. Longevity data in this table reach from several days in mayflies to decades in queen termites. In comparison with the 60-fold difference in the life-span of mammals adult insect's life-span differences are much more pronounced reaching an up to 5000-fold difference. Carey however emphasized how difficult it is to speak on life-span in general and particular on life-span in insects. In contrast to life expectancy and age specific mortality which are explicit measurable values "life-span" is weakly defined as time after which no member of a given species can survive even under the most favorable conditions. "Maximum life-span potential" is another description. It is reasonable therefore that maximum life-span determination depends on the number of individuals under determination being higher when large numbers are observed. On the other hand longevity data obtained under laboratory conditions may have nothing to do with the same data recorded under free living conditions. The problem in general is further strengthened when maximum life-spans of ectothermic organisms as insects have to determined. Insects are extremely sensitive against abiotic factors like temperature, humidity or light regime. It is therefore meaningless – as Carey pointed out – "to consider life-span for any species without considering environmental, ecological and evolutionary contexts."

H.D. Osiewacz (ed.), Aging of Organisms, 99–123.

Extensive studies on aging mechanisms are mostly restricted to dipterans. The fruit fly *Drosophila* of course is the most throughout studied insect even in this sense. Other more often studied dipterans are the house fly *Musca*, the blowflies *Phormia regina* and our own model organism *Phormia terraenovae* [3] and Mosquitoes – in most cases *Aedes*. The flour beetle *Tribolium* and the mealworm *Tenebrio* are some other examples.

Due to this limitation those biologists who were interested in comparative aging studies are in an unfavorable situation. The large differences in life history strategies of insects of the same genus would offer the opportunity to test the pylogenetically adopted causes of aging. As we will see these are the allocation of energy between body maintenance and reproduction and the influence of reproductive effort on the mortality character. In the last years special attention was given to social hymenopters as bees, wasps and ants with their remarkable long living queens. It has to be shown that their different casts are strongly dependent on epigenetic factors like nutrition or hormonal stimuli. This means that members with the same or a related genotype exhibit very different longevity.

The use of insects as model systems for aging mechanism studies in general is not new. The work of Loeb and Northrop [4] on metabolic rate and life-span in *Drosophila* marks the beginning of a comparative biology of aging. The authors took up an idea of Rubner [5]. Working with mammals he showed that the amount of life long metabolic energy per gram body weight is more or less constant. The later development of the rate of living theory [6] – accepted or not – is unthinkable without the insect model system. Beginning in the late seventies a number of studies especially on the housefly *Musca domestica* were undertaken by Sohal and co-workers to investigate this old problem with modern methods in greater detail. It were the results of these studies which Sohal leads to redefine the rate of living theory: "The rate of aging is directly related to the rate of unrepaired molecular damage inflicted by the by-products of oxygen metabolism and it is inversely related to the efficiency of antioxidant and repair mechanisms" [7]. Since that time insects have their true place in the study of the influence of oxidative stress, free radical damage and repair on aging and longevity. These factors are generally accepted as being of central importance for the understanding of aging mechanisms.

Insects, among other invertebrates, offer several advantages when chosen as models for aging studies: They can be raised in large quantities over short periods of time. They are in their adult stage composed of postmitotic cells – except for hemocytes and germ cells. Most of them are short living organisms. They share many basic metabolic similarities with other invertebrates as well as with vertebrates. These similarities in aging mechanisms however can result from different aging strategies when taking the life history of a species into consideration. The aging pattern doubtless is under genetic control, but this pattern – as mentioned above – at the same time is extremely flexible and depends on several biotic and abiotic factors. Life-span includes the longevity of all developmental stages an individual passes through during its life cycle. Individuals of insect species (as well as that of other species) experience a number of critical points in time during their life cycles which influence the life-span from egg over larvae up to pupa and adult stages [8]. In unawareness of such modifying factors the life-span of an insect species is unpredictable.

It is the aim of the following contribution to bring attention to the close connection between aging, environmental conditions and life-span of insects.

Evolution of life-span and life history trade-offs

It seems reasonable first to have a short look on the evolution of different life-spans in general before speaking of phylogenetically adopted or in an individuals life optional favored life history trade-offs. For further details see Kirkwood and Austad [9].

The celebrated German biologist August Weismann was the first to reflect basic considerations on the evolution of different life cycles. In a small publication "Über die Dauer des Lebens" (on the duration of life) [10] he stated: from my point of view death is in the last instance an adaptation. The life-span of an organism is therefore a species-specific trait. It is a constituent of the ecological niche of a species. "Ecological niche" does not denote any room compartment but has to be taken as an evolved answer to a network of interrelationships of biotic and abiotic factors. Life-span is therefore liable to selection. The evolution of different life-span arises from the necessity to obtain under different environmental conditions a high reproductive fitness. Genotypes with high fitness are more pronounced in the gene pool of the next generation against such with low fitness. With this life-span and reproduction in general are intimately interrelated. (It should be mentioned however that different mussels and crayfishes obviously does not fit in this scheme.) In these instances no correlation between reproductive output and aging could be detected. The same holds for organisms which are able to perfectly renew their body parts as it was described for the fresh water polyp *Hydra* [11, 12].

A young organism stands at the beginning of his reproductive phase. If there exist a mutation which acts in a life shortening manner, a strong selection pressure would act against this mutation. A mutation which happens later in life or exerts its negative influence later in life influences the reproductive success to a lesser extend. This is why the selection pressure on life-span is weakened with increasing chronological age.

The population biology introduced the term "reproductive value" which means that a trajectory of reproductive output exists. It culminates in a maximum in which the relative number of offspring which can arise from females aged to an age x is highest. Consequently the selection pressure against positive or negative traits declines with advancing reproductive age thereby favoring the accumulation of deleterious acting genes (mutation accumulation) or the acting of genes which were favorable early in life but deleterious later in life (pleiotropy) or both (Figure 1).

With this in mind we have an idea of the evolution of limited life-spans. The conditions however under which these different life-spans could evolve deserve further attention. Weismann's idea of life-span as an adapted trait will help again.

This idea offered the basic assumption of the "disposable soma theory" of Kirkwood [13, 14]. This theory is an integral part of all reflections on different life history strategies. The life history of a species in general is characterized by limiting

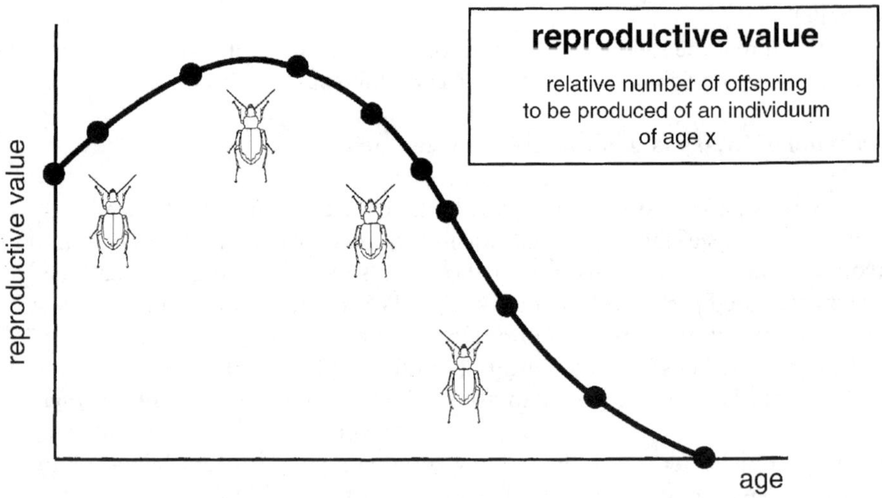

Figure 1. Reproductive value. Trajectory of reproductive output with a maximum after which the effect of selection declines. Declining selection favors the accumulation of deleterious acting genes or such genes which were favorable early in life but deleterious later in life or both.

factors and limited resources. These could be of very different nature e.g., nutrition deficiency, ephemeral substrates for egg deposition, climatic conditions and much more. Accordingly different reproductive strategies to obtain a high reproductive fitness were evolved. The mentioned limited resources ("constraints") enforce compromises ("trade-offs") between investment for reproduction and maintenance of body integrity. Given a high external risk it is the successful strategy to invest as much as possible into reproduction. Such organisms as a general rule are semelpar which means that they reproduce once in their life but in an explosive manner. This results in a short length of life. In the population biology this kind of life cycle is known as "r-strategy." Many insect species belong to this type of life history. "K-strategies" on the other hand are adopted by organisms living under low external risk factors. They invest more into body maintenance, reproduce several times – iteropar – during their life and live longer. Birds, bats and humans are typical examples. (The letters "r" – rapid – and "K" – capacity – denominate constants in population growing curves). As indicated in Figure 2 both life history strategies have to be seen relative to each other. To sum up this "life history theory on aging" it has to be stated that high extrinsic mortality rates caused by different environmental hazards led to the evolution of high intrinsic mortality rates in connection with early reproduction and vice versa. Experimental evidence to prove this central claim was given by Stearns and co-workers [15] and later in a more extensive form by Gasser *et al.* [16].

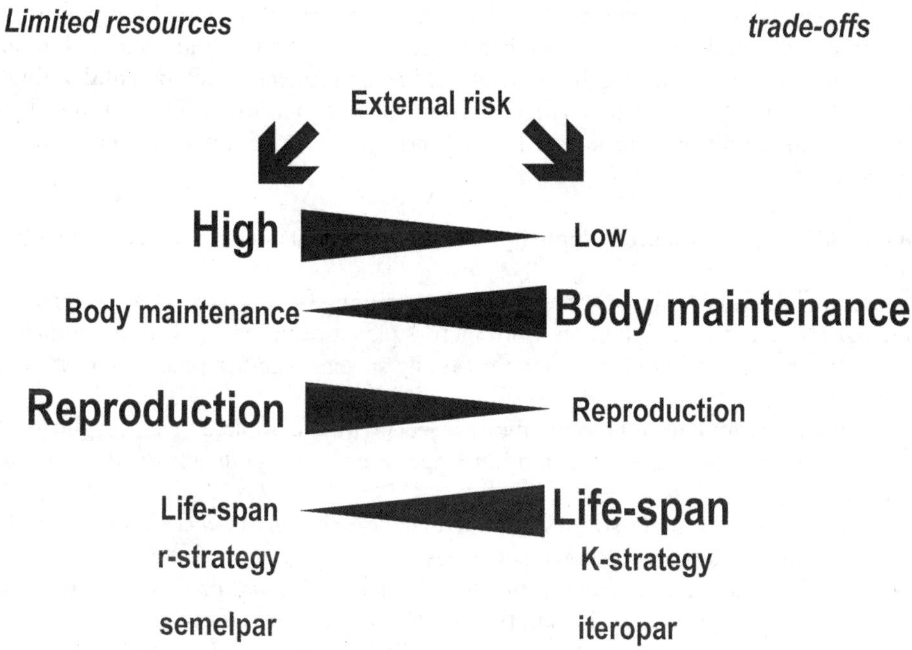

Figure 2. The "disposable-soma" theory. Allocation of limited resources between body maintenance and reproduction results in different life-span and reproductive output. The different life history strategies are denominated as "r"- or "K"-strategy.

Regarding this thoughts about the evolution of species-specific life-span it was consequent to attempt for laboratory selection on longevity in short living insects. This was often done successfully in the last decade especially in *Drosophila* [17].

It is however not the matter of this contribution to go in further details of the evolutionary aspect.

For our further purpose it is rather important to note that these trade-offs not only offer the explanation for evolved life strategies but were used in a very flexible manner depending on biotic or abiotic factors which act on different populations of a species living under variable conditions. The differences in mean life-spans in populations of a species are therefore not only genetic in nature but depend on a variety of environmental factors.

In investigating the longevity effects of a variety of environmental and genetic manipulations (temperature changes, dietary restriction, and laboratory selection for increased longevity) in rodents and *Drosophila* it could be shown that the underlying age-specific patterns of mortality are not always the same. Increases in longevity through laboratory selection result primarily from a reduction in baseline mortality whereas lower temperature and caloric restriction slow the rate of aging [18].

It is remarkable however that long-lived strains in *Drospophila* comparable to laboratory-selected strains have not been observed in the wild. Buck *et al.* [19] found that laboratory selected long lived strains exhibit a reduced developmental viability and a developmental lethality twice as high as that of normal lived strains. Thus preimaginal conditions are affected in a fitness-reducing manner in such selection experiments.

Which kind of factors are responsible for the modification of an insects life-span?

In our introductory remarks some life-span modifying factors as *ambient temperature* and *light regime* were touched on. Both factors play a prominent role in habitats with distinctive seasons where they decide for many species whether phases of dormancy occur or not.

During development in holometabolic insects with a complete metamorphosis the *preimaginal conditions* from egg to adult stages are not insignificant for the adult life course.

In most if not all insect species it is further the *reproductive success* which opt for the following course of life. Reproductive success not only depends on the availability of a mate but also on the presence or absence of *specific food*, protein in most cases, for the development of eggs. This in turn is also of great influence on longevity.

The acquisition of both mates and eventually a specific food naturally requires a variability in flight performance. In light of our critical points physiological trade-offs between flight capability, reproduction and longevity have to come into consideration. A prerequisite for a longer life in colonies of *eusocial insects* it to be in the right caste.

Finally *caloric restriction* in general can further act as a life prolonging strategy not only, as often reported in mammals, but also in insects. These factors now should be dealt with in more details.

Temperature and light

Life-span of insects and other ectotherms have been observed by many authors to be longer at lower than at higher temperatures. This was stated in the early work of Loeb and Northrop [4] as mentioned above. The interdependence of temperature and life-span was attributed to the varying metabolic rates of ectothermic organisms being higher at higher temperatures with the known longevity reducing consequences. The relationship however is in many cases not as simple as originally thought. Metabolic adaptations can alter the metabolic rate at different temperatures.

When we look at temperature as an abiotic factor for alternate life tactics its capacity to induce different forms of dormant or resting periods is more relevant.

Diapause

In insects diapause is a wide spread phenomenon. It is termed a prospective form of dormancy like hibernation in small mammals. A third form of this prospective resting period is aestivation or "summer diapause." All these forms represent an advanced mode to survive adverse climatic conditions such as low or high temperatures or draught. Diapause and aestivation are detectable in all developmental stages: eggs, larvae pupae and adults. It is evident that diapause and other resting stages have a profound influence on development, aging and life-span. Hereby it has to be taken into account that depending on the geographical position in which different populations of one species spend their life diapause can happen or not. In addition to the determination of the individual life program of the members in such a population diapause therefore affects the outcome of generations. This should be illustrated for two examples [8].

Depending on day length and temperature larvae of the cabbage moth *Mamestra brassicae* develop into normal, aestivating (drought period surviving) or overwintering pupae. The developmental time of unmodified pupae in a temperature range of 20 to 25°C takes 18 to 34 days. With increasing day length and high temperatures the pupae enter an aestivation diapause therefore prolonging their developmental life-span up to 65 days and more and with this their total life-span. A similar situation occurs under short photoperiods and low temperatures when pupae start into overwintering diapause. An aestivation diapause allows the emergence of a second generation during one year (polyvoltine or iteropar). Moths entering hibernation diapause in contrast are monovoltine (semelpar).

A second instance is the scorpion fly *Panorpa vulgaris*. This mecopter produces under prevailing environmental conditions two generations per year a spring and an autumn generation.

The last larval stage of the second generation enters an overwintering diapause. Under unfavorable dry periods *Panorpa* is able to enter in diapause as second larval stage of the first generation. In this case however it is impossible to produce a second generation. Corresponding to these different development abilities we find different longevity of the members of the different generations: In populations exhibiting no diapause larvae of the last stage live 20 days at maximum, overwintering larvae live at least 210 days. To these life-spans the life-span of about 80 days of the adult has to be added. This example evidently show that depending on the prevailing environmental conditions aging and life-span are modified in a complex manner.

In insects whose adults do not feed the option "larval-diapause" or "non-larval-diapause" decides directly on the adult's life-span. This is because larvae in diapause consume a great deal of their body energy reserves which then are not available for the adults. Larvae not experiencing a diapause transfer stored material through the metamorphosis into the adults which results in a life prolonging effect.

The larval stage in which diapause occurs can be variable within a species. The larval diapause of the beetle *Lagria hirta* for instance is characterized by additional moultings in advanced instars. The duration of each single stage lasts longer and moulting is desynchronized. *Lagria* therefore can overwinter in different larval stages. A chilling period which is often necessary to terminate diapause synchronize and accelerate the following development. This flexibility of larval development is the result of genetic determined polyphenism and depends not on the external conditions. It increases the flexibility of the species life cycle against unpredictable seasonal changes and occasionally the duration of life [20].

Concerning the extension of life-span the adult or reproductive diapause is the most fascinating form of age modification. Larvae and embryonic diapause interrupts the normal development for about half a year at maximum but thereafter development continues and eventually deteriorated cells or organs can replaced in the process of metamorphosis. Not so in reproductive diapause. In insects exhibiting reproductive diapause vitellogenesis (the synthesis of specific yolk proteins in the fat body) arrests as well as the production of secretes of male and female gonadal accessory glands. The male sexual behavior stops and a hypertrophy of the fat body in both sexes appear. Insects in reproductive diapause are otherwise by no means inactive as can be seen especially in dispersing insects.

In the extreme case of wandering butterflies such as the monarch *(Danaus plexippus)* adults travel in diapause thousands of kilometres to their overwintering places. "Monarch reproductive diapause provides escape in space and time" said Tatar and Yin [21] in their recent paper which addresses insect adult diapause to the slowing of aging processes. Life-span of diapausing and travelling adults can overweight that of reproductive adults more than five fold (Figure 3). As pointed out insects after adult diapause can never renew damaged tissues. We have only scarce understanding of what are the mechanisms leading to the eventually suspended aging process during diapause. All observations however point to the conclusion that these obviously enormous differences in the reduction of mortality are the result of intrinsic metabolic rearrangements and not the consequence of lower temperatures.

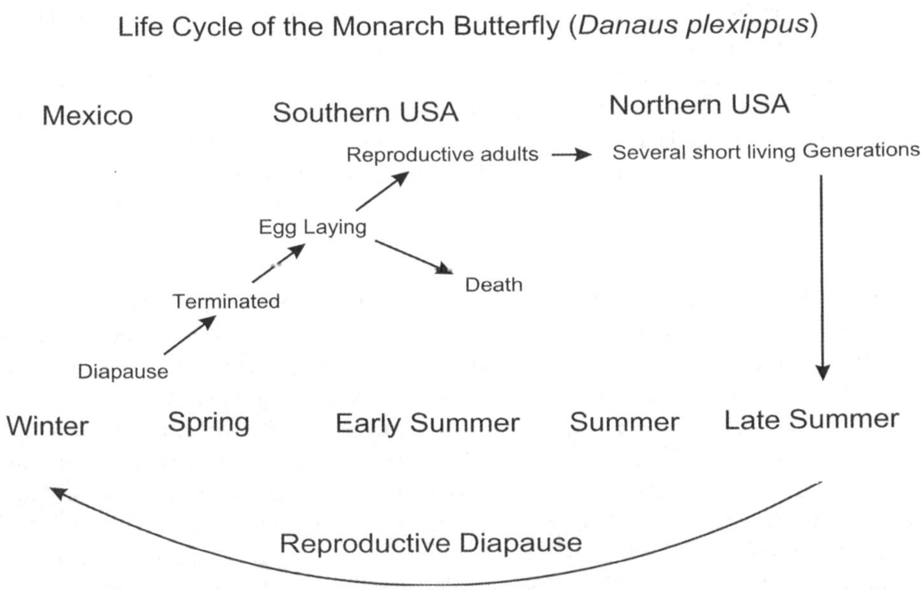

***Figure 3**. Life cycle of the monarch butterfly (Danaus plexippus). Adults travel thousands of kilometres in diapause from Northern USA to Mexico. Life-span of such travelling adults can overweight that of reproductive adults more than five fold. Data after Tatar and Yin [21].*

A down regulation of juvenile hormone titer is assumed to be the key step in reproductive diapause. But in which way this contributes to the change or persistence of somatic functions is a completely unresolved problem.

It has been proposed that a down regulation of juvenile hormone production reduces the metabolic rate by lowering the oxygen consumption. This however stands in sharp contrast to the inducible and voluntary high flight activity of juvenile hormone deficient adults.

For further studies on aging in diapause anautogenous insects like the blowfly *Phormia* should be favorable models see Tatar and Yin [21]. Under free-living conditions a short day photoperiod and temperatures below 15°C are the diapause inducing variables for *Phormia*. Adults can stay in diapause for 6 to 7 months. The life-span of reproducing *Phormia* flies lasts irrespectively of its flexibility depending on reproductive success (especially mates and protein) not more than two months at maximum. Induction of ovarian development in diapause free samples can easily manipulated by application of protein meals and a comparison can be made to the situation under diapause conditions. From the works of Stoffolano [22], and Yin and Stoffolano [23, 24], we have information about the connection between protein intake and the hormonal events in reproduction and diapause in *Phormia*. The metabolic consequences however remain obscure.

Regarding the signal transduction pathway which triggers such fundamental metabolic adjustments it is speculated if the dauer form of the nematode *Caenorhabditis elegans* could act as model. In some phenomenons (arrested development) dauer in *C. elegans* looks like larval diapause forms in insects but not in others (aphagy, small body size). Dauer and longevity of *C. elegans* is regulated over an receptor (DAF2) acting on a kinase in a way which on the one side have similarities with the insuline/IGF receptor pathway in mammals and on the other side is homolog to the InR (insuline-like receptor) of *Drosophila* [25, 26]. InR mutants produce dwarf forms which exhibit an up to 85% elongated life-span against the wild strain. Juvenile hormone treatment restores the life-span of the wild type. The InR pathway therefore regulates both the production of juvenile hormone and longevity. Males and females of the InR mutants are sterile and the ovaries represent that of diapausing flies. Two possibilities come onto consideration for the influence of the InR pathway. Either the sterility favors an allocation of metabolic reserves for maintenance of body integrity in the sense of the disposable soma theory or juvenile hormone induces diapause-specific metabolic pathways. In both InR mutants and flies under diapause one can find highly elevated superoxide dismutase (SOD) and lipid levels pointing to an improved resistance against oxidative stress and starvation conditions.

Different light regimes at constant temperature

Independent of the diapause state different light regimes also influence the adult life-span. This has been shown for *Drosophila* [27]. Living under constant light in contrast to light regimes of 12:12 light/dark or constant darkness has – as expected – a life shortening effect. In this situation the egg production of virgin and mated females increases dramatically and is made responsible for the observed effect. Reproducing males are affected in the same manner whereas the life-span of males without access to females was not significantly influenced by different light regimes. A different influence was observed when larvae were reared under constant light. They show a faster preadult development as compared to 12:12 light/dark or constant darkness. But there were no effect on preadult mortality. The preadult light regime however seems to be relevant for the longevity of the adults since *Drosophila* reared in an 12:12 light/dark cycle were found to have the highest median adult life-span even under this light regime compared to others [28]. This last example not only points to the next question after the influence of preimaginal conditions on life-span but emphasizes the importance of coordinated circadian rhythms for the well-being of organisms. A desynchronization of endogenous rhythms seems to be of major influence on aging and disease. Such a life shortening effect of circadian dyschronism could be clearly demonstrated in insects after keeping blowflies in non 24-h light-dark cycles or subjecting them to a jet lag [29]. To my knowledge except of two short publications on aging of a biochemical rhythm in *Phormia* [30, 31] this encouraging attempt to take insects as models for chronobiological aging studies was never haunted.

Preimaginal conditions

In holometabolous insects larval and adult stages are strictly separated by a pupa in which the dramatic process of metamorphosis takes place. It was therefore often questioned if the preimaginal conditions have any influence on the adult longevity. Recent studies of Blake [32] seem to confirm earlier results that at least in *Drosophila* the acceleration or slowing down of pre-adult development has little or no effect on adult life expectancy. A putative physiological clock appears to be reset upon eclosion. That this is at least questionable indicates the above-mentioned observation [28]. More relevant than life-span for the adult fitness however is the reproductive effort after different larval treatments. This question got only minor attention. In contrast to Lints and Lints [33] who could not detect any difference in total egg production when larvae of *Drosophila* were reared at temperatures of 16, 25, and 28°C own studies suggest that in *Phormia* such differences in fact exist. When larvae of blowflies were raised under temperatures of 13, 25, and 30°C respectively and adults held at 25°C the group witch emerged from the 13°C pretreated larvae exhibits a more than 90% reduction in egg production against the control group (larvae at 25°C). Egg production of the 30°C larval group was nearly unaffected. In addition food restriction in both 25°C and 30°C larval groups results in dwarf adult flies with minor egg production.

Very spectacular examples for the influence of preimaginal conditions on adult life can be found in social insects (see below). The queen of honeybees for instance develop faster, have a shorter pupal stage and emerge as an adult 5 days earlier than the workers. Different feeding regimes of the larvae account for this. Despite this faster development the queen live much longer than the workers.

Reproductive success

Reproduction shortens life-span in many species. This becomes especially intriguing if death happens immediately after reproduction. In addition to salmons, some other fishes, and a few mammals "reproductive death" is common in insects too. Many examples can be found in mayflies (*Ephemeroptera*) and saturniid moths. In contrast to vertebrates it is unlikely to assume specific hormonal or metabolic influences as life terminating incidences in insects. Many of them and mayflies in particular have reduced mouthparts, a gut which is filled with ambient air and they are unable to ingest food.

Despite their extremely short adult life-span mayflies are amenable for demographic studies. This was shown by Carey [34]. He conducted life table studies under laboratory conditions on both sexes of newly emerged subimagos of the mayfly species, *Hexagenia limbata* and *Ephemera simulans*, which have two- and one-year life cycles, respectively. The expectation of life at emergence for *H. limbata* was approximately 2.6 days for both sexes with the last individual (a female) dying on day 8. Expectation of life for *E. simulans* was 1.6 and 2.0 days for males and females, respectively, with the last individuals (both sexes) dying on day 3. He concluded that mortality does not occur in one large step in this group of rapidly senescing organisms. Rather deaths are spread over several days.

Mating, virginity and life-span

Many data exist on the effect of mating or virginity on the life-span of different insect species. In general mating or non-mating is not of great impact for longevity in males. Females however differ to a great extend in their life-span when virgin or mated.

Chapman *et al.* [35] computed the existing data of the effect of mating on female survival, egg production and sexual receptivity in 64 species of *Diptera*. He concluded that they are equivocal and do not show a clear pattern of increase or decrease. Observed differences can probably attributed to different experimental conditions. The measurement of the effect of mating on survival – as they pointed out – depends upon remating frequency itself, length of exposure to males and nutrition.

In their own studies on the interaction of mating, egg production and death rates in females of the Mediterranean fruit fly *Ceratitis capitata* they confirmed that virgin flies live longer than mated ones. This however is not due to the higher egg-laying rate of mated females as often described. At least in this system both mated and virgin females produce the same lifelong egg mass. The reason for the shorter life-span of mated females obviously lays in the cost of mating itself. Egg production demands an additional cost as is seen when X-irradiated (and therefore sterile) virgin females where compared with normal virgin females. But these costs are equal in mated and unmated females. The medfly is probably an exception in such a way that in the mentioned study. Chapman *et al.* [35] could not find any evidence that mating enhance the egg production via substance transfer from the males accessory glands. Otherwise this together with a decreased receptivity after the first mating is common in dipters. Since in agreement with the other data the receptivity of medflies is also reduced after mating they concluded that the uncoupling of seminal fluid mediated effects on egg production and receptivity may represent a novel pattern in *Diptera*. The transfer of accessory gland proteins from males to females during copulation is otherwise detrimental for females and one of the reasons for the accelerated mortality rate after mating. In contrast to other dipterans *Drosophila* females are capable of mating and storing sperm from different males. After Civetta and Clark [36] females mated to males from different lines showed significant variation in their average longevity. This indicates a genetic variance in the toxic effect of the accessory gland fluid. The mortality parameters in this case were significantly correlated with the proportion of progeny sired by the first mated male. This effect of sperm competition on postmating female mortality was discussed in more detail in the cited paper. As in other cases the increased mortality after mating is restricted to a short period so that the mean longevity not differ from virgin females.

At older ages the differences in longevity between virgin and non-virgin females become less clear than formerly thought. Carey and co-workers found a mortality cost of virginity in older ages when studied the mortality rates over the lifetime of huge numbers of the Mediterranean fruit fly *Ceratitis*. Despite a greater number of virgin females surviving to older ages the life expectancy of this very old females was shorter than that of the fewer number of mated females survived to this age. The authors suggest that there are two consequences of mating: a short-term mortality increase and a long term mortality decrease [37].

Trade-offs between flight capability, reproduction and longevity

The flight performance of many insects is subjected to accelerated aging changes after reproductive success [38].

Physiological trade-offs between flight capability, reproduction and longevity can best be studied in insects with wing dimorphism. The morph with well-developed flight muscles and wings is prepared for migration and dispersal. The other morph often shows flight muscle histolysis together with a shedding of their wings. Such dealation is common in crickets, termites, ants, and plant louses (aphids). In a recent paper Tanaka and Suzuki [39] described such trade-offs in a cricket *(Modicogryllus confirmatus)*. In the short wing morph the flight muscle mass of freshly emerged adults represents about 4% of wet body weight. In contrast the long wing morph has flight muscles sharing more than 11% of wet body weight. Both morphs did not show differences in their lipid energy reserves. The reproductive morph produced significantly more eggs than the migration morph and has a shorter life-span than the latter. The energy taken from nutrition ingested after adult development is obviously allocated on the one hand for reproduction at the expense of body maintenance and on the other hand to achieve stability of flight muscle and maintenance of body integrity. When experimentally dealated after emergence the long wing morphs histolyse their flight muscles and start to produce eggs instead.

Similar observations were made with other cricket species. In *Gryllus firmus* it could be shown that the allocation of energy is visible even on the organic level. The respiration rates of isolated thoraces of the dispersal morph of this species were up to fourfold higher than those of the short wing form. The activity of the mitochondrial citrate synthase – an indicator of flight capability – rose up to 70-fold [40].

Specific food

Regarding special food requirements as proteins insects and especially dipters can be divided into autogenous and anautogenous. Autogenous means that during metamorphosis larval protein is transferred into the adults. In female adults it is an indispensable requisite for the development of eggs. In male adults it often serves as source for the synthesis of accessory genital-gland fluids: a prerequisite for the insemination of females. Anautogenous insects rely on exogenic protein sources. This was proven for many blowflies and blood-sucking mosquitos. For these insects protein is an ephemeric substrate and the time of its acquisition influences the aging process.

Carey *et al.* [41] described dual modes of aging in females of the Mediterranean fruit fly *Ceratitis capitata.* When these medflies are deprived from protein and got only sugar they stay in a waiting mode in which mortality is low. Giving access to protein *Ceratitis* switches to the reproductive mode which low mortality at the beginning of egg laying and an increase of mortality thereafter. Depending on the time of protein access the life-span of medflies can be nearly duplicated. This strategy of prolonging survival while maintaining reproductive potential in periods of food scarcity as Carey *et al.* pointed out is widely adopted by animals.

Wilker in our laboratory was able to show that the anautogenic fly *Phormia terraenovae* behave in the same manner. It covers its energy demand solely by carbohydrates [3]. After the firs protein supply at the 17th day the sugar ingestion rises rapidly up to fivefold compared to the *ad libitum* with sugar and protein supplied flies. Corresponding to the delayed protein availability the egg deposition starts later in life together with an increase of life expectancy. When taking virgin females, which lay infertile eggs the life prolonging effect of delayed protein supply is much more pronounced. Its life expectancy increases up to 30% when giving protein at day 17. The age specific mortality rate reflects the prolonging effect of delayed protein supply in an impressive manner (Figure 4a,b). The enzyme arginine phosphokinasein (APK) in the flight muscle of the fly can be taken as metabolic marker for flight activity. Mated females show a maximum of activity which occurs earliest in life in the *ad libitum* group and is higher in protein deficient group or those which received their first protein meal at day 17. Virgin females in contrast exhibit elevated APK activities even in advanced ages being again highest in protein deficient groups (Figure 5a,b).

Special situation in parasitoids

Parasitoids are insects of different orders (*Hymenoptera, Coleoptera, Diptera, Lepidoptera, Neuroptera* and *Trichoptera*) whose larvae develop by feeding on arthropod hosts. Females of this group are able to paralyse hosts through an injection of venom. In context of aging studies they are of special interest since its fecundity and longevity depends not only on the kind of host but also on the host's age. The hymenoptera *Nasonia vitripennis* is a parasitoid on pupa of *Diptera* belonging to the group of *Muscidae.* When rearing on pupa of the housefly *Musca domestica* the highest longevity and fecundity was achieved with pupa being less than 48 h old [42]. The blowfly *Phaenicia sericata* served as a better host in comparison to *Musca* resulting in

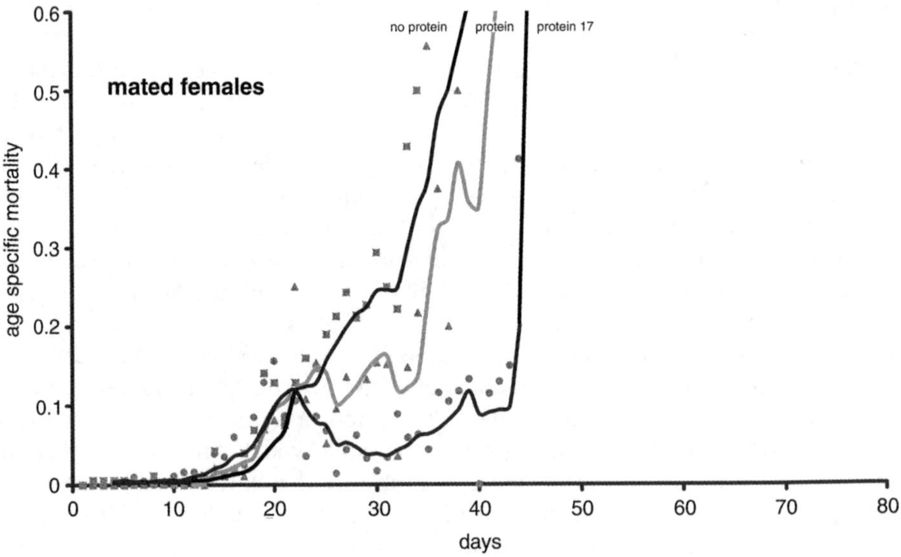

a

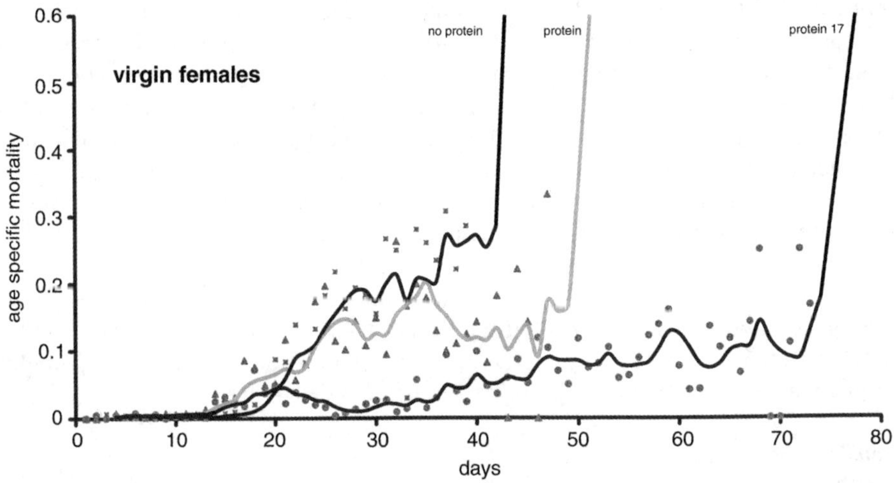

b

***Figure 4**. Age specific mortality trends (proportion of flies alive at the beginning of an age interval that die during that interval) of Phormia terraenovae with different protein supply based on populations of 400 flies each – a. mated females, b: virgin females. Wilker and Collatz own observations.*

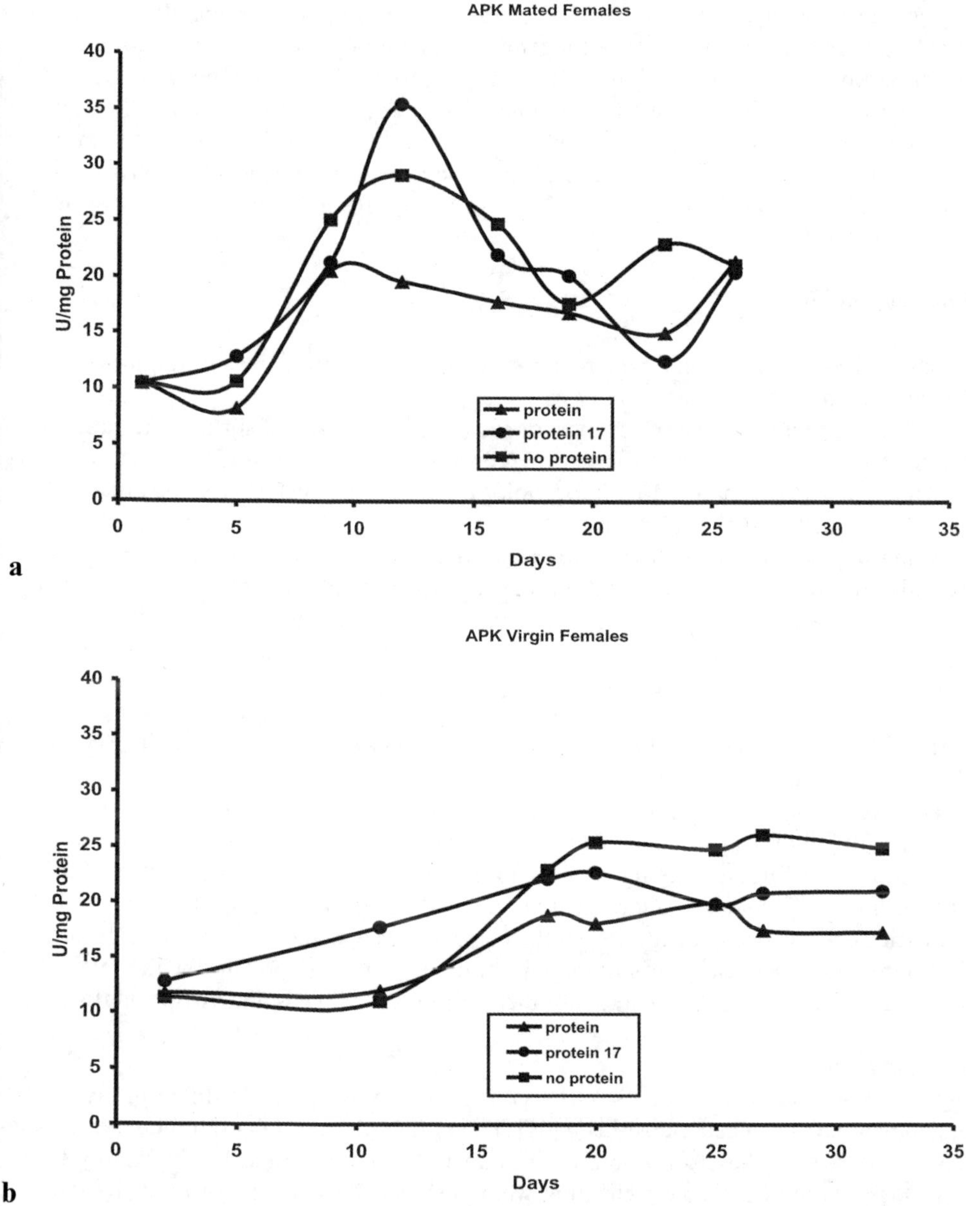

Figure 5. Age dependent specific activity of arginine phosphokinase as indicator of flight performance of Phormia terraenovae with different protein supply – a: mated females, b: virgin females. Mated females show a maximum of activity which occurs earliest in life in the ad libitum group and is higher in protein deficient group or those which received their frist protein meal at day 17. Virgin females in contrast exhibit elevated APK activities even in advanced ages being again highest in protein deficient grouprs. Wilker and Collatz own observations.

the greatest average longevity and progeny production of *Nasonia*. It was also reported that *Nasonia* females live longer at high host population densities [43]. This is especially remarkable and indicates that reproduction and longevity are not essentially correlated in a negative manner. As in the case of *Gryllus* and *Phormia* and other insects this parasitoid is able to stay in a kind of waiting mode to obtain successful reproduction. Maintaining on honey diet or sucrose alone has a life prolonging effect [44]. The survivorship curve drops after giving the pupa necessary for reproduction.

Eusocial insects

Different insect groups like wasps, bees, ants, and termites live in colonies with cooperating members of different casts.

The functional principle of colonies of social insects is that offspring take care for the next generation whereby their own reproduction is suppressed in favor of the queen. Albeit different in their morphology all different caste types share much genetic similarities with the queen.

Aging and longevity of the members of colonies of social insects are to a wide extend determined by extrinsic factors such as nutrition and kind of work. In all such colonies the queen has an extraordinary high life-span ranging from 4 to 8 years in wasps and up to 30 years in termites. A table containing maximal life-spans of different social insects can be found in Page and Peng [45].

Taking into consideration the normal life-span of a dog with about 12 years, a cat with say 20 years and most of the people in roman times with nor more than 30 years the duration of a termite queen's life remains still a mystery.

The evolution of this different life-spans of colony members is much better to conceive [46] than the proximate factors which enables the slow aging mechanisms of the queens and force the aging process in workers. We should not forget that this extraordinary life-span is achieved with a body consisting of postmitotic cells.

Much work has been done to reveal the mechanisms which lead to the development of different castes. Clearly it is the nutrition quality and composition which plays a significant role, but what special components are responsible is almost speculative.

Caste polyethism

In many insect societies workers of the same caste have to perform different tasks, a phenomenon called caste polyethism [47]. The kind and order of assignment which a member of a caste has to accomplish has an influence on longevity. Normally the safer tasks as nursing or nest cleaning were performed earlier in life than the risky ones as foraging or nest defending. After Calderone [48] in honeybees there is however only a weak causal link between the age of a worker and its duties and considerable variation in the age at which each performance were done. Instead it had been demonstrated repeatedly that heritable differences exist in the likelihood that individual workers from the same colony perform a specific task [49]. Together with the noted genetic variability in the age of performance this traits are selectable components and could affect colony development and organization. In more primitive organized social insects as the wasp *Ropalia marginata* a more definite correlation between age and performance of different tasks was described [50].

The shape of survivorship curves of colony members reflects the tasks individuals of the colony have to fulfil. If in worker castes there is no temporal division of labour the risk of dying is high even in younger workers. This was the case in bumblebee species where young workers starts foraging. The age specific mortality rate therefore is constant throughout the life and the shape of survivorship curves follows an exponential function. In contrast and uncommon for populations living under natural conditions the survivorship curve of honeybee workers resembles that of humans in developed countries (Figure 6). This is due to the more strictly regulated sequence in their tasks during adult life. Consequently the mortality in the protected environment in the hive which all workers enjoy during their first part of adult life is low in comparison to the second part when they suffer from predation and different accidents outside the nest.

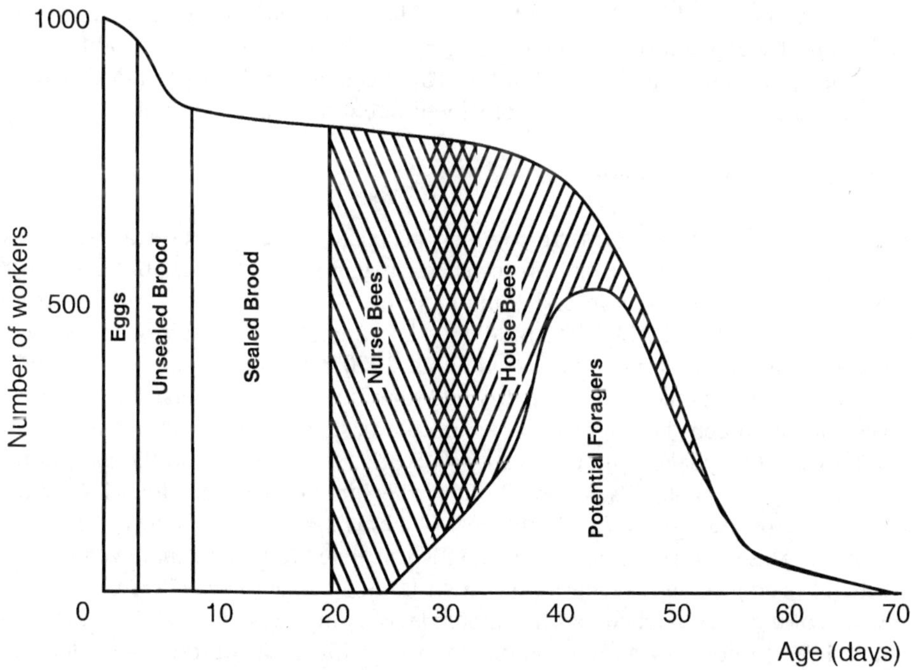

***Figure 6**. Survival curve of worker honey bees in different life stages. The death rate in the protected hive environment is low in comparison to the time outside the nest when bees suffer from predation and different accidents. Adapted from Wilson [47].*

It seems to be self evident that foraging and nest defending workers in social insects live shorter than their fellows inside the colony. The mentioned predators and other unpredictable accidents enhance the risk for dying. But the situation is not as simple as it seems. Here as in other cases we have to distinguish between extrinsic and intrinsic mortality.

Chapuisat and Keller [51] studied the life-span of weaver ant workers (*Oecophylla smaragdina*). The workers of this species are reported to be characterized by a clearly bimodal size distribution. Major workers length is about twice that of minor workers. This different size is associated with different duties. The minor workers take care for the brood inside the colony whereas the major workers have their working field outside the nest with foraging, defence and nest repair.

Taken into the laboratory without the many hazards one would expect the major workers to be longer living than the minor ones. At least the "rate of living theory" predicts that larger animals of the same species should have a greater life-span due to their lower metabolic rate per unit body mass. As Chapuisat and Keller [51] clearly show this does not apply to the weaver ant workers. Even under protected conditions smaller workers exhibit a longer life-span. The results are in agreement with the prediction of the evolutionary theory of aging that higher extrinsic mortality should also result in a higher intrinsic mortality. The mechanistic theory of metabolic rate and life-span covers only one aspect of why different life-spans do exist.

Intrinsic mortality in general

In general it is remarkable, that colonies of social insects sustain a big loss of individuals not only outside the protected environment of the nest but also inside the colony. One can find impressive numbers in the standard work of Edward O. Wilson on insect societies [47]. The following data are taken from this book. A colony of the army ant (*Eciton hamatum*) losts about 1000 workers per day comprising to approximately one percent. In another ant (Monomorium pharaonis) this loss is said to be twice as much. Different bumblebee colonies (*Bombus agrorum* and *Bombus humilis*) have to forsake four percent of its workers. Not better is the situation in honeybee or hornet colonies. From 60 000 workers in a honeybee colony 1800 dic off per day. The mortality rate inside the nest of *Vespa orientalis* amount to 9% and to 43% in foragers outside the nest. It would also a fallacy to believe that even in social insects all produced eggs reach maturity. In bumblebee colonies (*Bombus agrorum*) only 71% of the eggs hatched and from these larvae only 73% reached the pupal stage further 10% do not survive the metamorphosis yielding an reproductive success of only 47%. Similar results were obtained in ant colonies.

Honeybees

In honeybee colonies only one queen at the same time is tolerated by the workers. In comparison to the sterile female workers and the drones queens develop faster from egg to adult (see Table 1). The decision for development into queens or workers is made as short as three and a half day after hatching. Before that time the developmental pathway of a fertilized egg is not predetermined. The crucial role in

Table 1. Development and life-span in honey bee casts

	Egg	Larva	Prepupa pupa	Summer		Winter
Workers	3	6	12	15–18		140
Drones	3	6	15		20–40	
Queen	3	6	7		Up to 8 years	

Data from Page and Peng [45]

the process of queen differentiation lays in the quantity and quality of the secretions of the different food glands in head and thorax of the nurse bees [52]. In the L4 larval stage there is a dietary switch, which favor the prospective queens by continuous feeding of the "royal jelly." This results in a rapid weight gain so that the queen emerges with twice as much weight as the worker. The diet change further induces a sharp rise in juvenile hormone production which is together with ecdysteroids responsible for the final development into the queen. Fifty-six days after emergence queens attain sexual maturity. A short time after reaching maturity they start to up to three mating flies and mate during every flight with up to 13 males. The received sperm is stored over the whole period of queen's egg laying period without loosing its fertility. Up to now it is unknown what protective mechanism accounts for this.

Due to the different provided food workers grow slower and attain a lower weight. During larval to worker development apoptosis (programmed cell death) lead to a destruction of larval ovaries. The adult worker therefore has only rudimental ovaries and genital glands and in contrast to the queen a low juvenile hormone titer [45].

The life-span of workers depends to some extend on the development of larvae which has been shown considerable genetic variability. A much greater influence however exerts the season reaching from 1540 days life span in summer bees to about 140 days in winter bees. Live spans of more than 300 days for winter bees were also reported. The short life-span of summer bees was normally attributed to the high risk during foraging whereas the longer life-span of winter bees results from their prolonged hive period. Based on this well-known observation a study from Neukirch [53] on honeybee workers dealt with the dependence of life-span and energy consumption upon flight performance. Special attention was given to the foraging period of worker bees. The maximum flight performance during the foraging period was reported to be about 840 km. After reaching maximum flight performance the bees quickly died without a preceding drop of performance. The author suggested that probably enzymes of carbohydrate metabolism – the exclusive fuel for flight performance in bees – might be "exhausted" and concluded that the total flight performance of the individual seems to be fixed. A high flight performance on a per day basis leads to an earlier exhaustion of its "energy pool" and cause life shortening.

The chronological age at the onset of foraging was found to have no influence on flight duration and flight performance. Guzmán-Novoa *et al.* [49] however showed that individuals that begin foraging earlier in their life died sooner. On the other hand such individuals that begin foraging earlier in life forage for a longer time before they die. The authors suggested that there might be natural senescence in addition to mortality caused by the foraging activity itself. They could not found a correlation between foraging activity, as measured by the number of foraging trips, and foraging life or total length of life and suggest that fixed energetic reserves are not a major life-span determining factor. Whatever the actual cause of death of the old worker bee may be, the determination of life-span via the foraging efficiency is undoubtedly under hormonal regulation. Bees starting to forage were found to have greatest corpora allata volume and highest titers of juvenile hormone. This titer however was considerably lower than that of the queen (see above). Artificially applied high juvenile hormone titers result in an earlier start of foraging and a corresponding shorter life-span [53].

Drones require with 24 days the longest time for developing from egg into adult. Its body weight is indistinguishable from that of the queen.

The life-span of drones is dictated by several constraints. First the death is determined by the time of copulation itself since their endophallus as part of their sexual organ breaks off during mating which inevitably leads to death. Those drones which are still alive at the end of the mating season were killed by the workers. On the other hand it was claimed that drones which have not flown have rectums full of faeces which might directly affect their survival. Only the queen is allowed do defecate inside the nest [45]. To remove the faeces is one of the duties of workers. Workers itself as well as drones beside all other duties outside the nest have to fly in order to empty their recta.

For queens mating with older drones can be hazardous. This is because the seminal fluid of older drones contains material which blocks the queens oviducts. The queens become infertile and eventually die.

Ants

The most detailed summary of life-span of queens in ants (53 species) and termites (10 species) can be found in Keller [54]. Most life-span determinations however were done under laboratory conditions. Field data on the other hand are difficult to obtain and it could not ruled out that life-spans of queens were overestimated since in some cases queens were replaced thus prolonging the longevity of the colony and faking a greater life-span of one single queen [54].

In contrast to honeybees ants form colonies with – dependent on the species – one (monogyne) or several (polygyne) queens. In the same manner a queen alone can found the colonies themselves or the founding requires the help of workers. All these factors influence the longevity of the queen.

The better protected single queen in monogyne colonies live longer than each of the queens in the polygyne colonies. The "independent" founding queen (no workers are needed) life longer than the "dependent" founding queen. After Keller [54] the independent founding queen can produce sexuals only several years after the colony has been founded whereas queens returning to established nests after mating can do this sooner.

The latter was explained in evolutionary terms by Stearns [see Keller, ref. 54] claiming that the reproductive success of the independent founding queens increase with age against that of the dependent founding queens and therefore should selected for increasing life-span.

Caloric restriction

Beside all these critical points which modifies the pattern of aging in insects caloric restriction has been proven to be the best characterized and most reproducible method for extending life-span and retarding physiological aging at least in mammals. Beginning with the early works of McCay [55] numerous attempts have been made to study the underlying causes of this obviously general mechanism. An update of the discussion can be found in Masoro [56]. Up to now caloric restriction studies on insects are limited to a few examples. The data are not very convincing.

Evidence for enhanced survival time with reduced fertility under dietary restriction in insects among other invertebrates was provided by studies on waterstriders [57], carabid beetles [58] and *Drosophila.*

Dietary restriction in *Drosophila* has given inconsistent results. This might be attributable to different strains since inbreed strains which were often used in the laboratory do not show much phenotypic plasticity compared to hybrid strains. It could also be of influence if dietary restriction is achieved by a constant low caloric intake or intermittent starvation. Earlier studies failed to detect any enhancement of longevity under such regimen. David *et al.* [59] referred on a stimulation of egg production by high concentrated nutrition media but despite a life shortening effect with higher diluted food no influence on life-span could be detected. Le Bourg and Medioni [60] also did not found a connection between food restriction and longevity. In contrast after results of Partridge *et al.* [61], Chippindale *et al.* [62], Chapman and Partridge [63], and Clancy *et al.* [64], dietary restriction by food dilution to about 70% of the normal diet slows aging by increasing mean life-span in association with reduced fecundity in *Drosophila.* Using more diluted food results in increasing mortality due to starvation effects. In agreement with the earlier observations the age specific lifetime egg production increases with increasing food supply. The relation between nutriment use and life-span in *Drosophila* was further documented in selection experiments by Riha and Luckinbill [65] thereby showing that the evolutionary response and the phenotypically plastic response may share a common physiological base. Long-lived stocks are characterized by lower nutrition input. This was shown by the uptake of radiolabeled glucose and its incorporation into lipids and protein and by the *in vivo* metabolic flux though the pentose shunt. Even larvae of long-lived populations tend to feed less actively. Based on this and comparable studies Luckinbill and Foley [66] proposed a hypothesis describing the interaction between glucose level, damage by free oxygen radicals chaperonin proteins and the inevitable existing non enzymatic glycation process in glucose rich tissues. During ad libitum feeding high levels of glucose are available and the elevated metabolism produces excess amounts of free radicals damaging lipids, proteins and DNA. A side effect of the glycation process is the production of additional free radicals and the inactivation of protecting antioxidative enzymes such as superoxide-dismutase (SOD). If caloric restriction lowers available glucose as could be demonstrated this would result in increasing levels of antioxidant enzymes, reduced free radical production and a slowing of the rate of aging. Chaperonines play a role in the regulation of protein foldings and are found to be able to inhibit the production of advanced glycation end products (AGES). The relationship between oxidative stress, caloric restriction and aging in vertebrates and invertebrates was often discussed [see Sohal and Weindruch, ref. 67].

Caloric restriction obviously has no positive effect on reproduction. This was also reported from mammals where such treatment results in small animals with impaired fertility. In agreement with the disposable soma theory the increased longevity at the expense of fertility is the result of a shift of resources from reproductive activity to body maintenance. Under the aspect of a high reproductive fitness this energy allocation seems meaningless at first sight. So, what could be the meaning of the life prolonging effect of caloric restriction under the aspect of adaptivity? According to Holliday [68], Shanley and Kirkwood [69] assumed that animals have evolved a response to temporary changes in availability of resources. In this situation energy is allocated from reproduction to maintenance of somatic integrity resulting in impairment of survival and favoring the chance of successful reproduction when the famine is over. The authors developed a mathematical model to test this suggestion and found it to agree with the evolutionary hypothesis.

Conclusions

In the life history of a species numerous critical points exist which can modify the aging course in a shortening or delaying manner. The species answers to such abiotic or biotic factors are alternate life strategies. The ability to respond in the adequate manner is based on evolved characteristic traits and therefore under genetic control. It is an expression of the high genetic variability of life-span characters.

References

1. Rockstein M, Miquel J (1973). Aging in insects. In: Rockstein M, ed. *The Physiology of Insecta*. New York, London: Academic Press, pp. 371–478.
2. Carey JR (2000). Insect Biodemography. *Ann Rev Entomol*. 46: 79–110.
3. Collatz KG (1997). Fifteen years of *Phormia* – on the value of an insect for the study of ageing. *Arch Gerontol Geriatr*. 25: 83–90.
4. Loeb J, Northrop JH (1917). On the influence of food and temperature on the duration of life. *J Biol Chem* 32: 103–21.
5. Rubner M (1908). Das Problem der Lebensdauer und seine Beziehungen zu Wachstum und Ernährung, München, Berlin, Oldenburg.
6. Pearl R (1928). *The Rate of Living*. New York: Knopf.
7. Sohal RS (1986). The rate of living theory: a contemporary interpretation. In: Collatz KG, Sohal RS, eds. *Insect Ageing*. Berlin, Heidelberg, New York, Tokyo: Springer, pp. 23–44.

8. Sauer KP, Grüner C, Collatz KG (1986). Critical points in time and their influence on life cycle. In: Collatz KG, Sohal RS, eds. *Insect Ageing*. Berlin, Heidelberg, New York, Tokyo: Springer, pp. 9–22.
9. Kirkwood TBL, Austadt SN (2000). Why do we age? *Nature* 408: 233–8.
10. Weismann A (1882). *Über die Dauer des Lebens*. Jena: Fischer.
11. Finch CE (1990). *Longevity, Senescence, and the Genome*. Chigago and London: The University of Chicago Press.
12. Martinez DE (1998). Mortality patterns suggest lack of senescence in *Hydra*. *Exp Gerontol.* 33: 217–24.
13. Kirkwood TBL (1977). Evolution of ageing. *Nature* 270: 301–4.
14. Kirkwood TBL, Holliday R (1979). The evolution of ageing and longevity. *Proc R Soc Lond Ser B.* 205: 531–46.
15. Stearns SC, Ackermann M, Doebeli M (1998). The experimental evolution of ageing in fruitflies. *Exp Gerontol.* 33: 785–92.
16. Gasser M, Kaiser M, Berrigan D, Stearns SC (2000). Life-history correlates of evolution under high and low adult mortality. *Evolution* 54: 1260–72.
17. Luckinbill LS (1993). Prospective and retrospective tests of evolutionary theories of senescence. *Arch Gerontol Geriatr.* 16: 17–32.
18. Pletcher SD, Khazaeli AA, Curtsinger JW (2000). Why do life spans differ? Partitioning mean longevity differences in terms of age-specific mortality parameters. *J Gerontol.* 55: B381–9.
19. Buck S, Vettraino J, Force AG, Arking R (2000). Extended longevity in *Drosophila* is consistently associated with a decrease in developmental variability. *J Gerontol.* 55A: B292–301.
20. Zhou HZ, Topp W (2000). Diapause and polyphenism of life-history of *Lagria hirta*. *Entomol Exp Appl.* 94: 201–10.
21. Tatar M, Yin CM (2001). Slow ageing during insect reproductive diapause: why butterflies, grasshoppers and flies are like worms. *Exp Gerontol.* 36: 723–38.
22. Stoffolano JG (1974). Influence of diapause and diet on the development of the gonads and accessory reproductive glands of the black blowfly, *Phormia regina* (Meigen). *Can J Zool.* 52: 981–8.
23. Yin CM, Stoffolano Jr JG (1994). Endocrinology of vitellogenesis in blowflies. In: Davey KG, Peter RE, Tobe SS, eds. *Perspectives in Comparative Endocrinology*. Ottawa: National Research Council of Canada, pp. 291–8.
24. Yin CM, Stoffolano Jr JG (1997). Juvenile hormone regulation of reproduction in the Cyclorrhapuos Diptera with emphasis on oogenesis. *Arch Insect Biochem Physiol.* 35: 513–37.
25. Guarente L, Kenyon C (2000). Genetic pathways that regulate ageing in model organisms. *Nature* 408: 255–62.
26. Tatar M, Kopelman A, Epstein E, Tu MP, Yin CM, Garofalo RS (2001). A mutant *Drosophila* insulin receptor homolog that extends life-span and impairs neuroendocrine function. *Science* 292: 107–10.
27. Sheeba V, Sharma VK, Shubha K, Chandrashekaran MK, Joshi A (2000). The effect of different light regimes on adult life span in *Drosophila melanogaster* is partly mediated through reproductive output. *J Biol Rhythm.* 15: 380–92.
28. Pittendrigh CS, Minis DH (1972). Circadian systems: longevity as a function of circadian resonance in *Drosophila melanogaster*. *PNAS.* 69: 1537–9.
29. Saint Paul U (1978). Longevity among blowflies *Phormia terraenovae* R. D. kept in non-24-h light-dark cycle. *J Comp Physiol A* 127: 191–5.

30. Collatz KG, Horning M (1990). Age dependent changes of a biochemical rhythm – the glycolytic oscillator of the blowfly *Phormia terraenovae*. *Comp Biochem Physiol.* 96B: 771–4.
31. Horning M, Collatz KG (1990). First description of the glycolytic oscillator of an insect, the blowfly *Phormia terraenovae*. *Comp Biochem Physiol.* 95B: 613–18.
32. Blake KJ, Hoopengardner B, Centurion A, Helfland SL (1996). A molecular marker confirms that the rate of adult maturation is largely independent of the rate of pre-adult development in *Drosophila melanogaster*. *Dev Gen.* 18: 125–30.
33. Lints FA, Lints CV (1971). Influence of preimaginal environment on fecundity and ageing in *Drosophila melanogaster* hybrids – II. Preimaginal temperature. *Exp Gerontol.* 6: 417–26.
34. Carey JR (2002). Longevity minimalists: life tabel studies of two species of northern Michigan adult mayflies. *Exp Gerontol.* 37: 567–70.
35. Chapman T, Miyatake T, Smith HK, Partridge L (1998). Interaction of mating, egg production and death rates in females of the mediterrranean fruit fly, *Ceratitis capitata*. *Proc R Soc Lond Ser B* 265: 1879–94.
36. Civetta A, Clark A (2000). Correlated effects of sperm competition and postmating female mortality. *PNAS.* 97: 1362–5.
37. Carey JR, Liedo P, Harshman L, Zhang Y, Müller HG, Partridge L, Wang JL (2001). A mortality cost of virginity at older ages in female mediterranean fruit flies. *Exp Gerontol.* 37: 507–12.
38. Collatz KG, Wilps H (1986). Ageing of flight mechanism. In: Collatz KG, Sohal RS, eds. *Insect Ageing*. Berlin, Heidelberg, New York, Tokyo: Springer, pp. 55–72.
39. Tanaka S, Suzuki Y (1998). Physiological trade-offs between reproduction, flight capability and longevity in a wing-dimorphic cricket, *Modicogryllus confirmatus*. *J Insect Physiol.* 44: 121–9.
40. Zera A, Potts J, Kobus K (1998). The physiology of life-history trade-offs: experimental analysis of a hormonally induced life-history trade-off in *Gryllus assimilis*. *Am Nat.* 152: 7–23.
41. Carey JR, Liedo P, Müller HG, Wang JL, Vaupel W (1998). Dual modes of ageing in mediterranea fruit fly females. *Science* 281: 996–8.
42. Wylie HG (1962). An effect of host age on female longevitiy in *Nasonia vitripennis* (Walk.) (Hymenoptera, Pteromalidae). *Can Entomol.* 94: 990–3.
43. Wylie HG (1966). Survival and reproduction of *Nasonia vitripennis* (Walk.) at different host population densities. *Can Entomol.* 98: 275–81.
44. Davies I (1975). A study of the effect of diet on the life span of *Nasonia vitripennis* (Walk.) (Hymenoptera, Ptermoalidae). *J Gerontol.* 30: 294–8.
45. Page Jr. RE, Peng CYS (2001). Ageing and development in social insects with emphasis on the honey bee, *Apis mellifera*, L. *Exp Gerontol.* 36: 695–711.
46. Carey JR (2001). Demographic mechanisms for the evolution of long life in social insects. *Exp Gerontol.* 36: 713–22.
47. Wilson EO (1971). *The Insect Societies*. Cambridge, Mass: The Belknap Press of Harvard University Press
48. Calderone NW (1998). Proximate mechanisms of age polyethism in the honey bee, *Apis mellifera* L. *Apidologie* 29: 127–58.
49. Guzmán-Novoa E, Page Jr. RE, Gary NE (1994). Behavioural and life-history components of division of labor in honey bees (*Apis mellifera* L.). *Behav Ecol Sociobiol.* 34: 409–17.

50. Naugh D, Gadagkar R (1998). The role of age in temporal polyethism in a primitively eusocial wasp. *Behav Ecol Sociobiol.* 42: 37–47.
51. Chapuisat M, Keller L (2002). Division of labour influences the rate of ageing in weaver ant workers. *Proc R Soc Lond Ser B* 269: 909–13.
52. Moritz RFA (1994). Nourishment and sociality in honeybees. In: Hunt JH, Nalepa CA, eds. *Nourishment and Evolution in Insect Societies*. Boulder, San Francisco, Oxford: Westview Press, pp. 345–90.
53. Neukirch A (1982). Dependence of the life span of the honeybee (*Apis mellifica*) upon flight performance and energy consumption. *J Comp Physiol B.* 146: 35–40.
54. Keller L (1998). Queen lifespan and colony characteristics in ants and termites. *Insectes Soc.* 45: 235–46.
55. McCay CM, Maynard L (1935). The effect of retarded growth upon length of life and upon ultimate size. *J Nutr.* 10: 63–79.
56. Masoro EJ (2000). Caloric restriction and ageing: an update. *Exp Gerontol.* 35: 299–305.
57. Kaitala A (1991). Phenotypic plasticity in reproductive behaviour of waterstriders: trade-offs between reproduction an longevity during food stress. Funct Ecol, 5: 12–18.
58. Ernsting G, Isaaks JA (1991). Accerlerated ageing: a cost of reproduction in the carabid beetle *Notiophilius biguttatus* F. *Funct Ecol.* 5: 299–303.
59. David M, van Herrewege J, Fuillet P (1971). Quantitative under-feeding of *Drosophila:* effects on adult longevity and fecundity. *Exp Gerontol.* 6: 249–57.
60. Le Bourg E, Medioni J (1991). Food restriction and longevity in *Drosophila melanogaster. Age Nutr.* 2: 90–4.
61. Partridge L, Green A, Fowler K (1987). Effects of egg-production and of exposure to males on female survival in *Drosophila melanogaster. J Insect Physiol.* 33: 745–9.
62. Chippindale AK, Leroi AM, Kim SB, Rose MR (1993). Phenotypic plasticity and selection in *Drosophila* life history evolution. I. Nutrition and the cost of reproduction. *J Evol Biol.* 6: 171–93.
63. Chapman T, Partridge L (1996). Female fitness in *Drosophila melanogaste*r: an interaction between the effect of nutrition and of encounter rate with males. *Proc R Soc Lond Ser B.* 263: 755–9.
64. Clancy DJ, Gems D, Hafen E, Leevers SJ, Partridge L (2002). Dietary restriction in long lived dwarf flies. *Science* 296: 319.
65. Riha VF, Luckinbill LS (1996). Selection for longevity favours stringent metabolic control in *Drosophila melanogaster. J Gerontol A* 51: B284–49.
66. Luckinbill LS, Foley P (2000). The role of metabolism in ageing. *J Am Ageing Ass.* 23: 85–93.
67. Sohal RS, Weindruch R (1996). Oxidative stress, caloric restriction, and ageing. *Science* 273: 59–63.
68. Holliday R (1989). Food, reproduction and longevity: is the extended life span of caloric-restricted animals an evolutionary adaptation? *Bioessays* 10: 125–7.
69. Shanley DP, Kirkwood TBL (2000). Caloric restriction and ageing: a life-history analysis. *Evolution* 54: 740–50.

Genetics of Aging in *Drosophila*

Linda Partridge and Scott D. Pletcher
Department of Biology, University College London, Darwin Building, Gower Street, London, WC1E 6BT, United Kingdom

Introduction

Recent years have seen what can only be described as an all-out assault on the genetic mechanisms of aging in *Drosophila*. New technology and innovative experiments have led a two-pronged attack: one at the level of single-gene intervention and a second at the level of whole genome analysis. More than ten years elapsed between verification that the rate of aging was heritable in flies [1] and discovery of the first single mutant that extended lifespan [2]. In the four years since, 10 or so additional genes have been identified. Research in *Drosophila* has also embraced the genomics revolution, revealing key insights into the process of aging itself.

Invertebrate model organisms allow more rapid progress in genetic analysis of aging than do mammals, and they are therefore vital in the initial discovery of important genes and mechanisms. *Drosophila*, like the other major invertebrate models of aging, yeast (*Saccharomyces cerevisiae*) and the nematode worm (*Caenohabditis elegans*), are small and relatively short-lived. However, it is much easier to culture and monitor very large numbers of individual flies than it is yeast or worms, which is a great advantage for demographic analysis [3]. Demographic analyses use age-specific mortality rates to study the pattern of physiological deterioration with advancing age and to identify specific ages over which an experimental manipulation affects patterns of aging [4]. For instance, this approach has been used in *Drosophila* to demonstrate that, although male flies homozygous for the mutant *chico* (see below for further discussion of this mutant) have shortened mean and maximum lifespan, their mortality rate increases with age more slowly than does that of control flies. Their shortened lifespan is a result of an elevated death rate at all ages. In other words, *chico* males are sick, but they age slowly [5, 6]. Examination of the age-specific effects of genetic and environmental interventions will be important in looking beyond the identification of interventions that increase lifespan to an understanding of the nature of the underlying mechanisms through which they act [3].

H.D. Osiewacz (ed.), Aging of Organisms, 125–161.

If the findings in invertebrates are to be of relevance in mammals, particularly humans, mechanisms of aging must be conserved over the very large evolutionary distance involved [7, 8]. Early indications are that, indeed, there are specific genetic pathways that regulate aging in diverse species [8]. Comparative work in *Drosophila* on mechanisms discovered in *C. elegans* and in yeast is therefore a valuable step in the assessment of evolutionary conservation. This comparative approach led to the discovery that, as in *C. elegans*, reduction of the activity of the insulin/IGF-like signalling pathway can extend lifespan in *Drosophila* [5, 9]. In addition to conservation of specific genetic effects, recent advances in comparative genomics have identified candidate biological systems that are regulated similarly in aging *C. elegans* and *Drosophila* [9a, 9b].

Interventions, whether genetic or environmental, that slow aging are potentially of great value for understanding its mechanisms because they may specifically impact the normal aging process. Several genes potentially important in slowing aging in *Drosophila* have been identified from programmes of mutagenesis for increased lifespan [2, 10–13]. Mutant expression of known genes affecting candidate processes, such as defence against oxidative damage [14–17] and other forms of cellular stress and molecular damage e.g., [18, 19] has also been used. Mutant gene expression that results in extended lifespan or slowed aging identifies genes that may contribute directly either to the generation of aging-related damage or to defence against it. Genome level studies of gene expression have discovered candidate genes for slowed aging and have provided evidence that, at the level of the transcriptome, caloric restriction slows many of the normal changes that occur with age [20].

In this chapter we expand upon these approaches and findings and discuss modern genetic analysis of mechanisms of aging in the fruit fly *Drosophila melanogaster*. We begin on an historical note by illustrating some of the reasons why *Drosophila* can provide unique insights into the genetics of aging. We then review the methods that have recently been used to understand genetic mechanisms of aging in the fly. The details of the methods are often important, because they can determine the limits to what has been and, in some cases, can be, learned from particular approaches. We then consider what the experimental work has revealed. Finally, we discuss directions for future work. Throughout the article we define aging as the loss of reproductive potential with age, both through increasing vulnerability to death and through declining age-specific fecundity [21, 22].

An historical note

Drosophila as a model system for aging research is unique in that a body of established work has provided an evolutionary and ecological context for interpreting manipulations that extend lifespan. The fly has been used extensively in analysis of the effects of quantitative, polygenic, genetic variation on aging. Like humans, but unlike yeast and the nematode worm, the fly is an outbreeder. Its populations therefore harbour significant levels of natural genetic variation. This variation has been used to show that the rate of aging in *Drosophila* is heritable [1, 23–29], that lifespan and fecundity are both reduced by naturally-occurring new mutations [30–34] and that both traits show inbreeding depression [35–39].

Drosophila has also been used extensively in experimental work on the evolution of aging. Most of our current understanding of the population-genetic routes to the evolution of aging has come from the study of the properties of quantitative genetic variation in *Drosophila* [40, 41]. This work has shown that aging in the fly has evolved at least in part because of a delayed trade-off between reproductive rate and subsequent aging. Higher rates of reproduction lead to a more rapid subsequent increase in aging-related mortality [1, 24, 25, 28, 42, 43].

These evolutionary and ecological considerations have two important implications for molecular genetic studies of aging. First, there is considerable natural variation in lifespan and, depending on their respective culture and selection regimes in the laboratory, different strains may routinely exhibit very large differences in patterns of aging. Crosses between such strains may have unpredictable results. Second, the accumulation and fixation of deleterious mutations may lead to closely related stocks quickly becoming genetically differentiated, even in the absence of selection. The rate and average effects of new mutations that affect aging are substantial [32] and they may have a large influence on survival even at very old ages [33, 34]. The presence of deleterious mutations means that crosses between fly stocks often lead to extended longevity in the F1 flies through heterosis [44], a potentially dangerous confounding influence in studies using binary or tertiary transgenic systems (see below).

Despite these early insights, quantitative genetic variation has not proved useful for understanding the mechanisms by which *Drosophila* ages. None of the genes producing quantitative genetic variation for survival or fecundity have been identified. Nor are the mechanisms that cause the association between high reproductive rate and shortened lifespan understood. Although associations between increased lifespan and various types of stress-resistance have been reported [45–53], these associations are not always found [54, 55], suggesting that they may not indicate causality. This view is further supported by the finding that selection for stress-resistance *per se* does not lead to extended longevity [56–58], suggesting that stress resistance is not sufficient to slow aging. Resolving these issues requires more incisive experimental interventions.

Methods

The importance of genetic background

A powerful tool kit has been developed for genetic analysis in *Drosophila* (Table 1). Several methods of mutagenesis and transgenesis can be used to produce mutant over- or under-expression of genes. However, it is important that, in the course of pursuing identified genetic effects, we do not put the lights out behind us and forget essential lessons learned from quantitative genetic analysis of aging in the fly. Quantitative genetic effects on aging can be substantial, often much greater than those of mutants in single genes.

An important source of potentially uncontrolled variation in *Drosophila* comes from inbreeding depression [35–39, 59]. Adult survival and fecundity both show substantial effects of inbreeding depression, presumably because deleterious recessive mutations present in this outbreeding species are more likely to become homozygous,

Table 1. *Summary of various methods used to study genetic effects on aging in Drosophila*

Method	Advantages	Disadvantages	Aging genes discovered/ investigated
Quantitative genetics	Examines natural variation; provides genetic, evolutionary, and ecological context for experimental manipulations	Difficult to identify specific genes that affect aging	None
Mutagenesis			
Standard P-element mutagenesis	Relatively easy. Genomic position of insert is readily determined	Positional effects, size effects. Mutant may have effects in pre-adult life stages	*mth*, *chico*, *Inr*, *Rpd3*, *Indy*
Gal4-UAS elements/ EP and gene search	Allows mis-expression screens. Affected gene is usually identifiable. Tissue- (and to some extent stage-) specific mis-expression is possible	Binary system makes rigorous control of genetic background difficult	dPOSH, *hsp26*, nla, etc. See Seong *et al.* 2001
EP elements + adult induction (Flip-out, RU486 or tetracycline)	Temporal control of expression is better controlled: pre-adult deleterious effects are avoided	Binary system requires control of genetic background. Induction method may impact aging	α-mannosidase II
Targeted disruption/over-expression			
RNAi	Possible temporal control for loss-of-function manipulations	Sometimes difficult to implement. Especially susceptible to "leaky" expression	*Pgm*
Homologous recombination	Precise control of target gene and type of mutation.	Difficult and time-consuming to implement on a large scale	None
Promoter–gene fusion	Genes can be over-expressed in a "natural" way	Position and size effects of insert are difficult to control for	Cu/Zn-SOD, catalase, *hsp70*, Mn-SOD

Table 1. *(continued)*

Method	Advantages	Disadvantages	Aging genes discovered/ investigated
Gal4-UAS	Different combinations of tissue and stage specific expression can be rapidly examined. Many lines/genes are currently available with/as UAS constructs	Genetic background control difficult. Many different promoters may need to be tested	PCMT, MSRA
Gal4-RU486-UAS	Temporal control of transgene expression. Many lines/genes are currently available with/as UAS constructs.	Individual variation in exposure to inducible drug. Induction method may impact aging. Genetic background control difficult.	None
FLP/FRT "flip-out"	Constitutive expression is induced with transient heat-shock.	Possible cellular variation in flip-out efficiency. Heat pulse may have unwanted effects on aging.	Cu/Zn-SOD, catalase, Mn-SOD
Expression Monitoring			
Northern blots/Quantitative PCR	Can measure relative/absolute transcript levels. Direct measure of expression.	Difficult to do on a large scale (ie, for many genes)	*hsp70*, *hsp22*
Microarray	Can monitor many genes simultaneously with a small amount of starting material	Nearly always restricted to measures of relative transcript levels	Immunity genes, etc. See Pletcher *et al.* 2002.
Enhancer traps	Can precisely observe and measure cell, tissue and stage specific expression levels. Can determine absolute transcript levels	Difficult to do for very many genes. Expression patterns may be artifacts	Immunity genes. PGRP-LC. *Indy*

and hence to have deleterious phenotypic effects, in the offspring of matings between relatives. Inbreeding itself does not always abolish longevity differences among strains [59], but unequal effects of inbreeding depression in mutant and control strains must be avoided. Ideally, a mutant genetic effect that apparently slows aging in an inbred strain should also be tested in either outbred lines or in designed crosses among different strains [e.g., ref. 60] to ensure that it is not merely rescuing the effects of inbreeding depression. Inbreeding effects are most difficult to control when examining mutations in heterozygous form; the possibility that increased longevity is an artefact of heterosis at other loci must be ruled out.

A second problem is that adult survival and fecundity evolve rapidly in response to laboratory culture [61–64]. In a typical fly lab, the newly emerged adult flies are tipped into new culture vessels, allowed to lay eggs, and then discarded. This regime, when applied to flies collected from nature, leads to a rapid increase in early fecundity, which is selectively favoured under this culture regime, and also to a large decline in adult survival rates, as a correlated response to the increase in early fecundity. Furthermore, new mutations can also produce genetic differentiation between laboratory stocks [32]. As we mentioned above, this means that different strains of wild type flies found in *Drosophila* laboratories vary substantially in their adult survival and fecundity. It is therefore essential that the effects of mutant gene action be studied in a standard genetic background.

All *Drosophila* transgenic strains are produced by the insertion of transposons, nearly always the *P* element vector. The strains and crosses necessary to detect the presence of the newly inserted transposon, to determine into which chromosome(s) it has inserted, and to create a stable strain carrying the insertion often introduce genetic material from the other strains into the genetic background of the mutation. Judicious use of balancer chromosomes can place transgenes in an approximately matched genetic background to that of controls, although such matching is never perfect [65]. The alternative approach, backcrossing of the mutant or transgene to a standard, control genetic background, will always leave some linked variation at other loci. Other lines of evidence that the effect is due to the gene in question are therefore desirable. For instance more than one independently-derived mutant or transgene could be used [e.g., refs 15, 16]. Rescue of the mutant phenotype could be demonstrated by introduction of a wild type transgene [e.g., ref. 5] or a relevant environmental manipulation such as a ligand [e.g., ref. 66]. However, where "rescue" consists of taking slowed aging back to normal, the possibility of a non-specific deleterious effect of the intervention should be borne in mind and experimentally ruled out.

An additional issue with the use of transgenes comes from position effects, which manifest themselves in two ways. First, depending upon where in the genome a transgene is inserted, the extent and the spatial and temporal patterns of its expression will vary [65]. Second, transgenic insertions can produce substantial and highly variable mutagenic effects upon lifespan. These effects will depend upon their size, their position of insertion and the genetic background into which the insertion is made [67]. Random insertions tend to lower lifespan in *Drosophila* [67, 68]. In fact, mobilization and insertion of *P* element constructs is a standard method of mutagenesis in *Drosophila*.

One method for controlling for the mutagenic effect of insertion is to estimate the impact of the transgene on the *distribution* of lifespans in multiple control and insert lines, each with an independent transgenic insert [e.g., ref. 69]. However, because the size of an insert can have a bearing on its mutagenic effects, the control inserts should be of the same size as those containing the transgene, and should not consist only of the empty vector. This could be accomplished by inserting a mutant transgene that is not expressed [65], but this control has in general not been built into experimental studies. Data from a real experiment with transgenic *Drosophila* [70] have been used to estimate that 15 experimental and control lines would have been needed to detect a 10% difference in lifespan as a result of transgene expression [71]. Systems for working with a transgenic insert in the same genomic location in the experimental and control stocks have therefore been developed [72].

Mutagenesis

Mutagenesis has been used in *Drosophila* to induce new mutations and then to screen them for an effect on lifespan. Mobile genetic *P* elements have been used in preference to chemical mutagens, because the position of the insert can be located and multiple hits detected by *in situ* hybridization, and the flanking sequences around the *P* insert can be cloned and sequenced to identify the gene whose sequence has been disrupted. In general this type of procedure generates loss-of-function mutations, because it disrupts genes and their control sequences. The genes so identified, by analogous reasoning to the biology of oncogenes and tumor-suppressors, are most likely to be recessive and to encode elements of processes that contribute to the generation of aging-related damage. *Methuselah* [2] and *Indy* [10] were both identified by this type of procedure.

Methods of mutagenesis have also been developed to allow screens for effects of gene over- or mis-expression on lifespan. In general, the genes detected in over- and mis-expression screens are likely to be ones involved in defence against aging-related damage and they are likely to be dominant. In the gene search system [73], the *P* element vector used for mutagenesis contains Upstream Activating Sequences (UAS) that direct transcription outwards from the element when present in the same fly as a second *P* element vector encoding GAL4, the transcription factor for UAS (the "EP" element of ref. 74). Depending upon the characteristics of the GAL4 driver used, different temporal and spatial patterns of expression of the gene search vector are produced. Because of the crosses involved, the controls used in this type of screen usually have a different genetic background from the experimental flies. GAL4 inserts also differ in their genomic location, and their mutagenic effects must therefore be considered. In these cases, confirmatory evidence of a role of the identified gene in aging is therefore required.

When studying aging in *Drosophila* we are interested in events that happen during the adult stage of the life history. Gene action during pre-adult development may affect the rate or style of the subsequent aging of the adult and could be of interest for the aging process. However, mutant gene action during development may also be lethal or deleterious, and for the study of subsequent aging is best avoided. Mutagenesis screens where mutant gene action is confined to the adult period can be achieved by systems for conditional expression of the induced mutations [65]. In *Drosophila*, three types of inducible system have been used in mutagenesis programmes for lifespan. In the first, the temperature-sensitive *heat shock protein 70* (*hsp70*) promoter is used. This is placed upstream of the GAL4 driver element in the gene search system and is activated only at high temperature [11, 13] (Figure 1A). The flies are reared at normal temperature (25°C) to avoid pre-adult effects of gene mis-expression and then screened at 30°C, which activates the *hsp70* promoter. As with other GAL4–UAS systems, this one can lead to difficulties with genetic background differing between experimental and control strains, although by screening at both permissive and restrictive temperatures some control for genetic background can be built in. Another potential problem for inducible systems is that the inducer may have a direct effect on lifespan. For instance, heat shock itself can increase life expectancy in flies as a result of a medium-term increase in survival rates [75]. This kind of direct effect can be controlled for to some extent by exposing the control flies to the same thermal regime but, if they differ from the experimental flies in genetic background, complications could arise. Finally, inducible systems can show leaky expression in the absence of the inducer, and this possibility should be ruled out experimentally.

The second inducible technique is the UAS-linked RU486 ("GeneSwitch") system [76, 77]. This system takes advantage of the previously described "EP" and gene-search system, which are both based on UAS constructs, and extends it to allow for precise temporal control of expression (Figure 1B). The standard GAL4 driver is replaced with a fusion of GAL4, the human Progesterone-receptor-ligand-binding-domain, and the human protein p65. This chimeric molecule binds to a UAS only in the presence of the anitprogestin, RU486. This system can be used to drive temporally defined mis-expression of genes in many existing UAS fly stocks.

A third inducible system, as well as allowing age-specific mutant gene expression, also builds in a control for genetic background and for position and mutagenic effects of the insert. The system requires two *P* element constructs. In the first, the cytoplasmic actin (*actin5C*) promoter is used to drive constitutive expression of the *r*everse *t*etracycline *t*rans *a*ctivator in a construct called *rtTA* (Figure 1C). The second contains a tetracycline-inducible promoter upstream of the *hsp70* core promoter that directs transcription outwards from the element. This system is therefore inducible by the antibiotic deoxycycline (the "tet-on" promoter) [12]. In this type of screen, flies of the same genotype are maintained either in the presence or absence of deoxycycline and are compared for lifespan. Deoxycycline itself has been shown to be without effect on fly lifespan at the doses used [78]. This type of inducible system has led to the isolation of several mutants that increase fly lifespan [12].

Targeted gene disruption and over-expression

Candidate genes that have been examined for a role in aging in *Drosophila* have been both disrupted and over-expressed. Targeted gene disruption in *Drosophila* can be done by a variety of methods that take advantage of the fact that the gene is mapped and that other genetic reagents are available in the genomic region. For instance, several of the mutations in the insulin/IGF-like signalling (IIS) pathway that have

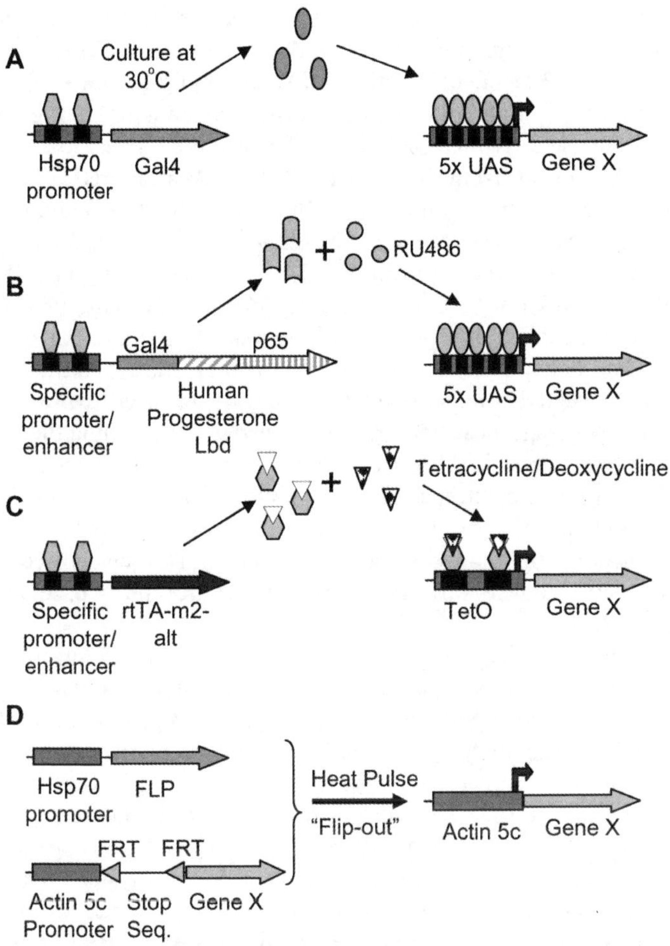

Figure 1. Illustrations describing four methods of inducing transgene expression in adult Drosophila. All are binary systems requiring two genetic constructs be present simultaneously in the genome of the adult fly. (A) hsp70-GAL4: Culturing adults at 30°C induces expression of GAL4 via the hsp70 promoter. The GAL4 then binds to its corresponding **u***pstream* **a***ctivating* **s***equence (UAS) to initiate expression of the target gene of interest (Gene X). Eggs and larvae are cultured at lower temperatures where the promoter is largely inactive. (B) GeneSwitch: Expression of a protein composed of the GAL4 DNA binding domain, human progesterone ligand binding domain, and human protein p65 is driven by a ubiquitous or tissue-specific promoter. This chimeric protein binds to UAS only in the presence of the anitprogestin, RU486. Expression of Gene X is therefore initiated when adult flies are fed RU486. (C) Tet-on: Similar to the RU486 system. A specific promoter drives expression of rtTA (***r***everse* **t***etracycline* **T***rans* **A***ctivator), which only binds to the tet activating sequence (TetO) in the presence of tetracycline or its analog, deoxycycline. (D) Flip-out: A short heat pulse is used to drive transient expression of the yeast FLP recombinase via the hsp70 promoter. FLP catalyses recombination at target FRT sites, which initially flank a transcriptional stop sequence separating Gene X from the Actin5c promoter. Transient expression of FLP stimulates excision of the FRT flanked sequence resulting in expression of Gene X.*

been examined for effects on lifespan in *Drosophila* were first induced to examine their effects on growth [e.g., refs. 79, 80]. In the case of the *chico* gene, the first mutant was the result of a *P* element insertion into the coding region of the gene while a second was induced by imprecise excision of the *P* element and hence excision of part of the gene [80]. The expression of specific, targeted genes can be reduced or abolished in *Drosophila* using double stranded RNA interference (RNAi) [81] and this interference can be made inducible both by heat shock [82] and by deoxycycline [83]. RNAi can be used to suppress the expression of an endogenous gene in order to test the effects of specific candidate alleles of the same gene. For instance, suppression of expression of the endogenous *phosphogluconate mutase* (*Pgm*) gene allowed a test of the effect of experimental expression of two alleles of the *Pgm* gene, one of which had been found to increase in frequency relative to the other in lines of flies artificially selected for increased lifespan [84]. Induced expression of the two alleles resulted in no significant difference in lifespan [85], leading the investigators to conclude that the change in allele frequency observed in the selection lines was not causal in the extension of lifespan. Homologous recombination [86, 87] can also be used for gene knock-out and for targeted mutagenesis of specific genes in *Drosophila* [88]. However, this procedure usually produce mutations with unconditionally altered expression and so far this approach has not been used for analysis of mechanisms of aging.

Several systems have been used to examine the effects of over- or mis-expression of specific genes on fly lifspan, all of them based on the use of transgenes in *P* element vectors [65]. In some studies the vector has contained the gene in question together with its 5′ and 3′ regulatory regions [69, 70, 89–91]. In these cases, the expected outcome is over-expression of the gene at the stages and in the tissues where it is normally expressed. Other systems allow the spatial and temporal pattern of expression of the transgene to be altered. One method is to use a heterologous promoter either in a single transgene or the GAL4–UAS system. The usual caveats about genetic background, position effects and insertional mutagenesis apply. Position effects can be ameliorated by using a system that induces germline recombination to produce transgene over-expression. This can be done using the yeast-derived FLP recombinase to catalyze recombination at its FRT target sites [15, 92, 93]. By putting the FLP recombinase under the control of the *hsp70* promoter, recombination can be induced by a heat pulse, and therefore confined to the adult period (Figure 1D). The transgene of interest is inserted into a second *P* element vector, downstream of a promoter such as actin 5C. Initially, transcription of the transgene is inactivated by the incorporation of a transcriptional "stop" sequence, flanked by FRT sites, between the promoter and the gene sequence. Upon a heat pulse, FLP is activated and catalyzes recombination between the FRT sites and hence the excision of the stop sequence; transcription of the transgene is hence activated. In addition to its other advantages, this system avoids leaky expression of the transgene until it is activated. However it does rely on heat shock activation, which introduces complications from its direct effects upon survival. Inducible systems such as tet-on [78] and the UAS-linked RU486 [76, 77] may provide improved systems for induction.

Changes in gene expression with age

Changes in gene expression with age are of interest because they can provide clues about how the organism is being affected by (or responding to) the aging process. For instance, the observation that many immune system genes increase in expression with age (see below) motivates a study of the role of infection in aging. Two fundamentally different techniques have been used to examine changes in gene expression with age: nucleic acid hybridization and "trapping" the expression pattern of a particular gene as it is being transcribed. Before discussing these methods, however, we highlight two important issues that should be kept in mind when considering the different methods.

When studying gene expression, there is always the possibility that changes in transcript representation are not mirrored where it is biologically important, at the protein level. This argument has been directed toward expression studies since their inception, and it has become increasingly prevalent with the rise of genomics [94, 95]. For studies of aging, one must not only be concerned with the relationship between RNA level and protein at a single age but also be wary of the possibility that the relationship changes with age. For example, a study looking at transcript representation, protein level, and protein activity for a specific gene, EF-1α, showed that these three measures of biological function may provide very different information depending on the age of the animal [96]. In one experiment, RNA transcript representation for EF-1α remained roughly constant (possibly even increasing) with age, while protein level decreased slightly only early in life, and protein activity decreased considerably throughout the lifespan [96]. Although measures of transcript representation have been shown to correlate with protein level and activity [97], the idea that some of the effects of aging are confined to the level of the proteome [98] suggests that we need to be especially vigilant about this connection.

A second issue derives from the growing evidence for a significant decline in both total and ribosomal RNA levels [96] (S.P. and L.P., unpublished data) in normally aging flies [20]. If this is true for mRNA levels as well, then normalized measures of gene expression (obtained by dividing the expression level of a target gene by that from a presumed "housekeeping" gene) are difficult to interpret. Normalized measures present data on a relative scale, where the expression level of a particular gene not only depends on the mRNA abundance for that gene but also on abundances of every other gene in the genome. If, for example, a small number of genes that are highly expressed early in life get turned-off at older ages, the relative contribution of genes whose rate of transcription is unaffected by age will increase [20]. Attention might be drawn to these genes in order to determine what aspect of aging is causing such an "induced response," when, in this example, the true effect of aging is being mediated through down-regulation of an entirely different set of genes. Authors seldom clarify whether expression is presented in absolute or relative terms.

The use of hybridization techniques to study gene expression and aging is manifest in three forms: northern blots, reverse transcription (RT) PCR, and DNA microarrays. Northern blots require a comparatively large amount of starting material and they are essentially a single-gene assay. This is somewhat restrictive for studying complicated processes like aging. An important advantage of northern blots is that the amount of RNA that is transferred to the membrane can be adjusted in such a way as to measure differences in either relative or absolute levels of a particular transcript. Unlike northern blots, RT-PCR requires only a very small amount of fly RNA for what is essentially a two step procedure [99]. This approach can be very reliable when real-time PCR methods are used to quantify the amount of transcript [99, 100]. Although it is sometimes possible to quantify differences in both relative and absolute amount of transcript for a particular gene, it can be difficult to obtain both types of information when a treatment has a large impact on global rates of transcription, as aging seems to do [99–101]. Over the last 3–5 years a third application of hybridization has gained general acceptance for measuring gene expression: DNA microarrays [102]. RNA is extracted from a group of flies, and non-specific primers are used to generate labelled cDNA or cRNA from virtually all of the transcripts in the RNA pool, which is then hybridized to the microarray. In this way quantitative measures of expression can be obtained simultaneously for every gene represented on the array using a single sample of RNA [20]. The use of DNA microarrays for studying a complex, multigenetic process, such as aging, is quite alluring. Global patterns of expression with advancing age can be observed and analyzed, and candidate genes with strong age-dependent regulation can be rapidly screened [20].

A second method of characterizing gene expression is fundamentally different from hybridization and seeks to capture, *in vivo*, the transcriptional activity in and around a gene of interest [103, 104]. In this technique, a *P*-element containing a transgene is allowed to integrate into the *Drosophila* genome. Depending on where the *P*-element ends up, regulatory sequences surrounding the integration site may stimulate transcription of the transgene. Specific properties of transgenes that make them more or less vulnerable to these position effects are known and have been used to engineer constructs called "enhancer traps," which are especially susceptible [104]. The hope is that the transgene will be expressed in a pattern similar to a nearby gene whose enhancers it has "trapped." Enhancer traps usually contain either a specific reporter gene, such as lacZ, or the GAL4 gene for use in combination with UAS transgene reporters. The expression of the reporter genes are then monitored using standard methods [105, 106].

Enhancer traps are useful in that they can be mobilized to new locations and transcription of the reporter genes they drive is easily measured and localized. As a result, detailed expression patterns in specific tissues and even specific cells can be characterized [106, 107]. Rapid screens for tissue- and stage-specific expression are possible and, once characterized, enhancer trap lines that encode GAL4 can be used to drive any gene of interest in a tissue- and age-specific manner [107]. There are however, certain limitations when using this technique to infer patterns of gene expression. Depending on the regulatory complexity of nearby genes, the enhancer trap may come under the influence of only a subset of the regulatory elements that are characteristic of any particular gene or may be subject to combinations of enhancers from multiple genes [108]. The resulting expression pattern may therefore not be representative of any specific gene. Of course, interesting expression patterns discovered using this method can be verified using northern blots, RT-PCR, or even microarrays. Because expression of a reporter construct can be quantified in an individual fly without any normalization, this is a powerful method for measuring changes in absolute levels of expression.

Mechanisms of aging in *Drosophila*

Genes identified by homology – the insulin/IGF-like signalling pathway and histone acetylation

The insulin/IGF-like signalling pathway

Genes in the insulin/IGF-like signalling (IIS) pathway in *Drosophila* were first examined for their effects on aging because their orthologues extend lifespan in *C. elegans*. The pathway in the worm was first discovered, not because of its effect on aging, but rather because single gene mutations caused an alternative pattern of pre-adult development. In wild-type *C. elegans*, under conditions of food scarcity and high population density, rather than developing directly to a reproductive adult, a dauer larva is formed. Dauers are long-lived, stress-resistant, lipid-storing and non-feeding, and they can resume development to reproductive adults if food levels increase [109]. A group of mutations was isolated in the worm that caused dauer larvae to be formed even in the presence of abundant food and the absence of crowding. Weak and temperature-sensitive mutations in these same genes gave rise to long-lived adults. The first of these genes to be identified, *age-1*, encoded a lipid kinase enzyme (phosphatidylinositol-3 kinase) [110]. Another gene, *daf-2*, encoded a homologue of the mammalian insulin/IGF receptor [111, 112]. This insulin/IGF-like signalling (IIS) pathway in the worm has since been shown to act in both dauer-formation and aging by down-regulating the activity of the DAF-16 forkhead transcription factor [113, 114]. A plausible guess is that DAF-2 responds to an insulin-like ligand. Bioinformatic analysis has shown that the *C. elegans* genome (which has been completely sequenced) contains only a single insulin/IGF receptor gene, but that there are an astonishing 37 *C. elegans* genes encoding insulin-like proteins. Of these, INS-1 most resembles human insulin. Using genetic tests, both INS-1 and human insulin have been shown to modulate the action of DAF-2 [115].

The *Drosophila* equivalents of *daf-2* (*Inr*), and *age-1* (*Dp110*), and also *chico*, a fruitfly insulin receptor substrate, were first identified by their homology to mammalian genes and because hypomorphic and null mutations in them reduce growth rate and eventual body size [79, 80, 116–119]. Using a heteroallelic combination of hypomorphic mutants in *Inr* and a null mutation in *chico*, back-crossed into the genetic background of the controls, it was demonstrated that a mild reduction of *Inr* function increased mean female lifespan by up to 85% [66], and loss of function of *chico* by up to 52% [5]. These fly genotypes also show slowed growth and give rise to dwarf adults. The dwarfism may not be causal in the extension of lifespan since females heterozygous for *chico* are also long-lived but normal in body size [5]. It thus appears that the effects of IIS on aging are conserved over large evolutionary distances, and that this is therefore a candidate mechanism for modulating aging in mammals [7, 8, 120].

IIS in the worm appears to be a two-step endocrine signalling process, with the cells that express the receptor *DAF-2* acting through the forkhead transcription factor *DAF-16* to produce an as yet unidentified second secreted ligand [121, 122], the receptor for which may be the steroid receptor DAF-12 [123–125]. A similar two-step endocrine signalling process may occur in *Drosophila*, because the rate of production of juvenile hormone by the *corpora allata* of *Inr* mutant flies has been reported to be reduced relative to that in controls. Moreover, administration of the juvenile hormone analogue methoprene to long-lived *Inr* genotypes reduces their lifespan to control values [9]. Only certain IIS mutations increase lifespan in the fly, and largely in females [5]. Males homozygous for *chico* are short-lived, although they also age slowly [5, 60]. Mutations in IIS that have so far been shown to increase lifespan in the fly also render the flies sterile or subfertile [5, 66, 80]. In contrast, it appears to be possible to extend lifespan in the worm without impairing fecundity [126], although resumption of fecundity after food shortage may be impaired [127]. Understanding the mechanisms by which this pathway extends lifespan and the reasons for the similarities and differences in its precise effects in different organisms will be a priority for future research.

The role of IIS in mediating the response of lifespan to caloric restriction (CR) has been investigated in *Drosophila*. CR can be carried out by food dilution [128]. As food concentration declines from a maximum, lifespan first increases to a peak at intermediate food concentration through CR and then declines through starvation. The daily and lifetime fecundity of the flies increases with food concentration throughout the concentration range. CR increases lifespan throughout the animal kingdom, by mechanisms that are not understood.

One method to identify genes that may mediate the response to CR is to measure the effect of CR on lifespan when candidate genes are mutated or over-expressed. If the response to CR is blocked or lowered, then the gene may have a role in the normal response. Some caution is required in the interpretation of such interaction experiments [129]. In particular, if lifespan is not maximized by either of the interventions, then the two treatments together may give a greater lifespan than either alone. Such a result cannot be taken to mean that the interventions act through different pathways, as has sometimes been suggested [e.g., refs. 130–133]. However, if lifespan is maximized by one intervention and is then not further increased by addition of the other, then the result suggests that they two act through a common pathway.

The *chico* mutation has been used to determine if reduced IIS and CR extend lifespan by common mechanisms in *Drosophila* [134]. The relationship between lifespan and food concentration was compared in *chico* homozygote and control females (Figure 2). At the food concentration that maximized lifespan in the controls, the *chico* flies were shorter-lived. The relationship between lifespan and food concentration was right-shifted in the *chico* flies relative to controls. These two interventions must therefore share common pathways and *chico* flies are calorically restricted by their genotype.

Histone acetylation

In addition to insulin/IGF signalling, modification of levels of histone acetylation are turning out to regulate lifespan across distant species. Chromosomal regions where histones are hyperacetylated are relatively "open" and characteristic of transcription-

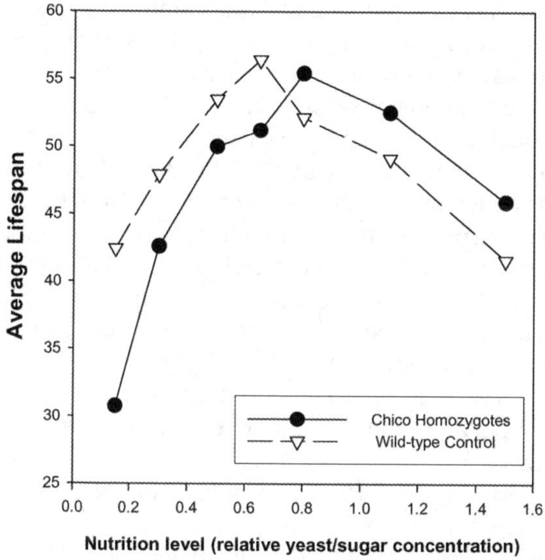

Figure 2. *Relationship between mean lifespan and nutrition level (× normal food used in Drosophila culture) for homozygous chico females and controls (redrawn from ref. 133). At the nutrition level that maximizes control lifespan chico females are shorter lived, and the whole relationship for chico females is right-shifted relative to controls.*

ally active domains [135]. Histones lacking acetyl groups condense the nucleosome structure and result in steric hindrance to the transcriptional machinery. Genes encoding histone acetyltransferases are involved in opening DNA through acetylation, while genes encoding histone deacetylases remove histone and thus are often involved in gene silencing through chromosomal condensation. Two histone deacetylase genes, SIR2 and RPD3, were first identified as important in determining replicative lifespan in yeast [136, 137].

SIR2 is an NAD-dependent histone deacetylase that is involved in silencing at mating type loci, at telomeres, and at the RDN1 locus in yeast, which consists of highly repeated copies of genes encoding ribosomal RNA [135, 138]. It also appears to function in DNA repair through non-homologous end-joining [139], but it is the silencing properties of SIR2 that have been shown to affect replicative aging in yeast. Deletion of SIR2 resulted in reduced lifespan [136]. A single extra copy of SIR2 increased lifespan [136], but high levels of expression were toxic [140]. It therefore appears that increased replicative lifespan requires a relatively precise amount of SIR2 protein. It has been argued that the mechanism responsible for increased longevity is the ability of SIR2 protein to silence activity at the rDNA locus, thereby reducing the formation of extrachromosomal rDNA circles (ERCs) [141]. However, whether the accumulation of ERCs is a cause or a consequence of aging in yeast has yet to be firmly established. There is evidence against a direct link. For example, petite yeast cells (cells that lack mitochondria) accumulate levels of ERCs comparable to wild-type cells, yet they are long-lived [142]. There is no evidence linking the formation of ECRs to aging in any other organism.

The closest nematode homolog of yeast SIR2 is *sir-2.1*, and a 50% increase in gene dosage, obtained through chromosomal duplication, increases lifespan in *C. elegans* by roughly 50% [143]. Lifespan increase by *sir-2.1* is dependent on a functional *daf-2* pathway, but the gene appears to act upstream of *daf-16* [143]. Extra copies of *sir-2.1* do not induce the pleiotropic effects that are characteristic of some mutants in the *daf-2* pathway: decreased brood size, increased larval lethality or increased dauer formation. However, *sir-2.1* does promote dauer formation when placed in a genetic background that has been sensitized to changes in insulin-like signalling [143]. These findings suggest that over-expression of *sir-2.1* weakly inhibits IIS.

Like SIR2, RPD3 is a histone deacetylase that affects replicative lifespan in yeast. There are, however, important differences between these genes. In contrast to SIR2, where increased levels of protein increase lifespan, it is a loss of function of RPD3 that makes yeast long-lived [137]. RPD3 does not require NAD for its deacetylation activity [135]. Gene expression studies revealed that deletion of RPD3 and SIR2 result in down- or up-regulation of several to many genes in addition to their known targets, suggesting that each regulate genes with distinct biological functions [145].

Several studies in yeast have identified a link between increased replicative lifespan by modification of histone deacetylase activity and by caloric restriction, but conflicting results among different laboratories make generalization difficult. Sir2p (the protein produced by SIR2) is required for extension of replicative lifespan caused by a putative genetic mimic of caloric restriction (the *cdc25-10* mutation) [146], but when caloric restriction is applied directly by limiting glucose or amino acid availability, lifespan extension via caloric restriction occurs in the absence of Sir2p. Deletion of RPD3 seems to block further extension by caloric restriction [137]. Indications are that there is some general interaction between histone deacetylase activity and the physiological effects of caloric restriction. However, the failure of the sir2 deletion to block the caloric restriction effect when modulated directly by glucose or amino acid concentration brings into question the importance of NAD availability as a mechanism of lifespan regulation by caloric restriction.

Investigations into the effects caused by manipulation of histone deacetylase activity in *Drosophila* are in their infancy. Currently there is no published evidence for modification of *Drosophila* lifespan by manipulation of Sir2. However, recent work has shown that moderate loss of function mutations in Rpd3 increase lifespan and block the caloric restriction effect [147]. Complete loss of function of this gene is lethal [148], suggesting that, as for the yeast SIR2, an intermediate amount of Rpd3p is optimal. Interestingly, Sir2 is upregulated in Rpd3 mutant flies [147]. In wild-type flies, caloric restriction resulted in significant down-regulation of Rpd3 throughout the lifespan. In contrast, Sir2 transcript representation increased with age but did not respond to caloric restriction [20]. Ingestion of the drug phenylbutyrate (PBA), which is a general inhibitor of histone deacetylase activity, can extend lifespan in flies [149], and the age-dependent progression of neuronal degeneration in *Drosophila* models of polyglutamine disease is halted upon treatment with histone deacetylase inhibitors [150]

Genes discovered in Drosophila – *mutagenesis*
Two mutations that increase lifespan in *Drosophila* have been discovered as a result of disruption of gene function, one of them fortuitously. The first, *methuselah* (*mth*), was discovered in a systematic screen for *P*-element insertions that increased lifespan [2]. The role of the *P*-element insertion in producing the increase in lifespan was confirmed by reversion to control lifespan when the *P*-element was precisely excised. The *mth* gene consisted of a single open reading frame and its sequence homology suggested that it encoded a G protein-coupled transmembrane receptor, of unknown function. The *mth* mutant appeared to be a partial loss of function as a result of insertion of the *P*-element in the third exon of the gene. Subsequent work has revealed that *mth* is a member of the secretin family of G protein-coupled receptors, and the structure of its extracellular domain and potential ligand-binding sites have been determined [151]. The receptor controls synaptic efficacy at glutaminergic neuromuscular junctions in the larva, and is required in the pre-synaptic motorneuron to acutely upregulate neurotransmitter exocytosis. In the mutant, synaptic transmission is reduced by 50%, apparently as a result of abnormalities in vesicle docking and clustering [152]. The mechanism by which *mth* extends fly lifespan is not understood. *mth* mutant flies show enhanced resistance to several forms of stress, including heat, starvation, and the free-radical generator paraquat [2], but the precise mechanisms at work await further analysis.

Mutants in a second gene, *Indy*, extend fly lifespan when heterozygous. Two long-lived mutants were initially found in a set of enhancer trap lines and others were subsequently obtained from other sources [10]. The role of the inserts in causing the increase in lifespan was confirmed by precise excision. The gene was found to encode a homologue of the rat and human sodium dicarboxylate cotransporters, which are responsible for the uptake or reuptake of Krebs cycle intermediates. The mechanisms by which *Indy* mutations extend lifespan are not clear. The homology of the gene and its pattern of expression in the adult fly suggests a role in absorption of metabolites and intermediary metabolism.

Two screens for extension of lifespan in *Drosophila* by conditional gene over-expression confined to adulthood have so far been reported. In the first, gene over-expression was driven by the tet-on system and the effect on lifespan of feeding with deoxycycline was determined for 13 insert lines [12]. In most of these, gene over-expression led to reduced lifespan but in one lifepsan was reproducibly increased by approximately 10%. The insert was found to be in the 5′ untranslated region of the α-*mannosidase II* gene, which is a Golgi apparatus enzyme involved in protein glycosylation. In the second screen the gene search system [73] was used, with heat shock being used to induce gene over-expression [11, 13]. Out of 646 lines screened, 23 that appeared to be long-lived relative to controls were subjected to further analysis. A total of 25 gene transcripts were overexpressed, 7 of these were associated with known *Drosophila* genes and 3 of these were related to stress response or redox balance. Among the remaining transcripts, one was homologous to the mouse gene POSH, and subsequent neural-specific over-expression of the *Drosophila* gene using the GAL4-UAS system was shown to be associated with a 14% increase in lifepsan

[153]. Further work will be needed to confirm the role of these genes in aging and to determine mechanisms at work.

Targeted gene manipulation – oxidative damage
Aging is caused by the accumulation of damage to macromolecules, cells and tissues. The free radical theory of aging [154] points to the key role of damage by Reactive Oxygen Species (ROS) as a public mechanism of aging [155–159]. Oxidative damage to macromolecules, membranes and tissues is a universal correlate of aging [155–157]. Furthermore, oxidative damage accumulates at a lower rate in long-lived bird and mammal species [e.g., refs. 160, 161]. ROS are produced as a toxic by-product of normal cellular metabolism, with mitochondria implicated as the main source [162, 163]. The mitochondrial electron transport chain generates the superoxide ($O_2^{\cdot-}$) anion as a consequence of leakage of electrons from the electron transport chain to molecular oxygen. Superoxide is generated mainly on the matrix side of the mitochondrial inner membrane, where it is converted to hydrogen peroxide (H_2O_2) by Mn-superoxide dismutase (Mn-SOD). Hydrogen peroxide can diffuse through the mitochondrial and other membranes. It may then undergo transition-metal-catalyzed, homolytic scission to generate the highly reactive hydroxyl free radical ($OH^{\cdot}$) and reactive nitrogen species, which can cause oxidative damage to a wide range of macromolecules [155–157, 164, 165].

Oxidative damage accumulates with age in *Drosophila* and other flies. The abundance of oxidized lipids [166], proteins [167–170] and DNA [171, 172] have all been reported to increase with age. Mitochondrial macromolecules such as aconitase [173, 174], the adenine nucleotide translocase [175] and cytochrome C oxidase [176] appear to be particularly vulnerable to aging-related oxidative damage. A comparison of 5 different species of fly, which ranged more than 2-fold in their lifespan potentials, showed that the longer-lived species accrued damage to proteins, as measured by protein carbonyls, at a slower rate. The longer-lived species also produced less superoxide and hydrogen peroxide from their isolated mitochondria [177]. Also consistent with the oxidative damage of aging are findings from work with the filamentous fungus *Podisma anserina* (see Chapter 2, this volume). Copper depletion leads to the induction of an alternative respiratory pathway that uses an alternative oxidase and also increases lifespan [178, 179].

The correlational evidence in general, and in *Drosophila* in particular, is thus consistent with a key role of oxidative damage in aging. To test the idea experimentally, it is necessary either to lower the rate at which ROS are generated, or to increase the rate at which they are removed, and to show that aging is slowed. In *Drosophila*, the second approach has been the one mainly adopted, by elevating endogenous or exogenous defences against ROS, and measuring the effect on lifespan.

Numerous studies have examined the effects of small molecule antioxidants added to the food on the lifespan of *Drosophila* [reviewed in ref. 180]. Extensions of lifespan have been reported for some chemicals, but other studies have sometimes found no effect or even shortening of lifespan with the same chemicals at the same doses. For instance, N-acetyl-cysteine [181] and propyl gallate [182] have both been suggested to increase lifespan in *Drosophila*, but these results were not replicated in an independent study (M. West and L. Partridge, unpublished data). It has recently been suggested that dietary melatonin [183] and pineal peptides [184–190] protect against oxidative damage and increase *Drosophila* lifespan. The robustness of these findings should be established by testing in other laboratories. In the nematode worm *C. elegans*, catalytic antioxidants fed to the worm in the diet have been reported to extend the lifespan of wild type adults [191]. However, independent experiments in a different laboratory have not replicated these findings [191a]. Administration of these catalytic antioxidants to house flies [192] and to *Drosophila* (M. West, D. Gems and L. Partridge, unpublished data) have similarly revealed only dose-dependent toxicity. The data from studies using exogenous antioxidants have hence not yet provided clear supporting evidence for the oxidative damage theory of aging, and have been beset by conflicting results. This could be explained in part by a publication bias towards studies that give positive results [180].

The inconsistencies of the anti-oxidant results are troubling. Even in circumstances where results are consistent, however, inferring a direct link between the physiological actions of a drug and extended lifespan is difficult. The interpretation of positive results is not straightforward. If a chemical that is expected to have antioxidant properties extends lifespan, it does not automatically follow that protection against oxidative damage is responsible. Studies of the effects of antioxidant administration do not in general measure the rate of accrual of oxidative damage or rule out the action of other mechanisms that could be at work. For instance, the drug could be aversive and hence induce the fly to eat less and become calorically restricted [128], which has been shown to be associated with reduced oxidative damage in other organisms [155]. The chemical could also reduce fecundity and in consequence extend lifespan [180]. Activity could also be reduced, which has been shown both to slow down the accrual of oxidative damage and to extend lifespan in houseflies [193]. Even if the drug both extends lifespan and reduces the impact of oxidative damage, it does not follow that the correlation is causal. Negative results are also hard to interpret. There are many reasons, including failure to reach the target, why a drug may fail to act. On the other hand, failure to slow aging by increasing exogenous defence against oxidative stress does not necessarily imply that oxidative damage is not a leading cause of aging. There may, for instance, be compensatory down-regulation of endogenous defence mechanisms.

Endogenous defence systems have also been manipulated in *Drosophila*. Insects in general lack glutatione peroxidase [194]. The enzymes superoxide dismutase (SOD) and catalase are therefore the major defences against superoxide and hydrogen peroxide respectively. In *Drosophila*, a single form of Mn-SOD is present in the matrix of the mitochondrion and of Cu/Zn-SOD in the cytosol. In most organisms, the disulphide-reducing enzymes glutathione reductase and thioredoxin reductase are highly conserved proteins that maintain intracellular redox homeostasis and protect against oxidative damage. *D. melanogaster* lacks glutathione reductase, and reduction of glutathione disulphide is carried out by a novel thioredoxin reductase [195, 196]. Interference with the activity of these defence mechanisms has been shown to have adverse effects on viability in *Drosophila*. For instance, null mutations in the gene *TrxR*, which encodes the novel thioredoxin reductase, are larval lethals and hypomorphic mutations drastically shorten adult lifespan [197, 198]. Null mutations in the gene encoding Cu/Zn-SOD both reduce resistance to oxidative stress and greatly reduce the lifespan of adult *Drosophila* [199, 200]. Catalase activity can be greatly reduced without any effect on fly lifespan [201] but homozygotes for null mutations do show reduced longevity [202].

The effects of over-expression of genes for mitochondrial (Mn-SOD) and cytoplasmic (Cu/Zn-SOD) superoxide dismutase and of catalase have been extensively studied in *Drosophila*. This series of experiments acted as a proving ground for different methods of gene over-expression. Initially, the effects of unconditional over expression of Cu/Zn-SOD and catalase was examined, using constructs containing the entire genomic gene sequences [70, 89, 90, 203–205]. Over-expression of either Cu/Zn-SOD or catalase alone was associated with insignificant effects on lifespan while joint expression of Cu/Zn-SOD and catalase gave an extension of lifespan of up to 33%. However, these experiments were subsequently criticized [65, 71, 206, 207]. The problems were lack of control for insertional mutagenic effects, different size of inserts in control and experimental lines, and uncontrolled genetic background.

Subsequent experiments addressed these issues. When over-expression of Cu/Zn-SOD was made ubiquitous using the *actin5C* promoter and confined to the adult period, by using the FLP/FRT system under the control of the heat shock promoter, mean lifespan was extended by up to 20% in females and 48% in males [15]. Over-expression of catalase alone or in conjunction with Cu/Zn-SOD had no detectable effect on lifespan. Under the conditions used, the heat shock inducer had neutral or negative effects on lifespan, so that the increase in lifespan in the transgenic lines over-expressing Cu/Zn-SOD after heat shock was probably underestimated. The effects of over-expression of MnSOD were examined in similar experiments, and mean lifespan was extended by up to 33%, with the degree of extension showing a strong correlation with the degree of Mn-SOD activity induced by the heat pulse [16]. This result contrasted with that of an earlier study [69], in which unconditional over-expression of Mn-SOD was induced under the native promoter and a slight (4–5%) decrease in lifespan was found in the over-expressing lines relative to the controls. Two possible explanations for the discrepancy are: (a) over-expression of Mn-SOD in the pre-adult period causes accelerated aging in the adult or (b) extension of lifespan by over-expression of Mn-SOD occurs only when the gene is over-expressed in tissues other than those that induce the native promoter. Some support for the latter interpretation comes from the finding that over-expression of Cu/Zn-SOD under a ubiquitous promoter [15] or under the GAL4–UAS system with over-expression targeted to motorneurons [14, 208] does extend fly lifespan while the unconditional expression in the earlier experiments did not [203].

Two other enzymes involved in defence against oxidative damage have been ubiquitously over-expressed, both using the GAL4–UAS system. Protein carboxyl methyltransferase (PCMT) is involved in the repair of modified isoaspartyl residues while peptide methionine sulphoxide reductase A (MSRA) catalyzes the repair of oxidized methionine. Over-expression of the gene encoding PCMT increased enzyme activity 3–7-fold at both 25°C and at 29°C but was associated with an extension of fly lifespan only at 29°C [19]. The authors concluded that elevated levels of PCMT were protective against higher levels of protein damage inflicted at 29°C. Over-expression of MSRA also increased both fly lifespan and resistance to the free radical generator paraquat [17]. In neither of these studies were there controls for effects of genetic background, heterosis or insertional mutagenesis, and independent confirmatory evidence for a role of these genes in aging is therefore desirable.

The picture that has emerged so far is one in which up-regulation of specific endogenous defence mechanisms, notably Cu/Zn- and Mn-SOD, does appear to be capable of increasing fly lifespan. However, it also appears that specific temporal and/or spatial patterns of up-regulation are required, and the reasons are unknown. The precise mechanisms by which extension of lifespan occurs are also not clear. Changes in redox status could interfere with cellular signalling [209–215] and with immune function [216, 217] and the effects on lifespan could be quite indirect.

Investigation of mechanisms of aging and its regulation through studies of gene expression

Changes in patterns of age-dependent gene expression in normal and experimentally manipulated flies provide important information about mechanisms of aging. Single gene assays, along with genomic approaches, such as enhancer traps and DNA microarrays, have provided fascinating glimpses of global regulation of gene expression in the aging fly and revealed new candidate genes involved in the regulation of the aging process.

Investigation of regulatory processes in the aging fly

An interesting perspective on the mechanisms that regulate biological processes in aging organisms has been developed based on age-dependent expression profiles. Using enhancer trap constructs, mostly in unknown genetic locations, Helfand and colleagues [105, 106, 218] characterized age-dependent expression profiles in the adult antenna. They demonstrated that expression patterns can be complicated and are highly regulated, even very late in life [219]. When temporal patterns were compared among strains of flies exhibiting significantly different lifespans, it was apparent that regulation of expression of certain genes was clearly linked to the rate of aging [105]. This was the first evidence that patterns of gene expression might serve as accurate biomarkers of physiological age.

Whole-genome expression profiling throughout the lifespan of control and experimentally manipulated flies has confirmed and extended these observations. The first microarray study of aging flies suggested that transcriptional changes in adult flies were rare and small [220], and a recent description of expression throughout the entire life-cycle seemed to verify this [221]. However, statistically powerful experiments, which focussed exclusively on the adult stage, identified many genes with age-dependent transcript representation [20], and examination of nearly 200 enhancer trap lines [107] conclusively showed over 80% with age- and tissue-dependent expression. It therefore appears that aging in this post-mitotic organism is highly dynamic at the level of the transcriptome, involving genes from a variety of different biological processes (see the next section).

Whole-genome expression profiling has also been used to characterize a transcriptional "signature of aging." By comparing the expression profiles of a large number of genes in control and calorically restricted flies, we identified a core set of nearly 1000 genes (roughly 7% of the fly genome) with characteristic age-dependent expression profiles [20]. Using this molecular signature, we showed that dietary restriction largely reduced the rate at which normal changes in gene expression occurred. This was the first evidence that, as it appears to do in mammals, CR caused flies to age more slowly. These signature profiles are reminiscent of the potential biomarkers of physiological age that were first identified using enhancer trap constructs [105].

Candidate mechanisms of aging identified through expression studies

Tower and colleagues [222, 223] investigated age- and tissue-dependent expression of a variety of heat shock protein genes (HSPs). HSPs are induced in response to protein damage, and increases in the abundance of inactive and/or damaged proteins with age have been observed in many species [21]. Although the results are occasionally conflicting, there is strong evidence for increased expression with age of both *hsp70* and *hsp22*. Reporter gene constructs and northern blots with fixed numbers of flies establish that the absolute amount of transcript for these genes is higher in old flies compared to young. Increased transcription of these genes in response to age-associated stress or reduced rates of degradation are possible explanations. *hsp70* seems predominantly induced in flight and leg muscle [222], while induction of *hsp22* is more broadly distributed [223]. Fly stocks that have been selected for increased lifespan exhibit higher levels of *hsp22* expression than the corresponding short-lived control [224]

Stimulated by this information, Tatar *et al.* [18] investigated the effect of over-expression of *hsp70* on adult survival. They used two matched strains of flies that contained either the normal quantity of constitutive and inducible *hsp70* genes or 12 extra copies of the inducible form of *hsp70*. Clever use of the FLP/FRT system [225] created two strains of flies each with a P-element in the same genomic location. In the extra-copy strain, this element contained the 12 additional *hsp70* genes, while only a remnant P-element remained in the control strain. This design controlled for position effects of the transgene, but the excision and extra-copy strains differed in the size of the transgenic insert. A brief heat pulse, sufficient to induce the heat shock protein only in the extra copy strain, resulted in a transient (approximately 10-day) increase in survival rates in the extra-copy strain but not in the controls. There are no published reports testing the effect of over-expression of *hsp22* on lifespan.

Recently, several large-scale examinations of age-dependent changes in gene expression have identified a variety of potential candidate genetic systems involved in *Drosophila* aging. Striking declines in the expression of genes localized to the mitochondrion, particularly those involved in electron transport, were observed with age. A recent comparative analysis of age-dependent expression profiles in the fly and the nematode identified this trend as one that is significantly conserved in both species (McCarroll *et al.*, in press). Clear signs of reproductive senescence are manifest as consistently reduced levels of chorion protein RNA in progressively older flies. Surprisingly, few genes involved in protein metabolism and modification were observed to exhibit age-dependent changes in expression, suggesting that these processes are highly regulated [20].

Both whole genome microarray [20] and enhancer trap [107] studies emphasize a large subset of genes that function in innate immunity that are dramatically up-regulated with age. Included in this group is the peptidoglycan recognition protein PGRP-LC, which responds to gram-negative bacterial infection by activating the Relish/imd antimicrobial pathway [226], along with well-known antimicrobial peptides such as Dipericin, various Attacins and Cecropins, and Metchnikowin. Using a very strict significance criterion, 23 of 69 genes identified by the Gene Ontology Consortium (http://www.geneontology.org) as having a role in immune response were shown to exhibit highly significant age-dependent transcript representation, and none of these expression profiles were significantly decreasing with age [20]. Examination of age-dependent transcript representation in all 69 genes reveals just how ubiquitous this effect is (Figure 3). Interestingly, the induction of roughly half of these genes was delayed in calorically restricted flies suggesting a link between the immune system and the rate of aging.

Conclusions

It is, perhaps, the end of the beginning for work on genetics of aging in *Drosophila*. The field is rapidly expanding and, thanks in part to results from other model organisms, significant advances have been made in the last few years. Continued success is virtually certain, and further discoveries about many different aspects of fly aging are on the horizon. We recognize three areas that would inordinately benefit from increased attention: (i) genetic analysis of demographic aging and more focus on temporal manipulation, (ii) precisely directed research to untangle the confusion that currently exists linking aging with stress resistance and oxidative damage, and (iii) continued exploitation of homologous genetic systems and their interaction with caloric restriction.

Demographic analyses focus on describing the age-pattern of physiological deterioration through changes in rates of age-specific mortality. Using these techniques it is possible to distinguish genetic manipulations that extend lifespan through a delay in the onset of age-dependent pathologies from those that slow their progression [3]. Because they can be easily cultured in very large numbers, *Drosophila* are more amenable to demographic analysis than are worms or yeast. Recently, such methods were used to show that short-lived male flies mutant for *chico* age more slowly [6], that long-lived flies derived from laboratory selection and from mutation in the gene *Indy* age in very different ways [4], and that various genetic and environmental manipulations that increase longevity can be classified by their effect on the pattern of age-specific mortality [3].

Progress in unravelling the physiological complexity of aging will require a sharper focus on characterizing the temporal aspects of genetic effects. To comprehend the role of candidate genes in the regulation of aging, we need to understand both where a gene is having an effect and *when during the lifespan* the important effects occur. Here again, *Drosophila* may prove advantageous. Age-specific mortality rates are a

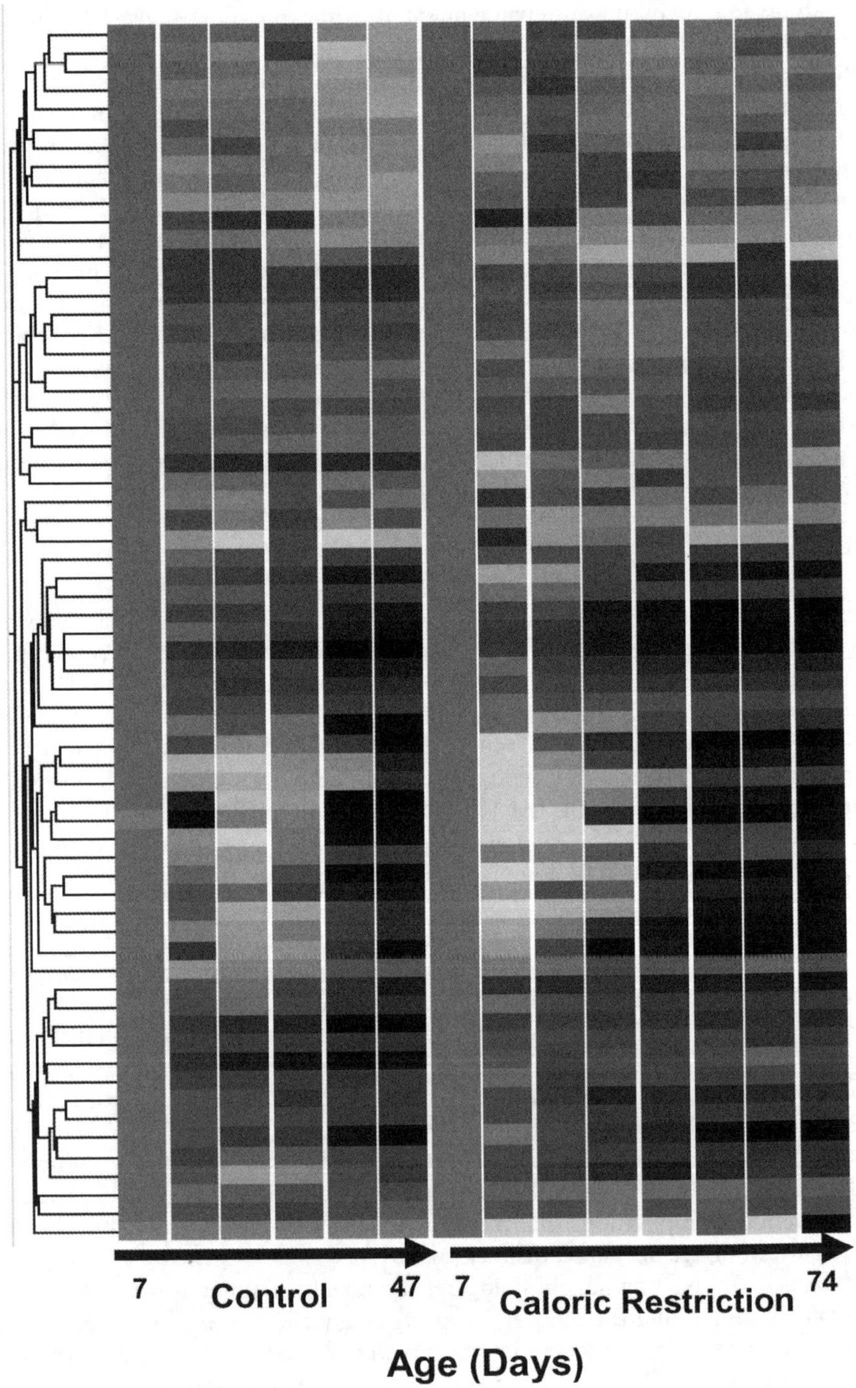
7
Control
47
7
Caloric Restriction
74
Age (Days)

unique measure of aging in that they identify specific ages over which an experimental manipulation affects patterns of aging [3]. New and efficient inducible expression systems in *Drosophila* [76–78, 83] provide us with the ability to execute genetic manipulation in an age-dependent manner, and detailed information on the age-dependent expression of nearly all genes in the genome is already available [20, 221, 227].

Although work in other model organisms suggests strong links between aging and resistance to various environmental stresses [e.g., refs. 228–234] and oxidative damage, these relationships are less robust in *Drosophila*. At least one mutation, *methuselah*, increases lifespan and stress resistance in the fly [2]. The *chico* mutation, however, increases lifespan while leaving stress resistance unaffected [5]. In addition, it is not clear why different studies of mutant over-expression of SOD and catalase give conflicting results. It is possible that unintended effects of transgenes in the larval or pupal stage are manifest or that tissue-specificity is important. Induction of a molecular chaperone increases adult survival rates over the ensuing days [18], and expression of two heat shock proteins increases with age [222, 223]. In total, the evidence seems to suggest that general stress resistance is capable of increasing adult survival in *Drosophila,* but it seems clear that increased stress resistance is not necessary for lifespan extension. Defence against damage to proteins during aging is probably important, but a causal link between aging and oxidative damage itself has not yet been established. Even if genes related to oxidative damage appear to show some effect on lifespan, it is far from clear how their effect is mediated. A great deal more work will be needed to establish the exact chain of causality producing any link between aging and oxidative damage.

Probably the most fruitful area of research thus far has been generated through homology. Our understanding of the role of IIS in the fly is increasing rapidly. The next few years is likely to see a detailed exploration of the mechanisms by which this pathway can extend lifespan, and on the links between the control of reproduction and of the rate of aging. The connections between histone acetylation, caloric restriction and lifespan are intriguing, although at present somewhat confusing. Sir2 over-expression increases lifespan, while Rpd3 loss-of-function does the same. Feeding of PBA (which causes hyperacetylation) increases lifespan. One immediate challenge is to understand how both deacetylation and acetylation of histones can increase lifespan. Gene expression studies show that genetic manipulation of Sir2 and Rpd3 (and feeding of PBA) cause many changes in gene expression. *Drosophila* is

---

Figure 3. *Age-dependent expression of the 69 genes identified by the Gene Ontology Consortium as being involved in Drosophila immune defense. For both control and caloric restriction treatments, expression levels are normalized to age 7 days (first column of both treatments, each shown in gray). Each row represents expression of a single gene and each column represents a different age (increasing in both treatments). Higher levels of expression are shown as darker rectangles; reduced levels of expression are lighter. It is clear that, although not all immune genes increase expression with age, the majority of them are dramatically upregulated in older flies.*

probably the best model organism to study this relationship at the moment. A robust system for caloric restriction has been developed. There are more candidate genes to investigate: in addition to Rpd3 and Sir2, the gene ontology recognizes 5 more histone deacetylases, one of them NAD dependent, and a variety of drugs are also available. There has been no more exciting time to be involved with aging research in *Drosophila* and we look forward to the unfolding of the story over the next few years.

Acknowledgments

We thank David Gems for comments on the manuscript and BBSRC and the Wellcome Trust for financial support.

References

1. Rose MR (1984). Laboratory evolution of postponed senescence in *Drosophila melanogaster*. *Evolution* 38: 1004–10.
2. Lin YJ, Seroude L, Benzer S (1998). Extended life-span and stress resistance in the Drosophila mutant methuselah. *Science* 282: 943–6
3. Pletcher SD, Khazaeli AA, Curtsinger JW (2000). Why do life spans differ? Partitioning mean longevity differences in terms of age-specific mortality parameters. *J Gerontol A Biol Sci Med Sci*. 55: B381–9.
4. Pletcher SD (2002). Mitigating the Tithonus error: genetic analysis of mortality phenotypes. Science SAGE KE, <http://sageke.sciencemag.org/cgi/content/full/sageke;2002/37/pe14>
5. Clancy DJ, Gems D, Harshman LG, et al. (2001). Extension of life-span by loss of CHICO, a *Drosophila* insulin receptor substrate protein. *Science* 292: 104–6.
6. Tu MP, Epstein D, Tatar M (2002). The demography of slow aging in male and female *Drosophila* mutants for the insulin-receptor substrate homologue *chico*. *Aging Cell* 1: 75–80.
7. Kenyon C (2001). A conserved regulatory system for aging. *Cell* 105: 165–8.
8. Partridge L, Gems D (2002). Mechanisms of ageing: public or private? *Nat Rev Genet*. 3: 165–75.
9. Tatar M, Kopelman A, Epstein D, Tu MP, Yin CM, Garofalo RS (2001). Mutations in the *Drosophila* insulin receptor homologue extend lifespan and impair neuroendocrine function. *Diabetes* 50: A79.

9a. Lee SS, Kennedy S, Tolonen AC, Ruvkun G (2003). DAF-16 target genes that control *C. elegans* life-span and metabolism. *Science* 300: 644–7.

9b. Murphy CT, McCarroll SA, Bargmann CI *et al.* (2003). Genes that act downstream of DAF-16 to influence the lifespan of *C. elegans*. *Nature* 424: 277–83.

10. Rogina B, Reenan RA, Nilsen SP, Helfand SL (2000). Extended life-span conferred by cotransporter gene mutations in *Drosophila*. *Science* 290: 2137–40.
11. Aigaki T, Ohsako T, Toba G, Seong KH, Matsuo T (2001). The gene search system: its application to functional genomics in *Drosophila melanogaster*. *J Neurogenet*. 15: 169–78.
12. Landis G, Bhole D, Lu L, Tower J (2001). High-frequency generation of conditional mutations affecting *Drosophila melanogaster* development and life span. *Genetics* 158: 1167–76.

13. Seong KH, Ogashiwa T, Matsuo T, Fuyama Y, Aigaki T (2001). Application of the gene search system to screen for longevity genes in Drosophila. *Biogerontology* 2: 209–17.
14. Parkes TL, Elia AJ, Dickinson D, Hilliker AJ, Phillips JP, Boulianne GL (1998). Extension of *Drosophila* lifespan by overexpression of human SOD1 in motorneurons. *Nat Genet.* 19: 171–4.
15. Sun J, Tower J (1999). FLP recombinase-mediated induction of Cu/Zn-superoxide dismutase transgene expression can extend the life span of adult *Drosophila melanogaster* flies. *Mol Cell Biol.* 19: 216–28.
16. Sun J, Folk D, Bradley TJ, Tower J (2002). Induced overexpression of mitochondrial Mn-superoxide dismutase extends the life span of adult *Drosophila melanogaster. Genetics* 161: 661–72.
17. Ruan H, Tang XD, Chen ML, *et al.* (2002). High-quality life extension by the enzyme peptide methionine sulfoxide reductase. *Proc Natl Acad Sci USA* 99: 2748–53.
18. Tatar M, Khazaeli AA, Curtsinger JW (1997). Chaperoning extended life. *Nature* 390: 30.
19. Chavous DA, Jackson FR, O'Connor CM (2001). Extension of the *Drosophila* lifespan by overexpression of a protein repair methyltransferase. *Proc Natl Acad Sci USA* 98: 14814–18.
20. Pletcher SD, Macdonald SJ, Marguerie R, *et al.* (2002). Genome-wide transcript profiles in aging and calorically restricted *Drosophila melanogaster. Curr Biol.* 12: 712–23.
21. Finch CE (1990). *Longevity, Senescence and the Genome.* Chicago: University of Chicago Press.
22. Partridge L, Barton NH (1996). On measuring the rate of ageing. *Proc R Soc Lond Ser B – Biol Sci.* 263: 1365–71.
23. Rose MR, Charlesworth B (1981). Genetics of life history in *Drosophila melanogaster.* I. Sib analysis of adult females. *Genetics* 97: 173–86.
24. Luckinbill LS, Arking R, Clare MJ, Cirocco WC, Buck SA (1984). Selection for delayed senescence in *Drosophila melanogaster. Evolution* 38: 996–1003.
25. Zwaan B, Bijlsma R, Hoekstra H (1996). Direct selection on life span in *Drosophila melanogaster. Evolution* 49: 649–59.
26. Tatar M, Promislow DE, Khazaeli AA, Curtsinger JW (1996). Age-specific patterns of genetic variance in *Drosophila melanogaster.* II. Fecundity and its genetic covariance with age-specific mortality. *Genetics* 143: 849–58.
27. Promislow DE, Tatar M, Khazaeli AA, Curtsinger JW (1996). Age-specific patterns of genetic variance in *Drosophila melanogaster.* I. Mortality. *Genetics* 143: 839–48.
28. Partridge L, Prowse N, Pignatelli P (1999). Another set of responses and correlated responses to selection on age at reproduction in *Drosophila melanogaster. Proc R Soc Lond B Biol Sci.* 266: 255–61.
29. Nuzhdin SV, Pasyukova EG, Dilda CL, Zeng ZB, Mackay TF (1997). Sex-specific quantitative trait loci affecting longevity in *Drosophila melanogaster. Proc Natl Acad Sci USA* 94: 9734–9.
30. Houle D, Hughes KA, Hoffmaster DK, *et al.* (1994). The effects of spontaneous mutation on quantitative traits. I. Variances and covariances of life history traits. *Genetics* 138: 773–85.
31. Houle D, Hughes KA, Assimacopoulos S, Charlesworth B (1997). The effects of spontaneous mutation on quantitative traits. II. Dominance of mutations with effects on life-history traits. *Genet Res.* 70: 27–34.
32. Pletcher SD, Houle D, Curtsinger JW (1998). Age-specific properties of spontaneous mutations affecting mortality in *Drosophila melanogaster. Genetics* 148: 287–303.

33. Yampolsky L, Pearse LE, Promislow DE (2000). Age-specific effects of novel mutations in *Drosophila melanogaster*. I. Mortality. *Genetica* 110: 11–29.
34. Mack PD, Lester VK, Promislow DE (2000). Age-specific effects of novel mutations in *Drosophila melanogaster*. II. Fecundity and male mating ability. *Genetica* 110: 31–41.
35. Johnson TE, Tedesco PM, Lithgow GJ (1993). Comparing mutants, selective breeding, and transgenics in the dissection of aging processes of *Caenorhabditis elegans*. *Genetica* 91: 65–77.
36. Roper C, Pignatelli P, Partridge L (1993). Evolutionary effects of selection on age at reproduction in larval and adult *Drosophila melanogaster*. *Evolution* 47: 445–55.
37. Hughes KA, Charlesworth B (1994). A genetic analysis of senescence in Drosophila. *Nature* 367: 64–6.
38. Hughes KA (1995). The inbreeding decline and average dominance of genes affecting male life-history characters in *Drosophila melanogaster*. *Genet Res.* 65: 41–52.
39. Charlesworth B, Hughes KA (1996). Age-specific inbreeding depression and components of genetic variance in relation to the evolution of senescence. *Proc Natl Acad Sci USA* 93: 6140–5.
40. Pletcher SD, Curtsinger JW (2000). The influence of environmentally induced heterogeneity on age-specific genetic variance for mortality rates. *Genet Res.* 75: 321–9.
41. Charlesworth B (2001). Patterns of age-specific means and genetic variances of mortality rates predicted by the mutation-accumulation theory of ageing. *J Theor Biol.* 210: 47–65.
42. Service PM, Fales AJ (1993). Evolution of delayed reproductive senescence in male fruit flies: sperm competition. *Genetica* 91: 111–25.
43. Sgro CM, Partridge L (1999). A delayed wave of death from reproduction in Drosophila. *Science* 286: 2521–4.
44. Rose MR (1991). *Evolutionary Biology of Aging*. Oxford: Oxford University Press.
45. Rose MR (1989). Genetics of increased lifespan in Drosophila. *Bioessays* 11: 132–5.
46. Chippindale AK, Chu TJF, Rose MR (1996). Complex trade-offs and the evolution of starvation resistance in *Drosophila melanogaster*. *Evolution* 50: 753–66.
47. Chippindale AK, Gibbs AG, Sheik M, *et al.* (1998). Resource acquisition and the evolution of stress resistance in *Drosophila melanogaster*. *Evolution* 52: 1342–52.
48. Djawdan M, Chippindale AK, Rose MR, Bradley TJ (1998). Metabolic reserves and evolved stress resistance in *Drosophila melanogaster*. *Physiol Zool.* 71: 584–94.
49. Arking R (1998). Molecular basis of extended longevity in selected *Drosophila* strains. *Curr Sci.* 74: 859–64.
50. Harshman LG, Haberer BA (2000). Oxidative stress resistance: a robust correlated response to selection in extended longevity lines of *Drosophila melanogaster*? *J Gerontol A Biol Sci Med Sci.* 55: B415–17.
51. Wang Y, Salmon AB, Harshman LG (2001). A cost of reproduction: oxidative stress susceptibility is associated with increased egg production in *Drosophila melanogaster*. Exp Gerontol, 36: 1349–59.
52. Arking R (2001). Gene expression and regulation in the extended longevity phenotypes of *Drosophila*. *Ann NY Acad Sci.* 928: 157–67.
53. Arking R, Buck S, Hwangbo DS, Lane M (2002). Metabolic alterations and shifts in energy allocations are corequisites for the expression of extended longevity genes in *Drosophila*. *Ann NY Acad Sci.* 959: 251–62.
54. Force AG, Staples T, Soliman S, Arking R (1995). Comparative biochemical and stress analysis of genetically selected *Drosophila* strains with different longevities. *Dev Genet.* 17: 340–51.

55. Arking R, Burde V, Graves K, *et al.* (2000). Identical longevity phenotypes are characterized by different patterns of gene expression and oxidative damage. *Exp Gerontol.* 35: 353–73.
56. Bradley TJ, Williams AE, Rose MR (1999). Physiological responses to selection for desiccation resistance in *Drosophila melanogaster. Am Zool.* 39: 337–45.
57. Harshman LG, Moore KM, Sty MA, Magwire MM (1999). Stress resistance and longevity in selected lines of *Drosophila melanogaster. Neurobiol Aging* 20: 521–9.
58. Vettraino J, Buck S, Arking R (2001). Direct selection for paraquat resistance in *Drosophila* results in a different extended longevity phenotype. *J Gerontol A Biol Sci Med Sci.* 56: B415–25.
59. Curtsinger JW, Khazaeli AA (2002). Lifespan, QTLs, age-specificity, and pleiotropy in *Drosophila. Mech Ageing Dev.* 123: 81–93.
60. Tu M-P, Epstein D, Tatar M (2002). The demography of slow ageing in male and female *Drosophila* mutant for the insulin-receptor substrate homologue *chico. Aging Cell* 1: 75–80.
61. Sgro CM, Partridge L (2000). Evolutionary responses of the life history of wild-caught *Drosophila melanogaster* to two standard methods of laboratory culture. *Am Nat.* 156: 341–53.
62. Sgro CM, Partridge L (2001). Laboratory adaptation of life history in *Drosophila. Am Nat.* 158: 657–8.
63. Promislow DE, Pletcher SD (2002). Advice to an aging scientist. *Mech Ageing Dev.* 123: 841–50.
64. Spencer CC, Promislow DEL (2002). Genes, Culture, and Aging Flies – What the Lab Can and Cannot Tell Us About Natural Genetic Variation for Senescence Science SAGE KE, <http://sageke.sciencemag.org/cgi/content/full/sageke>.
65. Tower J (2000). Transgenic methods for increasing *Drosophila* life span. *Mech Ageing Dev.* 118: 1–14.
66. Tatar M, Kopelman A, Epstein D, Tu MP, Yin CM, Garofalo RS (2001). A mutant *Drosophila* insulin receptor homolog that extends life-span and impairs neuroendocrine function. *Science* 292: 107–10.
67. Kaiser M, Gasser M, Ackermann R, Stearns SC (1997). P element inserts in transgenic flies: a cautionary tale. *Heredity* 78: 1–11.
68. Clark AG, Guadalupe RN (1995). Probing the evolution of senescence in *Drosophila melanogaster* with P-element tagging. *Genetica* 96: 225–34.
69. Mockett RJ, Orr WC, Rahmandar JJ, *et al.* (1999). Overexpression of Mn-containing superoxide dismutase in transgenic *Drosophila melanogaster. Arch Biochem Biophys.* 371: 260–9.
70. Orr WC, Sohal RS (1994). Extension of life-span by overexpression of superoxide dismutase and catalase in *Drosophila melanogaster. Science* 263: 1128–30.
71. Tatar M (1999). Transgenes in the analysis of life span and fitness. *Am Nat.* 154: S67–81.
72. Feder ME, Cartano NV, Milos L, Krebs RA, Lindquist SL (1996). Effect of engineering Hsp70 copy number on Hsp70 expression and tolerance of ecologically relevant heat shock in larvae and pupae of *Drosophila melanogaster. J Exp Biol.* 199: 1837–44.
73. Toba G, Ohsako T, Miyata N, Ohtsuka T, Seong KH, Aigaki T (1999). The gene search system: A method for efficient detection and rapid molecular identification of genes in *Drosophila melanogaster. Genetics* 151: 725–37.
74. Brand AH, Perrimon N (1993). Targeted gene-expression as a means of altering cell fates and generating dominant phenotypes. *Development* 118: 401–15.

75. Khazaeli AA, Tatar M, Pletcher SD, Curtsinger JW (1997). Heat-induced longevity extension in *Drosophila*. I. Heat treatment, mortality, and thermotolerance. *J Gerontol A Biol Sci Med Sci*. 52: B48–52.
76. Osterwalder T, Yoon KS, White BH, Keshishian H (2001). A conditional tissue-specific transgene expression system using inducible GAL4. *Proc Natl Acad Sci USA* 98: 12596–601.
77. Roman G, Endo K, Zong L, Davis RL (2001). P[Switch], a system for spatial and temporal control of gene expression in *Drosophila melanogaster*. *Proc Natl Acad Sci USA* 98: 12602–7.
78. Bieschke ET, Wheeler JC, Tower J (1998). Doxycycline-induced transgene expression during *Drosophila* development and aging. *Mol Gen Genet*. 258: 571–9.
79. Weinkove D, Neufeld TP, Twardzik T, Waterfield MD, Leevers SJ (1999). Regulation of imaginal disc cell size, cell number and organ size by *Drosophila* class I-A phosphoinositide 3-kinase and its adaptor. *Curr Biol*. 9: 1019–29.
80. Bohni R, Riesgo-Escovar J, Oldham S, *et al.* (1999). Autonomous control of cell and organ size by CHICO, a *Drosophila* homolog of vertebrate IRS1-4. *Cell* 97: 865–75.
81. Kennerdell JR, Carthew RW (2000). Heritable gene silencing in *Drosophila* using double-stranded RNA. *Nat Biotechnol*. 18: 896–8.
82. Lam G, Thummel CS (2000). Inducible expression of double-stranded RNA directs specific genetic interference in *Drosophila*. *Curr Biol*. 10: 957–63.
83. Allikian MJ, Deckert-Cruz D, Rose MR, Landis GN, Tower J (2002). Doxycycline-induced expression of sense and inverted-repeat constructs modulates phosphogluconate mutase (Pgm) gene expression in adult *Drosophila melanogaster*. *Genome Biol*. 3(5): research0021.1–0021.10.
84. DeckertCruz DJ, Tyler RH, Landmesser JE, Rose MR (1997). Allozymic differentiation in response to laboratory demographic selection of *Drosophila melanogaster*. *Evolution* 51: 865–72.
85. Allikian MJ, Deckert-Cruz D, Rose MR, Landis GN, Tower J (2002). Doxycycline-induced expression of sense and inverted-repeat constructs modulates phosphogluconate mutase (Pgm) gene expression in adult *Drosophila melanogaster*. *Genome Biol*. 3(5): research0021.1–0021.10.
86 . Rong YS, Golic KG (2000). Gene targeting by homologous recombination in *Drosophila*. *Science* 288: 2013–18.
87. Rong YS, Golic KG (2001). A targeted gene knockout in *Drosophila*. *Genetics* 157: 1307–12.
88. Rong YKS, Titen SW, Xie HB, *et al.* (2002). Targeted mutagenesis by homologous recombination in *D. melanogaster*. *Genes Dev*. 16: 1568–81.
89. Orr WC, Sohal RS (1993). Life-span extension in transgenic *Drosophila* containing extra copies of both catalase and Cu-Zn superoxide-dismutase. *Free Radic Biol Med*. 15: 546.
90. Sohal RS, Agarwal A, Agarwal S, Orr WC (1995). Simultaneous overexpression of copper- and zinc-containing superoxide dismutase and catalase retards age-related oxidative damage and increases metabolic potential in *Drosophila melanogaster*. *J Biol Chem*. 270: 15671–4.
91. Mockett RJ, Sohal RS, Orr WC (1999). Overexpression of glutathione reductase extends survival in transgenic *Drosophila melanogaster* under hyperoxia but not normoxia. *FASEB J*. 13: 1733–42.
92. Struhl G, Basler K (1993). Organizing activity of wingless protein in *Drosophila*. *Cell* 72: 527–40.

93. Basler K, Struhl G (1994). Compartment boundaries and the control of *Drosophila* limb pattern by hedgehog protein. *Nature* 368: 208–14.
94. Han ES, Hilsenbeck SG (2001). Array-based gene expression profiling to study aging. *Mech Ageing Dev.* 122: 999–1018.
95. Welle S (2002). Gene transcript profiling in aging research. *Exp Gerontol.* 37: 583–90.
96. Shikama N, Ackermann R, Brack C (1994). Protein synthesis elongation factor EF-1 alpha expression and longevity in *Drosophila melanogaster. Proc Natl Acad Sci USA* 91: 4199–203.
97. Gygi SP, Rochon Y, Franza BR, Aebersold R (1999). Correlation between protein and mRNA abundance in yeast. *Mol Cell Biol.* 19: 1720–30.
98. Tavernarakis N, Driscoll M (2002). Caloric restriction and lifespan: a role for protein turnover? *Mech Ageing Dev.* 123: 215–29.
99. Bustin SA (2000). Absolute quantification of mRNA using real-time reverse transcription polymerase chain reaction assays. *J Mol Endocrinol.* 25: 169–93.
100. Bustin SA (2002). Quantification of mRNA using real-time reverse transcription PCR (RT- PCR): trends and problems. *J Mol Endocrinol.* 29: 23–39.
101. Suzuki T, Higgins PJ, Crawford DR (2000). Control selection for RNA quantitation. *Biotechniques* 29: 332–7.
102. DeRisi JL, Iyer VR, Brown PO (1997). Exploring the metabolic and genetic control of gene expression on a genomic scale. *Science* 278: 680–6.
103. O'Kane CJ, Gehring WJ (1987). Detection *in situ* of genomic regulatory elements in *Drosophila. Proc Natl Acad Sci USA* 84: 9123–7.
104. Bellen HJ, O'Kane CJ, Wilson C, Grossniklaus U, Pearson RK, Gehring WJ (1989). P-element-mediated enhancer detection: a versatile method to study development in *Drosophila. Genes Dev.* 3: 1288–300.
105. Rogina B, Helfand SL (1995). Regulation of gene expression is linked to life span in adult *Drosophila. Genetics* 141: 1043–8.
106. Rogina B, Helfand SL (1996). Timing of expression of a gene in the adult *Drosophila* is regulated by mechanisms independent of temperature and metabolic rate. *Genetics* 143: 1643–51.
107. Seroude L, Brummel T, Kapahi P, Benzer S (2002). Spatio temporal analysis of gene expression during aging in *Drosophila melanogaster. Aging Cell* 1: 47–56.
108. Tower J (2002). Setting a trap for aging-related genes in *Drosophila*. Science SAGE KE, <http://sageke.sciencemag.org/cgi/content/full/sageke>.
109. Riddle DL, Albert PS (1997). Genetic and environmental regulation of dauer larva development. In: Riddle DL, Blumenthal T, Meyer BJ, Priess JR, eds. *C. Elegans II.* New York: Cold Spring Harbor Laboratory Press, pp. 739–68.
110. Friedman DB, Johnson TE (1988). A mutation in the age-1 gene in *Caenorhabditis elegans* lengthens life and reduces hermaphrodite fertility. *Genetics* 118: 75–86.
111. Kenyon C, Chang J, Gensch E, Rudner A, Tabtiang R (1993). A *C. elegans* mutant that lives twice as long as wild type. *Nature* 366: 461–4.
112. Kimura KD, Tissenbaum HA, Liu Y, Ruvkun G (1997). daf-2, an insulin receptor-like gene that regulates longevity and diapause in *Caenorhabditis elegans. Science* 277: 942–6.
113. Lin K, Dorman JB, Rodan A, Kenyon C (1997). daf-16: An HNF-3/forkhead family member that can function to double the life-span of *Caenorhabditis elegans. Science* 278: 1319–22.
114. Ogg S, Paradis S, Gottlieb S, *et al.* (1997). The fork head transcription factor DAF-16 transduces insulin-like metabolic and longevity signals in *C. elegans. Nature* 389: 994–9.

115. Pierce SB, Costa M, Wisotzkey R, *et al.* (2001). Regulation of DAF-2 receptor signaling by human insulin and ins-1, a member of the unusually large and diverse *C. elegans* insulin gene family. *Genes Dev.* 15: 672–86.
116. Ruan Y, Chen C, Cao Y, Garofalo RS (1995). The *Drosophila* insulin receptor contains a novel carboxyl-terminal extension likely to play an important role in signal transduction. *J Biol Chem.* 270: 4236–43.
117. Chen C, Jack J, Garofalo RS (1996). The *Drosophila* insulin receptor is required for normal growth. *Endocrinology* 137: 846–56.
118. Leevers SJ, Weinkove D, MacDougall LK, Hafen E, Waterfield MD (1996). The *Drosophila* phosphoinositide 3-kinase Dp110 promotes cell growth. *EMBO J.* 15: 6584–94.
119. Weinkove D, Leevers SJ, MacDougall LK, Waterfield MD (1997). p60 is an adaptor for the *Drosophila* phosphoinositide 3-kinase, Dp110. *J Biol Chem.* 272: 14606–10.
120. Gems D, Partridge L (2001). Insulin/IGF signalling and ageing: seeing the bigger picture. *Curr Opin Genet Dev.* 11: 287–92.
121. Apfeld J, Kenyon C (1998). Cell nonautonomy of *C. elegans* daf-2 function in the regulation of diapause and life span. *Cell* 95: 199–210.
122. Wolkow CA, Kimura KD, Lee MS, Ruvkun G (2000). Regulation of *C. elegans* life-span by insulinlike signaling in the nervous system. *Science* 290: 147–50.
123. Antebi A, Culotti JG, Hedgecock EM (1998). daf-12 regulates developmental age and the dauer alternative in *Caenorhabditis elegans. Development*, 125: 1191–205.
124. Antebi A, Yeh WH, Tait D, Hedgecock EM, Riddle DL (2000). daf-12 encodes a nuclear receptor that regulates the dauer diapause and developmental age in *C. elegans. Genes Dev.* 14: 1512–27.
125. Gerisch B, Weitzel C, Kober-Eisermann C, Rottiers V, Antebi A (2001). A hormonal signaling pathway influencing *C. elegans* metabolism, reproductive development, and life span. *Dev Cell* 1: 841–51.
126. Gems D, Sutton AJ, Sundermeyer ML, *et al.* (1998). Two pleiotropic classes of daf-2 mutation affect larval arrest, adult behavior, reproduction and longevity in *Caenorhabditis elegans. Genetics* 150: 129–55.
127. McColl G, Jenkins NL, Walker DW, Lithgow GJ (2000). Testing evolutionary theories of aging. *Ann NY Acad Sci.* 908: 319–20.
128. Chapman T, Partridge L (1996). Female fitness in *Drosophila melanogaster*: an interaction between the effect of nutrition and of encounter rate with males. *Proc R Soc Lond B Biol Sci.* 263: 755–9.
129. Gems D, Pletcher SD, Partridge L (2002). Interpreting interactions between treatments that slow aging. *Aging Cell* 1: 1–9.
130. Lakowski B, Hekimi S (1998). The genetics of caloric restriction in *Caenorhabditis elegans. Proc Natl Acad Sci USA* 95: 13091–6.
131 . Bartke A, Wright JC, Mattison JA, Ingram DK, Miller RA, Roth GS (2001). Extending the lifespan of long-lived mice. *Nature* 414: 412.
132 . Bartke A, Wright JC, Mattison JA, Ingram DK, Miller RA, Roth GS (2002). Dietary restriction and life-span. *Science* 296: 2141.
133. Clancy DJ, Gems D, Hafen E, Leevers SJ, Partridge L (2002). Dietary restriction and life-span – Response. *Science* 296: 2141–2.
134. Clancy DJ, Gems D, Hafen E, Leevers SJ, Partridge L (2002). Dietary restriction in long-lived dwarf flies. *Science* 296: 319.
135. Chang KT, Min KT (2002). Regulation of lifespan by histone deacetylase. *Ageing Res Rev.* 1: 313–26.

136. Kaeberlein M, McVey M, Guarente L (1999). The SIR2/3/4 complex and SIR2 alone promote longevity in *Saccharomyces cerevisiae* by two different mechanisms. *Genes Dev.* 13: 2570–80.
137. Kim S, Benguria A, Lai CY, Jazwinski SM (1999). Modulation of life-span by histone deacetylase genes in *Saccharomyces cerevisiae. Mol Biol Cell* 10: 3125–36.
138. Guarente L (2000). Sir2 links chromatin silencing, metabolism, and aging. *Genes Dev.* 14: 1021–6.
139. Martin GM (1999). Emerging paradigms and new perspectives for human geneticists interested in the biology of aging. *Age* 22: 63.
140. Holmes SG, Rose AB, Steuerle K, *et al.* (1997). Hyperactivation of the silencing proteins, Sir2p and Sir3p, causes chromosome loss. *Genetics* 145: 605–14.
141. Sinclair DA, Guarente L (1997). Extrachromosomal rDNA circles – a cause of aging in yeast. *Cell* 91: 1033–42.
142. Kirchman PA, Kim S, Lai CY, Jazwinski SM (1999). Interorganelle signaling is a determinant of longevity in *Saccharomyces cerevisiae. Genetics* 152: 179–90.
143. Tissenbaum HA, Guarente L (2001). Increased dosage of a sir-2 gene extends lifespan in *Caenorhabditis elegans. Nature* 410: 227–30.
144. Jiang JC, Wawryn J, Kumara HNCS, Jazwinski SM (2002). Distinct roles of processes modulated by histone deacetylasesRpd3p, Hdalp, and Sir2p in life extension by caloric restriction in yeast. *Exp Gerontol.* 37: 1023–30.
145. Bernstein BE, Tong JK, Schreiber SL (2000). Genimewide studies of histone deacetylasefunctuin in yeast. *Proc Natl Acad Sci USA* 97: 13708–13.
146. Lin SJ, Defossez PA, Guarente L (2000). Requirement of NAD and SIR2 for life-span extension by calorie restriction in *Saccharomyces cerevisiae. Science* 289: 2126–8.
147. Rogina B, Helfand SL, Frankel S (2002). Down-regulation of the rpd3 histone deacetylase in *Drosophila* is linked to extension of life-span and caloric restriction. *Science* in press:
148. Mannervik M, Levine M (1999). The Rpd3 histone deacetylase is required for segmentation of the *Drosophila* embryo. *Proc Natl Acad Sci USA* 96: 6797–801.
149. Kang HL, Benzer S, Min KT (2002). Life extension in *Drosophila* by feeding a drug. *Proc Natl Acad Sci USA* 99: 838–43.
150. Steffan JS, Bodai L, Pallos J, *et al.* (2001). Histone deacetylase inhibitors arrest polyglutamine-dependent neurodegeneration in *Drosophila. Nature* 413: 739–43.
151. West AP Jr, Llamas LL, Snow PM, Benzer S, Bjorkman PJ (2001). Crystal structure of the ectodomain of *Methuselah*, a *Drosophila* G protein-coupled receptor associated with extended lifespan. *Proc Natl Acad Sci USA* 98: 3744–9.
152. Song W, Ranjan R, Dawson-Scully K, *et al.* (2002). Presynaptic regulation of neurotransmission in *Drosophila* by the G protein-coupled receptor *Methuselah. Neuron* 36: 105–19.
153. Seong KH, Matsuo T, Fuyama Y, Aigaki T (2001). Neural-specific overexpression of drosophila plenty of SH3s (DPOSH) extends the longevity of adult flies. *Biogerontology* 2: 271–81.
154. Harman DJ (1956). Aging: a theory based on free radical and radiation chemistry. *J Gerontol.* 11: 298–300.
155. Sohal RS, Weindruch R (1996). Oxidative stress, caloric restriction, and aging. *Science* 273: 59–63.
156. Martin GM, Austad SN, Johnson TE (1996). Genetic analysis of ageing: role of oxidative damage and environmental stresses. *Nat Genet.*13: 25–34.

157. Beckman KB, Ames BN (1998). The free radical theory of aging matures. *Physiological Rev.* 78: 547–81.
158. Wallace DC (1999). Mitochondrial diseases in man and mouse. *Science* 283: 1482–8.
159. Finkel T, Holbrook NJ (2000). Oxidants, oxidative stress and the biology of ageing. *Nature* 408: 239–47.
160. Pamplona R, Portero-Otin M, Requena JR, Thorpe SR, Herrero A, Barja G (1999). A low degree of fatty acid unsaturation leads to lower lipid peroxidation and lipoxidation-derived protein modification in heart mitochondria of the longevous pigeon than in the short- lived rat. *Mech Ageing Dev.* 106: 283–96.
161. Barja G, Herrero A (2000). Oxidative damage to mitochondrial DNA is inversely related to maximum life span in the heart and brain of mammals. *FASEB J.* 14: 312–18.
162. Boveris A, Chance B (1973). The mitchondrial generation of hydrogen peroxide. General properties and hyperbaric effects of oxygen. *Biochemical J.* 134: 707–16.
163. Nohl H (1994). Generation of superoxide as a byproduct of cellular respiration. *Ann Biol Clin.* 52: 199–204.
164. Skulachev VP (1996). Role of uncoupled and non-coupled oxidations in maintenance of safely low levels of oxygen and its one-electron reductants. *Q Rev Biophys.* 29: 169–202.
165. Halliwell B, Gutteridge JMC (1999). *Free Radicals in Biology and Medicine.* Oxford: Oxford University Press.
166. Schwarze SR, Weindruch R, Aiken JM (1998). Oxidative stress and aging reduce COX I RNA and cytochrome oxidase activity in *Drosophila. Free Radic Biol Med.* 25: 740–7.
167. Agarwal S, Sohal RS (1993). Relationship between aging and susceptibility to protein oxidative damage. *Biochem Biophys Res Commun.* 194: 1203–6.
168. Agarwal S, Sohal RS (1994). Aging and protein oxidative damage. *Mech Ageing Dev.* 75: 11–19.
169. Agarwal S, Sohal RS (1995). Differential oxidative damage to mitochondrial proteins during aging. *Mech Ageing Dev.* 85: 55–63.
170. Agarwal S, Sohal RS (1996). Relationship between susceptibility to protein oxidation, aging, and maximum life span potential of different species. *Exp Gerontol.* 31: 365–72.
171. Agarwal S, Sohal RS (1994). DNA Oxidative damage and life expectancy in houseflies. *Proc Natl Acad Sci USA* 91: 12332–5.
172. Goddard KAB, Yu CE, Oshima J, *et al.* (1996). Toward localization of the Werner syndrome gene by linkage disequilibrium and ancestral haplotyping: lessons learned from analysis of 35 chromosome 8p11.1-21.1 markers. *Am J Hum Genet.* 58: 1286–302.
173. Yan LJ, Levine RL, Sohal RS (1997). Oxidative damage during aging targets mitochondrial aconitase. *Proc Natl Acad Sci USA* 94: 11168–72.
174. Das N, Levine RL, Orr WC, Sohal RS (2001). Selectivity of protein oxidative damage during aging in *Drosophila melanogaster. Biochem J.* 360: 209–16.
175. Yan LJ, Sohal RS (1998). Mitochondrial adenine nucleotide translocase is modified oxidatively during aging. *Proc Natl Acad Sci USA* 95: 12896–901.
176. Yan LJ, Levine RL, Sohal RS (2000). Effects of aging and hyperoxia on oxidative damage to cytochrome C in the housefly, *Musca domestica. Free Radic Biol Med.* 29: 90–7.
177. Sohal RS, Sohal BH, Orr WC (1995). Mitochondrial superoxide and hydrogen peroxide generation, protein oxidative damage, and longevity in different species of flies. *Free Radic Biol Med.* 19: 499–504.
178. Borghouts C, Werner A, Elthon T, Osiewacz HD (2001). Copper-modulated gene expression and senescence in the filamentous fungus *Podospoisma anserina. Mol Cell Biol.* 21: 390–9.

179. Osiewacz HD (2002). Genes, mitochondria and aging in filamentous fungi. *Ageing Res Rev.* 28: 1–18.
180. Le Bourg E (2001). Oxidative stress, aging and longevity in *Drosophila melanogaster. FEBS Lett.* 498: 183–6.
181. Brack C, Bechter-Thuring E, Labuhn M (1997). N-acetylcysteine slows down ageing and increases the life span of *Drosophila melanogaster. Cell Mol Life Sci.* 53: 960–6.
182. Ruddle DL, Yengoyan LS, Miquel J, Marcuson R, Fleming JE (1988). Propyl gallate delays senescence in *Drosophila melanogaster. Age* 11: 54–8.
183. Bonilla E, Medina-Leendertz S, Diaz S (2002). Extension of life span and stress resistance of *Drosophila melanogaster* by long-term supplementation with melatonin. *Exp Gerontol.* 37: 629–38.
184. Anisimov VN, Mylnikov SV, Oparina TI, Khavinson VK (1997). Effect of melatonin and pineal peptide preparation epithalamin on life span and free radical oxidation in *Drosophila melanogaster. Mech Ageing Dev.* 97: 81–91.
185. Anisimov VN, Mylnikov SV, Khavinson VK (1998). Pineal peptide preparation epithalamin increases the lifespan of fruit flies, mice and rats. *Mech Ageing Dev.* 103: 123–32.
186. Anisimov VN, Mylnikov SV, Khavinson VK (1998). Pineal peptide preparation epithalamin increases the lifespan of fruit flies, mice and rats. *Mech Ageing Dev.* 103: 123–32.
187. Khavinson VK, Izmaylov DM, Obukhova LK, Malinin VV (2000). Effect of epitalon on the lifespan increase in *Drosophila melanogaster. Mech Ageing Dev.* 120: 141–9.
188. Khavinson VK, Myl'nikov SV (2000). Effect of pineal tetrapeptide on antioxidant defense in *Drosophila melanogaster. Bull Exp Biol Med.* 129: 355–6.
189. Khavinson VK, Myl'nikov SV (2000). Effect of epithalone on the age-specific changes in the time course of lipid peroxidation in *Drosophila melanogaster. Bull Exp Biol Med.* 130: 1116–19.
190. Khavinson VK, Myl'nikov SV, Oparina TI, Arutyunyan AV (2001). Effects of peptides on generation of reactive oxygen species in subcellular fractions of *Drosophila melanogaster. Bull Exp Biol Med.* 132: 682–5.
191. Melov S, Ravenscroft J, Malik S, *et al.* (2000). Extension of life-span with superoxide dismutase/catalase mimetics. *Science* 289: 1567–9.
191a. Keaney M, Gems D (2003). No increase in lifespan in *Caenorhabditis elegans* upon treatment with the superoxide dismutase mimetic EUK-8. *Free Radic Biol Med.* 34: 277–82.
192. Bayne A-CV, Sohal RS (2002). Effects of superoxide dismutase/catalase mimetics on life span and oxidagtive stress resistance in the housefly, *Musca domestica. Free Radic Biol Med.* 11: 1229–34.
193. Yan LJ, Sohal RS (2000). Prevention of flight activity prolongs the life span of the housefly, *Musca domestica*, and attenuates the age-associated oxidative damage to specific mitochondrial proteins. *Free Radic Biol Med.* 29: 1143–50.
194. Felton GW, Summers CB (1995). Antioxidant systems in insects. *Arch Insect Biochem Physiol.* 29: 187–97.
195. Kanzok SM, Fechner A, Bauer H, *et al.* (2001). Substitution of the thioredoxin system for glutathione reductase in *Drosophila melanogaster. Science* 291: 643–6.
196. Bauer H, Kanzok SM, Schirmer RH (2002). Thioredoxin-2 but not thioredoxin-1 is a substrate of thioredoxin peroxidase-1 from *Drosophila melanogaster* – isolation and characterization of a second thioredoxin in *D. melanogaster* and evidence for distinct biological functions of Trx-1 and Trx-2. *J Biol Chem.* 277: 17457–63.

197. Missirlis F, Phillips JP, Jackle H (2001). Cooperative action of antioxidant defense systems in *Drosophila. Curr Biol.* 11: 1272–7.
198. Missirlis F, Ulschmid JK, Hirosawa-Takamori M, *et al.* (2002). Mitochondrial and cytoplasmic thioredoxin reductase variants encoded by a single *Drosophila* gene are both essential for viability. *J Biol Chem.* 277: 11521–6.
199. Phillips JP, Campbell SD, Michaud D, Charbonneau M, Hilliker AJ (1989). Null mutation of copper-zinc superoxide-dismutase in *Drosophila* confers hypersensitivity to paraquat and reduced longevity. *Proc Natl Acad Sci USA* 86: 2761–5.
200. Parkes TL, Kirby K, Phillips JP, Hilliker AJ (1998). Transgenic analysis of the cSOD-null phenotypic syndrome in *Drosophila. Genome* 41: 642–51.
201. Orr WC, Arnold LA, Sohal RS (1992). Relationship between catalase activity, life span and some parameters associated with antioxidant defenses in *Drosophila melanogaster. Mech Ageing Dev.* 63: 287–96.
202. Mackay WJ, Bewley GC (1989). The genetics of catalase in *Drosophila melanogaster* – isolation and characterization of acatalasemic mutants. *Genetics* 122: 643–52.
203. Orr WC, Sohal RS (1993). Effects of Cu-Zn superoxide dismutase overexpression of life span and resistance to oxidative stress in transgenic *Drosophila melanogaster. Arch Biochem Biophys.* 301: 34–40.
204. Orr WC, Sohal RS (1992). The effects of catalase gene overexpression on life-span and resistance to oxidative stress in transgenic *Drosophila melanogaster. Arch Biochem Biophys.* 297: 35–41.
205. Orr WC, Sohal RS (1993). Effects of Cu-Zn superoxide-dismutase overexpression on life- span and resistance to oxidative stress in transgenic *Drosophila melanogaster. Arch Biochem Biophys.* 301: 34–40.
206. Tower J (1996). Aging mechanisms in fruit files. *Bioessays* 18: 799–807.
207. Mockett RJ, Orr WC, Sohal RS (2002). Overexpression of Cu, ZnSOD and MnSOD in transgenic *Drosophila. Methods Enzymol.* 349: 213–20.
208. Phillips JP, Parkes TL, Hilliker AJ (2000). Targeted neuronal gene expression and longevity in *Drosophila. Exp Gerontol.* 35: 1157–64.
209. Wolin MS, Gupte SA, Oeckler RA (2002). Superoxide in the vascular system. *J Vasc Res.* 39: 191–207.
210. Guzik TJ, West NEJ, Pillal R, Taggart DP, Channon KM (2002) Nitric oxide modulates superoxide release and peroxynitrite formation in human blood vessels. *Hypertension* 39: 1088–94.
211. Martindale JL, Holbrook NJ (2002). Cellular response to oxidative stress: signaling for suicide and survival. *J Cell Physiol.* 192: 1–15.
212. Irani K, Xia Y, Zweier JL, *et al.* (1997). Mitogenic signaling mediated by oxidants in ras-transformed fibroblasts. *Science* 275: 1649–52.
213. Riley JCM, Behrman HR (1991). Oxygen radicals and reactive oxygen species in reproduction. *Proc Soc Exp Biol Med.* 198: 781–91.
214. Schreck R, Rieber P, Baeuerle PA (1991). Reactive oxygen intermediates as apparently widely used messengers in the activation of the Nf-kappa-B transcription factor and Hiv-1. *EMBO J.* 10: 2247–58.
215. Parkes TL, Hilliker AJ, Phillips JP (1999). Motorneurons, reactive oxygen, and life span in *Drosophila. Neurobiol Aging* 20: 531–5.
216. Grimm M, Spiecker M, De Caterina R, Shin WS, Liao JK (2002). Inhibition of major histocompatibility complex class II gene transcription by nitric oxide and antioxidants. *J Biol Chem.* 277: 26460–7.

217. Pavlick KP, Laroux FS, Fuseler J, *et al.* (2002). Role of reactive metabolites of oxygen and nitrogen in inflammatory bowel disease. *Free Radic Biol Med.* 33: 311–22.
218. Rogina B, Helfand SL (1997). Spatial and temporal pattern of expression of the wingless and engrailed genes in the adult antenna is regulated by age-dependent mechanisms. *Mech Dev.* 63: 89–97.
219. Rogina B, Vaupel JW, Partridge L, Helfand SL (1998). Regulation of gene expression is preserved in aging *Drosophila melanogaster. Curr Biol.* 8: 475–8.
220. Zou S, Meadows S, Sharp L, Jan LY, Jan YN (2000). Genome-wide study of aging and oxidative stress response in *Drosophila melanogaster. Proc Natl Acad Sci USA* 97: 13726–31.
221. Arbeitman MN, Furlong EE, Imam F, *et al.* (2002). Gene expression during the life cycle of *Drosophila melanogaster. Science* 297: 2270–5.
222. Wheeler JC, Bieschke ET, Tower J (1995). Muscle-specific expression of *Drosophila* hsp70 in response to aging and oxidative stress. *Proc Natl Acad Sci USA* 92: 10408–12.
223. King V, Tower J (1999). Aging-specific expression of *Drosophila* hsp22. *Dev Biol.* 207: 107–18.
224. Kurapati R, Passananti HB, Rose MR, Tower J (2000). Increased hsp22 RNA levels in *Drosophila* lines genetically selected for increased longevity. *J Gerontol A Biol Sci Med Sci.* 55: B552–9.
225. Welte MA, Tetrault JM, Dellavalle RP, Lindquist SL (1993). A new method for manipulating transgenes: engineering heat tolerance in a complex, multicellular organism. *Curr Biol.* 13: 842–53.
226. Ramet M, Manfruelli P, Pearson A, Mathey-Prevot B, Ezekowitz RA (2002). Functional genomic analysis of phagocytosis and identification of a *Drosophila* receptor for *E. coli. Nature* 416: 644–8.
227. Zou S, Meadows S, Sharp L, Jan LY, Jan YN (2000). Genom*e-wide study of aging and oxidative stress response in* Drosophila melanogaster. *Proc Nat Acad Sci USA* 97: 13726–31.
228. Braeckman BP, Houthoofd K, Vanfleteren JR (2001). Insulin-like signaling, metabolism, stress resistance and aging in *Caenorhabditis elegans. Mech Ageing Dev.* 122: 673–93.
229. Fabrizio P, Pozza F, Pletcher SD, Gendron CM, Longo VD (2001). Regulation of longevity and stress resistance by Sch9 in yeast. *Science* 292: 288–90.
230. Guarente L (1999). Mutant mice live longer. *Nature* 402: 243, 245.
231. Jazwinski SM (1999). Molecular mechanisms of yeast longevity. *Trends Microbiol.* 7: 247–52.
232. Lithgow GJ (2000). Stress response and aging in *Caenorhabditis elegans. Results Probl Cell Differ.* 29: 131–48.
233. Longo VD (1999). Mutations in signal transduction proteins increase stress resistance and longevity in yeast, nematodes, fruit flies, and mammalian neuronal cells. *Neurobiol Aging* 20: 479–86.
234. Walker GA, Walker DW, Lithgow GJ (1998). A relationship between thermotolerance and longevity in *Caenorhabditis elegans. J Investig Dermatol Symp Proc.* 3: 6–10.

Aging in *C. elegans*

Anders Olsen, James N. Sampayo and Gordon J. Lithgow
Buck Institute, NOVATO, CA 94945, USA

The contribution of the worm to aging research

The genetic dissection of lifespan
In the last decade of the 20th century one model organism has received more attention from gerontologists than any other, namely the free-living soil nematode *Caenorhabditis elegans* (*C. elegans*). We owe much of our general view of the genetic determination of lifespan to this organism. *C. elegans* became a popular organism for the analysis of complex biological problems following the pioneering work of Sidney Brenner and Sir John Sulston who determined the developmental cellular fate map [1]. It is also noted for the early work of Robert Horvitz and colleagues who used the cell lineage map to begin a genetic dissection of programmed cell death [2].

The use of *C. elegans* in aging research has a long history but its true strength emerged when aging studies were combined with its excellent characteristics as a genetic system. Two classic publications from the University of Colorado, Boulder, heralded the worm's arrival. In the first Thomas Johnson and William Wood demonstrated, perhaps for the first time, experimental evidence for the genetic determination of longevity [3]. In the second publication, Michael Klass demonstrated that genetic variants with large increases in lifespan could be isolated following mutagenesis of a population [4]. This was of great significance because it suggested that longevity might be studied using the same molecular genetic techniques that had proved highly successful in the study of development. Tom Johnson was later to show that mutation of only a single gene was responsible for longevity in Klass' mutants and that the mutation slowed the overall population aging rate [5–7]. He named this gene *age-1* and "molecular gerontology" became a reality [8]. The gene encoding the AGE-1 protein was said to confer the age phenotype (extended lifespan) and was the first in a long line of "gerontogenes."

H.D. Osiewacz (ed.), Aging of Organisms, 163–199.

A short life history and a long lived larval variant

There are several practical reasons for the worm's popularity amongst biologists including: (a) the easy and inexpensive culture techniques, (b) the relatively short generation time, (c) the simplicity of genetic manipulation, a result of its self fertilizing hermaphroditic mode of reproduction (males occur infrequently), (d) the ease of mutagenesis and the down-regulation of gene expression by RNA interference (RNAi), and finally, (e) the fully sequenced genome [9–11].

Most adults are self-fertilizing hermaphrodites with a mean lifespan of approximately 18–20 days under normal laboratory conditions (20°C with *E. coli* as a food source). This short lifespan is undoubtedly a major advantage for aging studies allowing relatively rapid analysis of survival phenotypes. Males that arise by chromosomal non-disjunction at a frequency of 1 in 500 have a lifespan a few days shorter than the hermaphrodites [12].

C. elegans has a three-day lifecycle with four larval stages (L1 to L4) before the final moult into the reproducing adult (Figure 1A). Another reason the worm attracted the interest of those researchers interested in aging, is the existence of a long-lived larval variant, called the dauer [13]. When nutrition is low or the population density is high, the worms activate the alternative developmental program that leads to formation of the dauer larvae [14]. The dauer larvae is a diapause stage that does not feed or reproduce and has a lifespan of approximately 60 days [13]. It is also highly stress resistant [15].

Loss of function mutations in dauer formation genes (Daf genes) cause constitutive dauer formation even in the presence of food. In general, strong loss-of-function alleles cause larvae to develop into dauers rather than reproductive adults (dauer formation constitutive – Daf-c) [14]. Weaker Daf-c alleles develop into adults at permissive temperatures and have a variety of altered phenotypes including reduced fertility, stress resistance, fat accumulation and are frequently long-lived (see section on "Insulin/IGF-1-like signaling"). The genes that determine the decision to progress through normal development or become a dauer were the subject of intense study [14], and many of these genes were later shown to also affect aging.

Beyond genetics

C. elegans is more than just a genetic model system. In an aging context, nematodes have also proven useful in screening for drugs that extend lifespan, for investigating metabolic changes with age, for studies on the effects of stress and there are even examples of studies on the evolutionary biology of aging and the pathology of aging.

The nematode has also proved useful for studying environmental influences on aging such as dietary restriction and hormesis (the life extending effects of mild stress) [16, 17]. This approach has, to date, not yielded the kind of detailed information on the determinants of aging that the genetic approach has provided. However, environmental manipulation in consort with genetic analysis is likely to be a very powerful approach. The investigation of the action of a gene under a single environmental condition can lead to misleading interpretations about its contribution to an animal's life history. The true picture of gene action in aging can only be

Aging in *C. elegans*

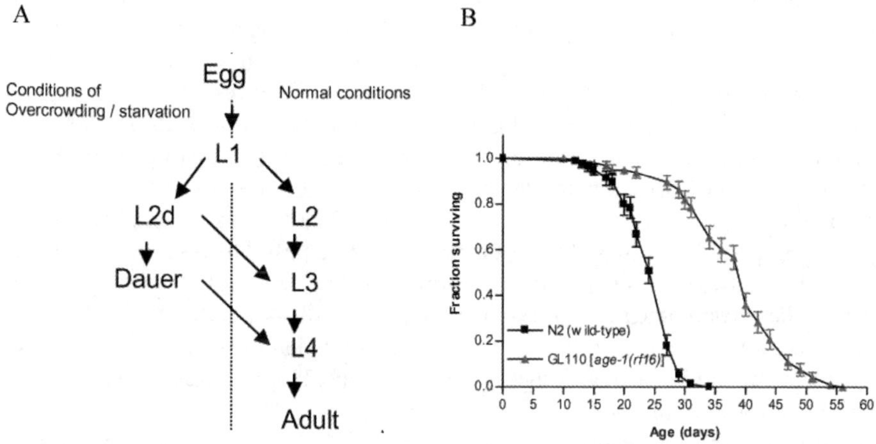

Figure 1. *The C. elegans lifecycle and lifespan. A: Under favorable conditions the nematode develops into a reproducing adult, after the egg has hatched, via 4 different larval stages (L1–L4). Under conditions of starvation or overcrowding the L1 larvae can develop into a pre-dauer L2d and then into a dauer. If the conditions change, the nematode can exit the dauer stage and develop into a normal reproducing adult. In contrast, the decision to activate the dauer formation program must be made before the L2 stage otherwise the nematode is committed to normal development. B: age-1(rf16) mutant displays enhanced longevity. Synchronously ageing populations of hermaphrodites (n >80) were scored for touch provoked movement every two or three days, at 20°C. Fraction surviving ± SEM was calculated for each time-point. Mean survival ± SD were as follows: N2 (wildtype) 24.4 ± 4.4 days, GL110 [age-1(rf16)] 38.3 ± 8.2 (David W. Walker (2000). PhD thesis, University of Manchester).*

established by examining the role of the gene in various environmental conditions and in conjunction with environmental treatments known to effect aging rate.

As Age genes have been identified in their hundreds, the challenge is to understand the mechanisms by which these mutations extend lifespan. In the future, genetic strategies will be combined with other experimental approaches to build a comprehensive understanding of aging at the systems level. Here we will review the literature on Age mutants but broaden the analysis to the wide variety of experimental approaches that are being undertaken with *C. elegans* to study aging. We wish to give a broad view of worm aging and consider the diverse roles and activities now ascribed to aging genes.

When worms die

Biodemography

One of the most important breakthroughs in aging research was the widespread adoption of "lifespan" or life expectancy at birth as a surrogate of aging itself (Figure 1B). However, the determination of a lifespan is much less informative than measurements of age-specific mortality that reveal features of aging essentially hidden by the traditional "survival curve" determination. For example, large scale analysis of a genetic sterile strain of *C. elegans* exhibited two distinct rates of aging with an apparent reduction in the rate at day ten of life [18, 19]. No adequate physiological explanation for this aging rate change has been forthcoming although it has been suggested that two modes of aging are acting [20]. The response of mortality rate to lifespan-extending hormetic treatment is the focus of studies designed to measure the physiological variance within clonal worm populations [21, 22]. This is an example of an extensive body of literature that demonstrates that the rate of increase in mortality rate with age is not constant [23–26]; a discovery with great significance for human populations.

Combining large scale biodemographic studies with genetic and environmental modulation of aging is a likely important route to the understanding of the large degree of survival heterogeneity present in even populations of inbred or clonal animals such as *C. elegans*. In general such studies require large numbers of animals and whilst the mass culturing (100 000+ worms) of *C. elegans* is trivial [27], the manual maintenance and analysis of such a large culture is highly labour intensive. The measurement of mortality rates may only become commonplace in *C. elegans* with the widespread use of automated worm handling technology.

Aging pathology

One of the traditional weaknesses of invertebrate aging, and *C. elegans* aging experiments in particular, is the lack of a detailed description of the pathology of the aging worm. Occasionally, a cause of death can be attributed to a worm such as the occurrence of a major trauma or rupture resulting in a massive loss of turgor pressure. However, most worms die without any specific cause of death being assigned. Despite this, a great range of gross anatomical and behavioral changes are apparent upon casual observation such as slowed movement, slowed eating and apparent loss of some cellular structure [28–30]. Comprehensive analysis of the cellular changes during aging using Green Fluorescent Protein (GFP) under the control of neuronal transcriptional promoters reveals that neurones are not lost during aging [31]. In contrast, very significant changes are apparent in the muscle groups of the worm. At the subcellular level, the nucleolus increases in size relative to the nucleus which itself becomes increasingly mis-shapen. With increasing age, the sarcomeres become disorganized and less dense, and muscle cells shrink and accumulate lipid [31].

Both these behavioral and cellular changes are highly unpredictable and vary considerably in their age of onset and severity between individuals [31]. Many of the changes are poor predictors of longevity although some changes are clearly delayed in long-lived genetic variants. This situation is not unique to the nematode as the search for good "biomarkers" of biological age is notoriously difficult. Despite this, it is important to continue the studies of the pathology of aging as this will provide an important link to studies of age-related disease.

The age mutant paradigm

More than 100 genes have been shown to influence adult *C. elegans* lifespan [32]. Most of these genes have been defined initially by mutation and many share identity to genes of known function or biochemical activity in other species [33]. The Age genes are diverse and have a wide variety of functions but genes encoding intracellular signaling and neuroendocrine processes figure prominently (Table 1).

Insulin/IGF-1-like signaling

The most significant discovery in invertebrate aging research to date is the realization that an insulin/IGF-1-like signaling pathway (Figure 2) has a large influence on aging rate in a range of diverse species [67, 105–109]. It is one of only a few examples of factors that appear to have a conserved function in lifespan determination in different invertebrate species [102]. As insulin/IGF-1 signaling is important in both worm and fruit fly aging, one can infer that it may also be important in mammalian aging. Consequently, the influence of these signaling pathways on aging and age-related disease in humans is a highly important area.

In *C. elegans*, the insulin-signaling pathway is initiated by the *daf-2* gene product, an insulin/IGF-1 like receptor [67]. The ligand or ligands for this receptor have not yet been identified, but 37 insulin-like genes have been noted [103]. In response to insulin receptor activation AGE-1, a phosphoinositide-3-OH kinase (PI3K), generates 3-phosphoinositides, including phosphatidylinositol-3,4-biphosphate ($PI(3,4)P_2$) and phosphatidylinositol-3,4,5-triphosphate ($PI(3,4,5)P_3$) [110]. These second messengers activate the PDK-1 and AKT-1 and AKT-2 kinases. *pdk-1* encodes a homologue of the mammalian AKT/PKB kinase [34]. It's activity has been shown to be both necessary and sufficient to propagate PI3K signals in this pathway. However, since an activating mutation in *pdk-1* requires *akt-1* and *akt-2* gene activity in order to suppress the *age-1* phenotype, the major function of *pdk-1*, appears to be, the transduction of signals from AGE-1 to AKT-1 and AKT-2 [111]. AKT-1 and AKT-2 are homologs of the mammalian AKT/PKB [111]. Activating mutations in these genes can bypass the requirement of AGE-1 PI3K signaling but not DAF-2 signaling. This indicates they are the major outputs of PI3K signaling but not the only output of DAF-2.

DAF-16, a forkhead transcription factor, is a downstream target of the AKT signaling and is a key molecule for the influence of insulin/IGF-1-like metabolism on longevity [112, 113]. All lifespan extending mutations identified in the insulin/IGF-1 pathway require functional DAF-16 activity including mutations in *age-1* and *daf-2*.

Mutation of the gene encoding the lipid phosphatase DAF-18, a homologue of the mammalian PTEN tumor suppressor, bypasses the normal requirement for AGE-1 [114] and DAF-2 signaling. It is likely that DAF-18 normally limits AKT-1 and

Table 1. *Genes that when modulated increase the lifespan of the nematode*

Gene	Function
Recessive mutations	
age-1	PI-3-kinase (catalytic subunit-p110) [5–7, 17, 34–38]
che-2	WD40 repeat protein [39, 40]
che-3	Cytosolic dynein heavy chain [40–44]
clk-1	Strong similarity to the human ubiquinone biosynthesis protein COQ7 [45–54]
clk-2	Similar to yeast telomere binding protein Tel2p [55–57]
daf-12	Predicted steroid hormone receptor, Age phenotype only in a *daf-2* mutant background [58–61]
daf-19	RFX-type transcription factor [40, 62, 63]
daf-2	Insulin-like / IGF-1 tyrosine kinase receptor [36, 64–70]
eat-2	Nicotinic acetylcholine receptor subunit [71, 72]
glp-1	Receptor for a germline-proliferation signal that is produced by the distal tip cells of the somatic gonad [73–76]
isp-1	Iron-sulfur protein of mitochodrial complex III [77]
mec-8	RNA-binding protein splicing factor [40, 78, 79]
mes-1	Tyrosine-protein kinase [73, 80, 81]
osm-3	Kinesin motor domain [40, 42, 82]
osm-5	Member of the TPR family [38, 40, 42, 63, 83, 85]
pdk-1	Serine/threonine kinase [34]
pgl-1	Component of germ line P-granules [73, 86]
spe-26	Kelch family (actin binding) protein [87, 88]
tax-4	Forms heteromeric cyclic nucleotide-gated channel along with Tax-2 [40, 83, 89]
unc-13	Synaptic transmission [9, 12, 90]
unc-31	Homologue of vertebrate CAPS, a calcium binding protein [9, 12, 91–94]
unc-32	Vesicular ATPase alpha subunit [9, 12]
unc-4	Neuronal homeodomain transcription factor [9, 12]
unc-64	Syntaxin homologue [9, 12, 92, 95, 96]
unc-76	FEZ family protein [12, 97]
RNAi	
ins-18	Insulin ligand predicted to bind DAF-2 [98, 99]
Overexpression	
tkr-1	Tyrosine kinase receptor [100, 101]
sir-2.1	Homology to NAD-dependent histone deacetylases [102]
ins-1	Insulin like protein [103]
INS	Human insulin [103]
Hsp70F	Possible heat shock protein [104]
Hsp16	Heat shock protein (Walker and Lithgow, in press)

Adapted from SAGE KE's Genes/Interventions Database.
(URL http://sageke.sciencemag.org/cgi/genesdb)

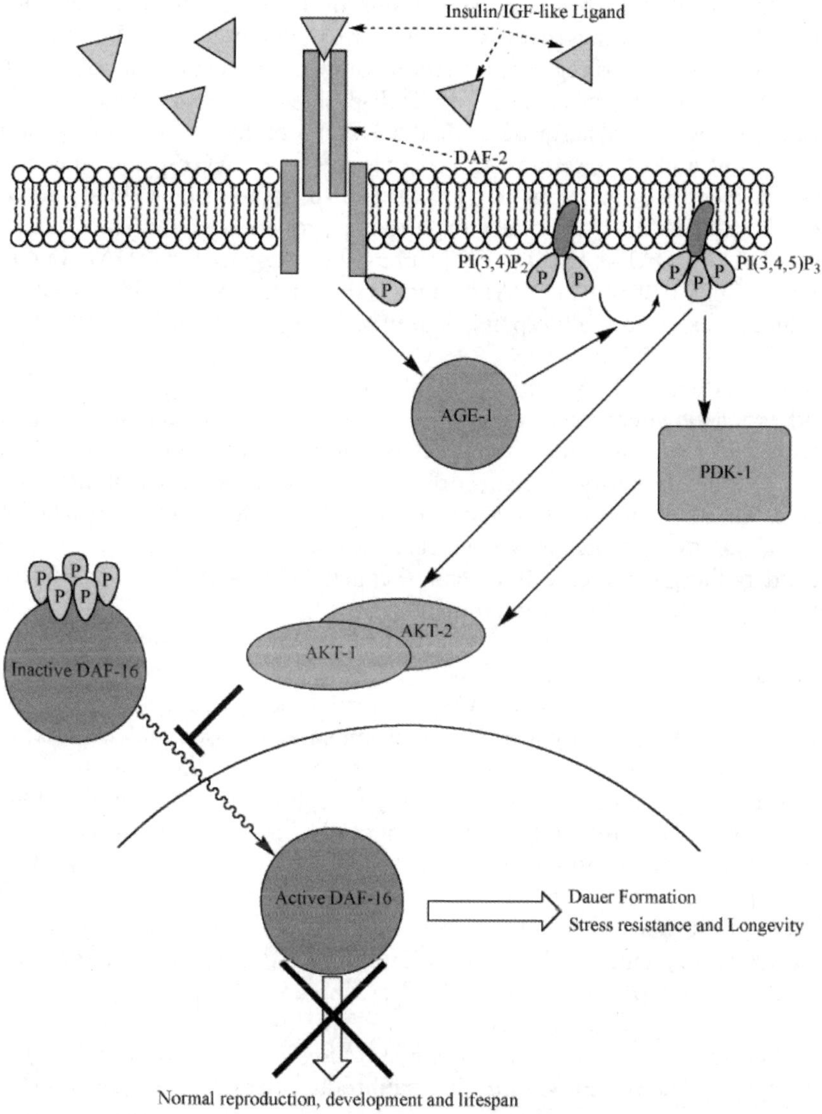

Figure 2. *The insulin/IGF-1-like signaling pathway. See text (see section on "Insulin/IGF-1-like signaling") for the description of the components of the pathway and their interactions. P signifies phosphate groups.*

AKT-2 kinase activity by decreasing levels of PI(3,4,5)P_3. This mechanism is supported by the finding that AGE-1 suppression requires AKT-1 and AKT-2 activity. How DAF-18 is regulated is still unknown, but in cultured cells PTEN has been shown to be down-regulated by TGF-β signaling. In *C. elegans*, a TGF-β signaling pathway is synergistic with the DAF-2 pathway for some phenotypic outcomes, but not all. Several of the components of the TGF-β signaling pathway have been identified, including DAF-7, a TGF-β homologue, DAF-1, a receptor protein kinase, DAF-4, a bone morphogenic protein receptor, and the SMAD proteins DAF-3, DAF-8, and DAF-14. It has been suggested that DAF-18 could be one of the links between the DAF-2 and TGF-β pathways [14, 115]. It should be noted that no extension in lifespan has been reported for worms carrying mutations of the TGF-β signaling pathway.

Lifespan extension by manipulation of sensory neurones, gonadal development and other genetic alterations all require DAF-16 activity (see section on "Endocrine signals are essential for life history determination"). The potentially detrimental effects of some mutations often complicate epistatic analysis of lifespan. For example, the dependence of *daf-16* for longevity may be due to a non-specific lowered fitness of *daf-16* mutants. However, there are a number of instances of *daf-16* independent lifespan increase and it can be concluded that DAF-16 does indeed play a pivotal role in lifespan determination.

At least one other intracellular signaling pathway is implicated in lifespan determination. Over-expression of the *old-1* gene, which encodes a tyrosine kinase receptor, is sufficient to extend the lifespan of wildtype worms [100, 101]. Expression of *old-1* is up-regulated by various environmental stresses and over-expression confers resistance to heat shock, ultraviolet radiation and oxidative stress, a characteristic of all long-lived worm lines (see section on "Long-lived mutants are resistant to extrinsic stress"). The expression of *old-1* is also under the control of *daf-16* such that the stress inducible expression is compromised in a *daf-16* mutant background. This suggests that *old-1* acts downstream of *daf-16* and perhaps is part of a mechanism that relays a signal from neuronal *daf-16* to other cells. Consistent with this idea, reducing *old-1* expression by RNAi suppresses the lifespan extension conferred by an *age-1* mutation [100, 101].

Insulin/IGF-1-like signaling acts cell non-autonomously to determine lifespan

Excluding germline cells, *C. elegans* hermaphrodites possess 959 differentiated cells. The position of each cell and the fate of each cell during development are known. A range of techniques is available to study tissue- and cell-specific events and the DAF-2 pathway has been investigated utilizing these techniques. Expression analysis using GFP fusion proteins indicates that *akt-1* and *akt-2* are expressed in the majority of head neurones (including sensory neurones), muscle cells of the pharynx, hypodermis, intestine, muscle, and in the excretory canal [34]. A similar expression pattern was observed for *pdk-1* consistent with a model where these molecules act in the same cell [34]. Mosaic analysis [66] and expression of wildtype *daf-2* in otherwise mutant worms [116] revealed that dauer formation and lifespan are regulated by DAF-2 function in a cell non-autonomous manner i.e., DAF-2 acts in groups of signaling cells to control the production or activity of other secondary signals that regulate development and lifespan of individual tissues [66]. Exposure to RNAi at various ages results in clear demonstration that the pathway acts in adults to confer longevity [117]. This is consistent with the prior observation that drug inhibition of AGE-1 in adults confers thermotolerance [118]. Together, these observations demonstrate that the insulin/IGF-1 effects are not the consequence of events that occur during development.

The importance of neurones

The discovery that DAF-2 can act in neurones to influence lifespan suggests that some neurones themselves are critical for a normal life history. There are two ways in which this may occur.

First, some neurones may be very sensitive to aging such that the rate of neuronal death may limit lifespan. This would be the case if life were not sustainable following the death of specific life-limiting neurones. Protection of these critical neurones from aging would therefore extend organismal lifespan [116].

The second possibility is that neurones produce endocrine signals that influence the rate of aging of other tissue types [119, 120]. Manipulation of the aging of these neurones would have very different effects on organismal aging depending on whether these neurones produced lifespan enhancing or lifespan shortening signals. The two hypotheses are not mutually exclusive particularly if critical neurones produce lifespan-enhancing signals. Indeed, the finding that the nervous system retains its cellular integrity during aging suggests that the death of neurones is one of the last pathological events prior to death [31].

Genetic evidence for neuronal determination of lifespan includes the effects discovered for mutations of *unc-64* and *unc-31* which both extend the nematode lifespan. *unc-64* encodes a syntaxin homolog [95] and *unc-31* encodes a CAPS (**c**alcium-dependent **a**ctivator **p**rotein for **s**ecretion) homologue [93] and both gene products are involved in Ca^{2+} signaling. *unc-64* has a neuronal site of action and the lifespan extension is fully suppressed by a *daf-16* mutation [91]. Double *daf-2; unc-64* mutants do not have a stronger phenotype than the single mutations indicating that these genes are involved in DAF-2 signaling. This is consistent with the suggestion that the site of action could be secretion of a ligand for the DAF-2 insulin receptor.

Sensory cilia are found in 60 of the 302 neurones of the hermaphrodite [121]. It has been shown that defects in sensory cilia and their support cells can increase the lifespan of the nematode [66]. Various defects in sensory cilia caused by mutations in nine different genes all resulted in significant lifespan extension. Although a *daf-16* mutation slightly reduced the lifespan increase in these mutants, complete reversion was not observed. The mutants also seem to act in a *daf-2* dependent pathway since no synergy was observed in *daf-2* double mutants, in fact, the double mutants were slightly shorter lived. These mutants need to be analyzed carefully in order to understand the molecular mechanisms involved but the study does illustrate that environmental cues are likely to influence lifespan through neuronal processes.

Alteration in the timing of life history

Clk mutations define an interesting class of genes that influence the timing of a range of biological events. Clk stands for "abnormal function of biological clocks." Mutation of Clk genes cause apparent deregulation of life history events resulting in slowed metabolism, slower development, and extended lifespan [45, 47]. The Clk phenotype should not be confused with circadian or other biological clocks. *C. elegans* does possess such a circadian timing mechanism, but its role in aging has not been investigated. The components of such a circadian clock act in concert to bring about specific cyclical events. The notion of an "aging clock" is difficult to reconcile with the evolutionary origins of aging (see section on "Why aging is not regulated") and further investigations of the Clk mutations demonstrated little evidence for an integrated age-related timing mechanism. Indeed, the Clk phenotype is a misnomer, as Clk related longevity is dependent on culture conditions (see section on "Alterations to the rate of ROS production and ROS-associated damage).

CLK-1 is a homolog of the yeast metabolic regulator Coq7p [51]. In yeast, the CLK-1 homolog is required for coenzyme Q (Q) production [49]. This function turned out to be conserved in the nematode and is discussed in a later section focusing on oxidative stress.

The *clk-2* gene mutation results in a modest increase in mean and maximum lifespan [45]. However, no functional relationship has been discovered between *clk-2* and *clk-1* and it seems likely that they influence lifespan by independent mechanisms. Three research groups independently discovered that *clk-2* encodes a protein homologous to yeast Tel2p [55–57]. Interestingly, *clk-2* and *rad-5* turned out to be the same gene [55]. In yeast, a mutation in the Tel2p protein results in shortened telomeres and a similar function was sought in the nematode. The results, however, were controversial since shortened telomeres [57], longer telomeres [56] and unaltered telomeres [55] were found in *clk-2* mutants. The discrepancy has been attributed to the fact that different *C. elegans* strains generally have very different telomere lengths [55]. Whereas telomere length has been shown to be critical for maintaining an immortal germline [122] the role of telomere length in *C. elegans* aging remains ambiguous.

Feeding behavior and lifespan

Caloric restriction (CR) is a well-documented way of extending lifespan reported in many different species [123]. The mechanisms of action of CR are not fully understood but many lines of evidence suggest that the effect may be due to a reduction of oxidative stress in CR animals. There is considerable hope that by combining CR with a genetic approach, the critical physiological alterations that occur during CR and are causal for lifespan extension can be identified. Mutant *C. elegans* that have altered feeding behaviors (Eat mutants) were examined for lifespan alterations. Some Eat mutants are indeed long-lived (Table 1) and do not require the activity of DAF-16. *daf-2*; *eat-2* double mutants live longer than single *daf-2* mutants [71]. In contrast, *clk-1*; *eat-2* double mutants do not have increased lifespan compared to *clk-1* mutants suggesting that *clk-1* and Eat mutations affect similar processes [71]. It should be noted that these experiments were performed under conditions where *clk-1* mutants utilize bacterial Q.

CR may exert some of its effects on lifespan by alterations to protein turnover [124]. The rate of protein turnover is known to decrease with age, which is believed to account for the accumulation of damaged proteins over time. There is a decrease in protein synthesis, which would lead to less efficient defence mechanisms (e.g., abundance of antioxidant enzymes and molecular chaperones) and a lack of newly synthesized, fully functional proteins to replace the damaged ones. There is also a decline in protein degradation, which leads to functionally compromized proteins accumulating within the cell. However, under the influence of CR, both protein degradation and synthesis are increased.

Surprisingly, no standard CR methodology has yet been established in the nematode. Pioneering work by Klass (1977) demonstrated that decreasing the food concentration did lead to an increase in lifespan [16]. However, a short period of starvation also results in increased stress resistance and extended lifespan (Lithgow, unpublished results). As discussed in the section on "Hormesis," short periods of mild stress often result in extended lifespan (sometimes known as hormesis) therefore the relationship between CR and hormesis may have some mechanistic commonality [125].

Stress resistance determines *C. elegans* lifespan

Long-lived mutants are resistant to extrinsic stress

A causal relationship between organismal resistance to stress and longevity has been established [126]. A range of manipulations that enhance stress tolerance in *C. elegans* also result in lifespan extension. Evidence for this relationship stemmed initially from the observation that all known Age mutants exhibit increased resistance to environmental stress, such as oxidative stress [127, 128], heat shock [17, 129], UV irradiation [87] and heavy metal stress [36]. A number of mutations that confer stress tolerance in young animals [130, 131] do not extend lifespan suggesting that some mutations also confer detrimental effects.

Organisms respond to acute stress by preferentially synthesizing stress proteins. For example, heat shock results in the up-regulation of genes encoding the heat shock proteins (hsps) many of which exhibit chaperone functions that protect against thermal stress [132]. Other forms of stress, including oxidative stress, also induce the heat shock response. It is thought that long-lived worms are stress resistant due to an up-regulation of hsp genes and other stress protein genes [36, 68, 133].

The optimal growth temperature for reproduction in *C. elegans* is 20°C. When worms are subjected to 35°C they stop laying eggs, stop feeding and death occurs within approximately eight hours. Mutation of *age-1* or *daf-2* results in a 40% increase in the mean survival time in these heat shock conditions [17, 129]. Mutation of *daf-2* or *age-1* also results in increased cadmium and copper ion resistance [36]. In a 24 h toxicity assay, lethal concentration (LC_{50}) was significantly higher for these mutants. The genes encoding metallothionein (MTL), which are important for metal detoxification, are up-regulated in these insulin-signaling mutants. *mtl-1* mRNA levels were significantly higher in *daf-2* mutants compared to *age-1* mutants and wildtype *C. elegans*. Even in the absence of elevated heavy metal concentrations and in the presence of toxic levels of cadmium, induction of *mtl-1* and *mtl-2* mRNA was 3 and 2-fold higher, respectively, in *daf-2* mutant worms than in wildtype [36]. As heavy metals are involved in many redox reactions and in the generation of Reactive Oxygen Species (ROS) [134], this elevated expression may result in a reduction of oxidative damage.

When worms are subjected to stress the DAF-16 protein moves from the cytoplasm to the nucleus [135]. When extra-copies of the wild type *daf-16* gene were introduced, the resulting worms were resistant to stress consistent with a model in which the insulin/IGF-1 pathway represses DAF-16-associated stress resistance [135, 136].

Hormesis

In an aging context, hormesis can be described as the beneficial effects resulting from the cellular response to mild repeated stress [137]. Hormesis has been shown to increase the lifespan of many different species including yeast [138], mice [139], flies [140], and nematodes [17]. Multiple stressors, such as increased heat, high oxygen [141] and gamma irradiation [142] can induce such a hormetic response in *C. elegans* and a single pre-treatment of heat early in life is sufficient to significantly extend lifespan. Repeated mild heat treatment throughout the lifespan of the nematode results in a small beneficial effect compared to single heat treatments early in life (Olsen and Lithgow, unpublished results). The relevance of hormesis in aging studies has been debated extensively [143]. It is clear from this discussion that a simple model organism such as the nematode, could prove very useful in order to fully understand the mechanisms involved, which are currently unknown.

Oxidative stress and aging in *C. elegans*

ROS and the worm

In the mid-1950s, Denham Harman proposed that the accumulation of damage caused by free radicals was the underlying mechanism by which organisms age [144, 145]. Since then, there has been an ever-growing body of literature refining and testing this theory, and it has now been transformed into a more general premise which highlights the importance of ROS in the aging process [146]. Although little direct evidence exists in support of the theory, it is still accepted by most biological gerontologists as the most likely mechanistic cause of aging [147]. Work with *C. elegans* provides some of the best direct evidence that ROS plays a role in lifespan determination [127, 128, 148, 149].

The majority of the endogenous production of ROS occurs in the mitochondrial matrix of aerobic organisms [150]. During respiration, cells transform carbohydrates and fat into adenosine triphosphate (ATP), the energy source in many essential biochemical reactions. Some ATP production occurs in the cytoplasm (with the conversion of sugars to pyruvate), but the majority is produced in the mitochondria if oxygen is available via the electron transport chain.

The electron transport chain is made up of four large protein complexes, which are situated within the inner membrane of the mitochondria. These complexes establish a proton gradient by pumping H^+ ions across the inner mitochondrial membrane utilizing the energy released from the transfer of electrons. ATP synthase (Complex V) then allows protons back into the matrix of the mitochondria, and the energy this flux generates is used to produce ATP. In a perfect system, the electrons would be passed along the chain and oxygen would be the final electron acceptor, yielding water. However, the system is not perfect and due to inappropriate electron donation (usually involving ubiquinone) ROS can be produced. Ubiquinone (also called Coenzyme Q or simply Q) accepts electrons from NADH in complex I and also from succinate in complex II. This reduced form of Q (now carrying 2 electrons) normally transfers its electrons, one at a time, to complex III. However, if the unstable intermediate, called ubisemiquinone (carrying 1 electron) comes into contact with oxygen, the reaction can give rise to ROS, in this case, the superoxide anion [151]. Coenzyme Q consists of a quinone head attached to a tail of isoprene units, and the number of attached units varies slightly between species. Coenzyme Q from *C. elegans* contains 9 isoprene units [49], compared to 8 in *E. coli* [49] and 9 or 10 in various mammalian species [152]. Manipulation of the levels or type of Coenzyme Q can have profound effects on *C. elegans* life history (see below).

ROS can also be generated from non-metabolic sources, such as ultraviolet light, ionizing radiation and extremes of temperature, although these sources yield less ROS than respiration [153].

The three most common ROS are the superoxide anion ($^{\cdot}O_2^-$), hydrogen peroxide (H_2O_2), and probably the most destructive, the hydroxyl radical ($^{\cdot}OH$). Superoxide carries a negative charge, which means it cannot traverse the mitochondrial membrane. In contrast, H_2O_2 can diffuse through membranes, and although less reactive, can give rise to the hydroxyl radical if it comes into contact with metal ions (via the Fenton reaction). These ROS can cause damage within the mitochondria, within the cytosol and even outside the cell [147].

All of the main biological macromolecules are at risk from ROS-mediated damage, and the accumulation of this damage increases as a function of age in a wide variety of tissues of many species [123]. ROS can react with the polyunsaturated fatty acids of lipids leading to decreased membrane fluidity. This, in turn, can have deleterious effects on the function of proteins (e.g., receptors) and enzyme complexes located in the membrane, including those of the electron transport chain [154].

Proteins can also be directly damaged by ROS, one of the most common results being the formation of carbonyl groups either by peptide cleavage or by modification of the amino acid side chains [155]. These alterations can have profound effects on protein function and can also lead to protein aggregates being formed within the cell. The products of protein and lipid oxidation can also form undegradable, highly cross-linked molecules, such as lipofuscin, another oxidation product that accumulates within the cell as it ages [156]. It has been shown that certain proteins are more at risk than others, and it has been suggested that specific damage to key enzymes may explain the age related decline in cellular function [145, 157, 158]. Aconitase and adenine nucleotide transferase, two key enzymes involved in mitochondrial function, were shown to accumulate carbonyl content distinctly more than other proteins in the housefly. The link between carbonyl accumulation and lifespan has also been demonstrated in *C. elegans* [37], and an egg yolk protein (Vitellogenin-6), has been shown to undergo extensive carbonylation with age [159].

Nucleic acids are also subject to ROS attack. DNA bases are often modified by ROS and deletions may be induced [147, 160]. When the DNA is replicated, these modifications can cause errors to be propagated. This may occur in both nuclear and mitochondrial DNA, but as the mitochondria is the site of the majority of ROS production, the mitochondrial DNA suffers from a higher rate of damage [161–163]. The frequency of deletions in the mitochondrial genome of *C. elegans* has been shown to increase with age, and a long-lived strain carrying an *age-1* mutation was shown to accumulate these deletions less rapidly than wildtype strains [161–163].

Defences against ROS

A number of protective enzymes have evolved to help organisms cope with oxidative stress. These include superoxide dismutases (SODs), catalases and enzymes involved with the redox status of glutathione. Most eukaryotes possess three types of SODs, which reside in distinct areas in or around the cell. There is a cytosolic CuZn-SOD, an extracellular CuZn-SOD and a Mn-SOD which is located in the mitochondria [164]. In *C. elegans* there appear to be 4 different SODs and these differ in the metal ions found at their active sites and their locations. SOD-1 (CuZn-SOD) is present in the cytoplasm [127], SOD-2 and SOD-3 (both Mn-SODs) are found in the mitochondria [165, 166], and SOD-4 (CuZn-SOD) is located in the extracellular space [167]. There are two types of catalase in the worm, which are encoded by the genes *ctl-1* and *ctl-2*. They too reside in different locations; *ctl-1* in the cytosol and *ctl-2* in the peroxisome of the cell [168]. Although important in many vertebrates, the levels of intracellular glutathione peroxidase, and hence its significance, have proved difficult to measure in *C. elegans* [128, 169, 170], however glutathione itself has been shown to be important in ROS scavenging in the worm [148].

There are also a number of nonenzymatic antioxidants that help protect cells against oxidative damage. Caretenoids, thiols and tocopherols are some of these compounds that scavenge ROS and enhance the overall effectiveness of the antioxidant defence system [171]. Metallothioneins also play an important role in protection against ROS. They display *in-vitro* scavenging capacity and may also help reduce ROS production by complexing metal ions involved in Fenton reactions [172].

As an animal gets older, not only does mitochondrial function decrease (resulting in less ATP production), but also the accumulation of ROS-mediated damage increases [173]. Coupled with that is a decline in the efficiency of the antioxidant enzymes leading to an elevation of oxidative stress in aging tissues [151, 163]. It has been demonstrated that worms become more sensitive to oxidative stress as they age [170]. Young *C. elegans* were shown to adapt more readily to oxidative stress, compared to older worms, by increasing their levels of SOD.

Alterations to the rate of ROS production and ROS-associated damage
As described previously, insulin/IGF-1-like pathway mutants are resistant to different stressors [174–176]. *daf-2* and *age-1* mutants show increased levels of catalase and SOD-3 and these elevated levels are maintained for a prolonged period compared to wildtype worms [127, 128, 168]. *daf-2* mutants also have elevated levels of SOD-2 [68]. It has also been shown that the increased resistance conferred by these enzymes is critical to the lifespan extension observed in these mutants. Interestingly, the positive correlation between stress resistance and longevity does not hold for certain *daf-2* alleles and resistance towards hypoxia [69]. This uncoupling of stress resistance and lifespan extension is at the moment not very well understood. However, understanding this exception to the rule could prove very informative for understanding the interplay between stress response and longevity.

Historically, the lifespan extension exhibited by mutants of the insulin/IGF-1-like pathway was believed to be due to the inappropriate expression of dauer formation genes. There are phenotypic characteristics shared by dauers and long-lived adult worms. For example, the dauer larval stage is also resistant to numerous stresses, and it has been demonstrated that dauers too show elevated levels of SOD-3 and catalase [177]. The metabolism in dauer larvae is very different to that of the developing worm, and it has been suggested that these metabolic changes actually slow aging in the dauer larvae [177]. Dauer larvae consume substantially less oxygen than other larval stages and exhibit a marked reduction in metabolic capacity. A simple interpretation of these observations would be that the dauer produces less ROS and the ROS that is produced is more efficiently detoxified, thus decreasing the rate of oxidative damage, and improving the overall vitality of the worm. However, a reduction in flux through the electron transport chain does not necessarily result in a reduced rate of ROS production. On exit from the dauer stage, the metabolic capacity and oxygen consumption return to normal and typical age-related changes are observed. However, dauers do seem to age, but the slowed rate of decline and more gradual accumulation of damage are somehow reset on entry to normal development [177].

C. elegans is proving to be a useful organism to investigate the relationship between electron transport and aging. Some surprising alterations in aging rate result from mutations in genes encoding components of the electron transport chain.

A mutation in the *C. elegans* iron sulphur protein (*isp-1*) of mitochondrial complex III has been identified, which results in low oxygen consumption, decreased sensitivity to ROS, and increased life span [77]. ISP-1 is known to be involved in the transfer of electrons from cytochrome b to cytochrome c in the electron transport chain, and is also capable of superoxide anion production during this process. The combination of *isp-1* (qm150) and *daf-2* (known to increase resistance to ROS) did not result in any significant further increase in adult life span. These findings suggest that both *isp-1* and *daf-2* mutations increase lifespan by lowering oxidative stress, and that perhaps there is an upper limit to lifespan extension by the reduction of damage caused by ROS [77].

The *C. elegans* gene *clk-1* (see section on "Alteration in the timing of life history") encodes a di-iron carboxylate enzyme responsible for the final hydroxylase step in the synthesis of ubiquinone (Q) [178]. *clk-1* mutants lack the endogenously synthesized isoform Q_9 (the 9 denotes the number of isoprene units in the polyisoprene tail) and instead rely on the *E-coli*-derived isoform, Q_8, obtained from their diet. Mutations in *clk-1* result in the accumulation of the intermediate demethoxy-Q_9 (DMQ_9) [179] and increased levels of RQ_9 (denoting a shift toward anaerobic fermentation and a decrease in ROS production) [49]. This change in quinone species is thought to be responsible for lifespan extension in these mutants, perhaps by altering metabolism and the rate of ROS generation in the mitochondria [179]. The *clk-1* mutation is not suppressed by *daf-16* mutations and in a *daf-2*; *clk-1* double mutant an increase in lifespan of approximately 500% relative to wildtype has been observed [45]. An alternative hypothesis for the role of *clk-1* is in communication between the mitochondria and nucleus on cellular activity levels, and this signal is disrupted in the *clk-1* mutants resulting in lifespan extension by the uncoupling of energy production from energy consumption [47].

Further investigation into the role of Q in lifespan determination was undertaken by feeding worms a diet lacking Q [54]. The removal of Q from the diet of wildtype worms resulted in a ~60% increase in lifespan, and *clk-1* and other mutants were also shown to live longer on the Q-less diet. The authors suggest that a dietary source of Q is important in development but has a detrimental effect late in life, and that dietary Q and insulin-like signaling are parallel pathways that converge on the mitochondria to influence the production of ROS [54].

Mutants with increased sensitivity to ROS

An alternative strategy to investigate the role of oxidative stress in aging is the study of strains with mutations that result in increased ROS production and subsequent shortening of lifespan. A mutation in the *mev-1* gene was identified due to the mutants' hypersensitivity to methyl viologen (paraquat) [180]. This strain also exhibits hypersensitivity to elevated oxygen levels, a more rapid accumulation of fluorescent material and protein carbonyls, and a shortened lifespan. The *mev-1* mutant is defective in the cytochrome b large subunit (Cyt-1/ceSDHC) in complex II of the mitochondrial electron transport chain, and this gene has been shown to be important in both energy metabolism and superoxide anion production [181, 182].

The gene *gas-1* has also been shown to be critical to mitochondrial function in *C. elegans*. *gas-1* encodes a homologue of the 49kDa iron protein subunit of the mitochondrial NADH-ubiquinone-oxidoreductase found in complex I [183, 184]. Mutations in this gene render the worm more sensitive to volatile anaesthetics, reduces complex-I-dependent metabolism, and increases complex-II-dependent metabolism [184]. These mutants therefore have an increased sensitivity to oxidative stress, and display signs of rapid aging and a shortened lifespan.

Inducers of oxidative stress

The rate of ROS production can be increased in a variety of ways. One of the most common is the use of chemicals that promote ROS production (prooxidants).

Prooxidants, such as H_2O_2 and methyl viologen (paraquat), have been used in studies to demonstrate the increased protection against ROS conferred by mutations that also extend lifespan [127, 128]. H_2O_2, as previously mentioned, can give rise to the destructive hydroxyl radical. Paraquat is an example of a redox cycling agent, and interferes with the electron transport chain in mitochondria leading to the production of the superoxide radical [185]. Juglone, a redox quinone, has also been used to induce oxidative stress [141]. It too, generates superoxide anions from molecular oxygen during metabolism.

The oxygen concentration that worms are cultured in also has an effect on the rate of accumulation of protein damage and lifespan [37]. When the oxygen concentration was increased (hyperoxia), an increase in protein carbonyls was observed as well as a decrease in lifespan. In contrast, a modest lifespan increase was also demonstrated by lowering the oxygen concentration.

Antioxidant treatments

A number of antioxidant compounds have been tested on worms (e.g., Vitamin E [186], and tocotrienols [187]). The Ginkgo biloba extract, EGb 761, has been shown to increase resistance to heat and oxidative stress and extend lifespan in the worm [188]. In another notable study, the administration of small, synthetic superoxide dismutase/catalase mimetics (SCMs), resulted in a large increase in lifespan (both mean and maximum) in treated worms, with no apparent cost to development time or fertility [149]. The SCMs tested (Euk 8 and Euk 134) also rescued the reduced *mev-1* lifespan. Further studies have also shown that these drugs do indeed augment natural antioxidant defences and confer increased resistance to paraquat in treated young adult hermaphrodite worms (Sampayo, Olsen and Lithgow, unpublished results). The compounds only extend lifespan when the worms are grown in liquid culture (Golden and Melov, personal communication) and with no worm pharmacokinetics available, it is not known how the drug is taken up or to what tissues. The SCMs have been tested in a number of mammalian disease models in which oxidative stress is thought to play a part [189]. Hence these experiments suggest that endogenous oxidative stress is a major determinant of the rate of aging and profoundly important in many pathological conditions.

Other roles for ROS

ROS, at first glance, appear to be just noxious by-products of aerobic metabolism, but growing evidence implicates ROS in signal transduction and cellular communication [190, 191]. Increases in intracellular oxidants have two profound effects: (1) damage to cellular components, and (2) activation of signaling pathways. Both of these influence many processes and result in either normal cell function or cell death. Typically, low doses of ROS are mitogenic and promote cell proliferation. As the dose increases, growth can be arrested either temporarily or permanently depending on the duration of the stress. A high dose usually causes cell death via either apoptosis or necrosis. The large number of signaling pathways involved in these decisions in response to ROS levels are discussed in Finkel and Holbrook [192] and Martindale and Holbrook [193].

ROS appear to have a role in worm aging
As with the majority of experiments investigating ROS-mediated damage in relation to aging, experiments designed to test the ROS hypothesis are mainly correlative. However, the alterations in ROS defences in long-lived mutants and the extension of lifespan by catalytic antioxidants indicate that ROS contributes to aging in the worm. Further analysis is required to elucidate the mechanisms by which ROS and aging are connected in the nematode, and a range of techniques for manipulating ROS levels and making accurate measures of production and damage are also vital.

The endocrinology of longevity

Endocrine signals are essential for life history determination
It is widely accepted that a neuro-endocrine axis is, in part, responsible for aging in the nematode, *C. elegans*. This axis, involving an insulin/IGF-1-like pathway and a secondary steroid hormone mediated signal, has been the subject of much investigation and is involved in a number of different metabolic and physiological changes the worm undergoes through its lifetime.

As outlined earlier in this chapter and the subject of many studies, alterations to the signaling through the insulin/IGF-1-like pathway can result in the wildtype worm either undergoing normal growth, development, reproduction and aging, or transforming into the diapausal dauer larvae. The environmental cues that affect this developmental decision are detected via sensory neurones in the head of the worm. Worms are sensitive to temperature and a wide variety of soluble and volatile compounds. Some of these are food signals as well as a worm-produced pheromone. The pheromone signal indicates the density of the surrounding worm population, and coupled with signals emanating from food sources, give the individual worm an important insight into the availability of nutrients. This mechanism helps protect both the individual, and also its progeny, from starvation.

The sensory neurones transmit the environmental information via a serotonin signal that controls the release of a TGF-β-like ligand, an insulin-like ligand and is also involved in a cGMP signal. As described previously, the secretion of the insulin-like ligand, when food is abundant (most probably Ceinsulin-1) [98], is modulated by the gene products of the *unc-64* and *unc-31* genes [91]. The TGF-β pathway and its influence on dauer formation have been described in some comprehensive reviews by Patterson and Padgett [194] and Savage-Dunn [195]. The cGMP component of the dauer formation pathway is described in Riddle and Albert [14].

The insulin-like ligand is thought to bind to the insulin-like receptor, DAF-2 [67]. This receptor is present in many or all cell types throughout the worm, but exerts its' most profound effects in the neurones [116]. On binding of ligand, the DAF-2 receptor initiates a signaling cascade very similar to insulin signaling in mammals, that ultimately controls the phosphorylation state of the forkhead transcription factor, DAF-16. By the phosphorylation of DAF-16, the insulin-like signaling pathway inhibits its migration to the nucleus, and thereby represses its activity [136]. This results in normal growth and development and a bypassing of the dauer larvae stage.

If there are insufficient nutrients, the insulin-like ligand is not released and the signaling cascade is not activated. This results in newly formed DAF-16 remaining unphosphorylated, migrating to the nucleus of the neuronal cells and initiating the expression of genes that transform the L2 larvae into dauers. The downstream targets of the DAF-16 transcription factor remain largely unknown, although some have been identified in relation to the DAF-2 signaling pathway [196].

There is growing evidence that the activation state of DAF-16 and subsequent expression of specific genes in the neurones exerts its' effects at the whole organism level via a secondary signal, most probably a steroid hormone or hormones. These are as yet undetermined, however, a number of genes have been identified in the worm that may facilitate this global signal. These include DAF-9, a cytochrome P450 related to vertebrate enzymes involved in steroid ligand synthesis [119], and at least 3 other genes whose expression is upregulated in dauers and may be involved in the propagation of a steroid hormone signal (Poly (A) binding protein, Hsp90 and ASC-1-like protein) [197]. Consistent with this model, the protein encoded by *daf-12* is a member of the nuclear hormone receptor superfamily and has been identified as acting downstream of both *daf-16* and *daf-9* [58, 60]. In addition, *C. elegans* requires an exogenous source of cholesterol, the basic substrate for steroid biosynthesis, for normal growth and development.

Some genetic perturbations in insulin-like signaling can extend lifespan by over 400% [198]. All of these disruptions, such as inhibition of secretion (*unc-64* and *unc-31* mutants), decreased production of insulin-like ligand (Ceinsulin-1 mutant), reduced levels of functional insulin-like receptors (DAF-2), and reduced activity of specific kinases (AGE-1, AKT-1/2, PDK-1) have been shown to be DAF-16 dependent, and exert their influence on lifespan in neuronal cells. DAF-16 activation in neuronal cells results in, among other things, increased expression of a catalase (CTL-1) and a superoxide dismutase (SOD-3), two key enzymes involved in the detoxification of ROS. It has been alluded to, that neuronal protection against oxidative damage may be sufficient to extend lifespan in the whole organism.

The signals downstream of DAF-16 that exert their effect on the whole animal can be illustrated by a variety of models [Gill, personal communication]. If DAF-16 is inactive, (i.e., the environment favors reproduction) genes involved in steroid hormone synthesis are expressed. This signal leaves the neurones and diffuses into the cells throughout the whole animal. The steroid hormone then binds to the DAF-12 receptor and represses it's action, this complex may also repress DAF-16's action in the somatic cells. With DAF-12 (and DAF-16) repressed, the somatic cell expresses genes required for normal growth and development, and normal aging. If DAF-16 is activated (i.e., the environment favors dauer formation), the level of the steroid hormone secreted from the neurones would be decreased or abolished. Subsequently, DAF-12 (and DAF-16) would undergo no repression and the expression of dauer specific genes would result.

Alternatively, two or more separate steroid hormones may be involved that both bind to the DAF-12 receptor, one being a repressor of DAF-12 mediated gene expression, the other an activator. For normal development, the levels of the repressor would be high and the levels of the activator would be low. If the levels were reversed (repressor – low, activator – high), then dauer formation would ensue.

The finely balanced and strictly controlled levels of this signal/signals may determine the gene expression profiles in the somatic tissue of the worm, ultimately determining organismal fate. When this balance is severely upset in the developing worm, either the worm will become a dauer irrespective of most environmental cues (dauer formation constitutive, daf-c), or the worm, even under unfavorable reproductive conditions, will not enter dauer and become an egg-laying adult (dauer formation defective, daf-d). However, when this hormonal balance is less severely altered and the worm manages to bypass the dauer stage (e.g., in certain mutants, by being raised at a particular temperature), an extension in the lifespan of the adult worm is observed.

Laser ablation of specific cells in the developing gonad of the worm has uncovered another level of hormonal complexity. These studies point to the existence of a hormonal signal emanating from proliferating germline stem cells and also an insulin-like signal being secreted from the somatic gonad [61, 73]. The effects on lifespan of these have, interestingly, been shown to be antagonistic to one another. The removal of the entire gonad (germline and somatic) and subsequent elimination of both signals has no net effect on lifespan.

The signal from the germline stem cells has been shown to be *daf-9*, *daf-12* and *daf-16* dependent, but *daf-2* independent, therefore it is believed to be steroidal in nature. By removing the germline cells, thereby removing this signal, an increase in lifespan results and so it is thought that this hormone acts to limit lifespan, perhaps by repressing both DAF-12's and DAF-16's actions on dauer formation gene expression. The removal of the germline cells can be achieved without the destruction of the neighbouring somatic gonad cells. Somatic gonad cells also release a hormonal signal, although this is thought to be an insulin-like molecule as its action is DAF-2 dependent. This insulin-like signal is thought to promote long life by an alternative DAF-16-independent DAF-2 signaling pathway [136].

These studies have shown that DAF-2 and DAF-16, although involved together in insulin-like signaling, can also exert different effects independently of one another and play important roles in lifespan determination. They also indicate that hormonal signaling and reproductive status could determine how long a worm lives.

The signal from the germline stem cells could be the same as the steroid hormone secreted by the neurones, which promotes reproductive growth. These two sources of the same hormone may set up a positive feedback loop, which drives development towards reproduction and, by the allocation of limited resources, shortens lifespan. The second possible hormone (the DAF-12 activator) could exert its effect on the gonad by down regulating production of germline derived steroid hormone, and promoting production of the insulin-like hormone from the somatic gonad, thereby inhibiting reproductive growth and indirectly extending lifespan.

By coordinating the onset of reproduction with the rate of aging, the worm may age more slowly if reproduction is delayed. The advantage being that if reproduction is delayed, perhaps by starvation, the animal would be in a better, more robust condition when reproductive maturity was finally reached. Again, this example highlights the importance of the worm's response to environmental conditions and how this response is elicited in relation to aging.

Finding novel age genes

Surprisingly, most of the early discoveries of Age genes were serendipitous as they were unearthed whilst investigating other phenotypes. Very few Age genes were first identified due to their explicit effects on aging. More recently, genes have been uncovered by screening for specific associated phenotypes such as stress resistance [130, 131] and the number of genes identified from targeted approaches will undoubtedly increase dramatically with many groups actively undertaking such research. Since we have only a rudimentary understanding of how the known Age mutations extend lifespan, one may question the importance of screens for additional genes. Further genetic analysis often clarifies the order in which genes act and the identity of the genes tagged by suppressor mutations is likely to yield additional important information.

New genes are identified not only by traditional genetic screens but through the use of microarray analysis, RNAi [11, 199], by drug screening and by researching specific candidate genes. The different approaches have their advantages and disadvantages. Traditional screening is labour intensive and no guarantees can be made regarding success in identifying the mutated gene. In this respect, the RNAi screens clearly have the major advantage of prior knowledge of gene identity. However, for reasons not fully understood at present time, RNAi is not effective for all genes thus one is likely to miss interesting genes.

Transcriptional profiling and proteome analysis are among the new exciting tools for gene discovery and for defining physiological influences on aging. However, these are also major challenges for the bioinformatician, as one must determine from a large dataset the physiological importance of the gene expression or protein profile being observed. Microarray technology allows for the parallel determination of mRNA levels and facilitates novel comparisons between, for example, young versus old and mutant versus wildtype worms. The transcriptional profiles for nematodes of different ages have been published [200, 201]. In the more recent study, Lund *et al.* used microarrays, which contained about 90% of all the predicted genes in the nematode genome and the expression profiles at six different ages were examined. They found that less than 1% of the genes changed with aging based on stringent statistical criteria. The study revealed several gene changes consistent with the existing literature and aging hypotheses. For example, they found an increase in *hsp16* mRNA and *hsp70* mRNA expression during midlife, followed by a significant down-regulation very late in life. The study also revealed increased levels of *ins-2*, *ins-17* and *ins-18* but lower levels of *ins-7* with age. These insulin-like molecules could be the ligands for the *daf-2* insulin receptor and thus they could be involved with the endocrine regulation of lifespan. A comparison of the gene expression profile of dauer larvae and that of old worms revealed a significant overlap.

C. elegans and human age-related disease

As the factors that determine lifespan for invertebrates are identified, one is prompted to ask whether any of these genetic promoters of aging are also responsible for any aspects of age-related disease. It has long been argued that aging is a highly significant risk factor for human disease but little research has gone into establishing a direct mechanistic relationship. The worm is now being used as a model for human age-related disease and this offers a unique opportunity to bring basic biology of aging together with studies on the etiology of age-related disease.

Alzheimer's disease

Perhaps the prototype for all worm disease models was that constructed for studying Alzheimer's disease (AD) [202]. Transgenic animals were engineered to express human β-amyloid peptide (Aβ). These worms form intracellular β amyloid [203, 204] and develop a paralysis phenotype. These transgenic animals were used to identify proteins that specifically co-immunoprecipitated with Aβ. All proteins identified were chaperones, including two members of the HSP70 family, three HSP16s, and finally a putative orthologue of a mammalian protein, a small glutamine-rich tetratricopeptide-repeat-containing protein (SGT), known to negatively regulate HSP70 function [205]. RNAi of the *C. elegans* orthologue of SGT significantly suppressed the toxicity caused by Aβ expression.

Triplet repeat disease

Another very elegant example of disease modelling in the worm is the Huntington's disease model that addresses the effects of polyglutamine expansions. Initially, it was reported that expression of a huntingtin fragment containing a polyQ tract of 150 residues in ASH neurones lead to protein aggregate formation and neuronal dysfunction [206]. An increase in the number of ASH neurones containing aggregates was reported with age. Functional decline was observed in both dye filling assays, which address the integrity of sensory neurons, and motor function by touch provoked movement assays.

Aggregate formation of polyglutamine repeats expressed in a GFP fusion protein in body wall muscle cells have also been used [207]. The authors looked at expression of both Q19-GFP and Q82-GFP and found that only the latter resulted in aggregate formation. The presence of aggregates correlated with retarded larval growth. It was also found that the Q82-GFP aggregates activated the stress response as observed by elevated expression of a *hsp-16::lacZ* reporter construct. Interestingly, expression of the yeast chaperone HSP104 reduced the aggregate formation and, to some degree, alleviated the growth retardation. This protective effect of HSP104 was consistent with previous findings in yeast [208]. In order to test if a critical threshold of repeats exists beyond which aggregates are formed, Morley *et al.* investigated the effects of nine repeat lengths ranging from Q0 to Q82 [209].

They found appearance of aggregates did indeed correlate with the polyQ length and that the critical threshold was around 35–40 glutamines. For all aggregate lengths, a strong correlation was found between age and number of aggregates.

Moreover, it was also found that *age-1* mutants have a very significant delay in the onset of polyQ toxicity and appearance of protein aggregates. This could be explained by the fact that the *age-1* mutation increases lifespan in part by up-regulating chaperone capacity as discussed previously. Such a hypothesis is consistent with the protective effect observed by HSP104.

In a different study, expression of a huntingtin fragment with expanded polyQ tracts (88 or 128 residues) and a GFP fusion in touch receptor neurones was found to cause neuronal dysfunction but no cell death [210]. Neuronal function was measured by means of a mechanosensory assay which simply involved touching the worms with a hair and monitoring the response. Wildtype worms respond to touch by backing away but expression of polyQ expansions resulted in touch insensitivity, also called a Mec phenotype. The construct with more glutamine residues resulted in a more severe Mec phenotype. No significant correlation was found between protein aggregates and polyQ length. However, this could be because only two relatively large polyQ expansions were investigated.

Parkinson's disease

Parkinson's disease (PD) is another example of a neurodegenerative disease studied using the nematode. A transgenic nematode with great potential for studying dopamine neurone degeneration has been constructed [211]. Expressing GFP under the promoter of the presynaptic dopamine transporter (DAT), Nass *et al.* found GFP expression exclusively in the eight dopamine neurones present in the nematode. Importantly, they found GFP expression could be used to measure the integrity of these neurones. Specific killing of the dopamine neurones with the neurotoxin 6-hydroxydopamine resulted in complete absence of any fluorescent signal. Their transgenic strain thus provides a very simple system where viability of dopamine neurones can be monitored via presence of GFP. The potential of using this strain in studying the genetic and molecular basis of dopamine neurone degeneration is very promising. The authors point out that screening for mutants that protect the dopamine neurones could reveal some of the genetic components involved in this process. Also, that pharmacologically, it is very easy to utilize this nematode model in screening for novel compounds that protect the dopamine neurones.

These specific examples represent studies where the researchers have been able to create a model for a human pathology with varying degrees of success. In addition, using that particular model they have been able to intervene with the progression and toxicity of the pathology. Being able to do so in simple model systems provides valuable information for understanding the human pathologies. In the future, we expect to see more disease models in the nematode and the use of *C. elegans* in drug screens will no doubt become much more popular.

The nematode and the evolutionary origins of aging

Why aging is not regulated

A complete understanding of aging requires not only a description of the various biochemical and physiological processes that influence it but also an understanding of the evolutionary origins of these processes and their influence on Darwinian fitness. Over the years many researchers have expressed a wide variety of views on this issue and it is common to hear that aging is "regulated" in some fashion. In this chapter we have often described a gene as influencing or determining an aging rate but it is important to note that we are not implying that the gene affects aging in *C. elegans*' natural environment. Since many of the factors that influence aging are themselves subject to full selective pressure (e.g., stress response, reproduction and development), the false impression of a regulatory network for aging itself is readily formed from the genetic approaches described above. One might think that since mutation of a gene extends lifespan it must follow that the selected function of that gene must be to kill the animal. This is a teleological argument that does not fully explore the cause for the selection for the wildtype form of that gene. The appearance of the "regulation" of aging originates in the tight regulatory control over processes that "influence" aging and should not be mistaken for direct control.

Most researchers adhere to the general view that aging is non-adaptive – an epiphenomenon, rare in the wild hence not subject to the force of natural selection [212, 213]. This view, formalized by population genetic theory [214], is based on the observation that extrinsic mortality pressures (disease, predation, accident) tend to result in age-structured populations where older animals are less common that younger ones. Age structure arises from extrinsic hazard where each individual is subject to a finite probability of death in a given time period. In this situation, even if all age groups exhibited identical reproductive success, more progeny would arise from the more populous, younger age groups. As the younger age groups have a greater influence on the genetic makeup of subsequent generations, the force of natural selection is maximal here. Older age groups have less influence on the genetic composition of the next generation and so the force of natural selection can be shown to decline with age [215].

Based on the notion of a decline in the force of natural selection with advancing age, three general types of genes may have influence on the aging rate. The first are alleles that arise that have no effect on early life fitness but have detrimental effects in late life. Since late life selection is weak, such late-life detrimental alleles are maintained in the population and accumulate [216]. This notion is known as "mutation accumulation" [217]. The second class of genes is comprised of allelic forms that have beneficial effects in early life and have late life detrimental effects. Such alleles would be subject to high early life selective force and would be selected due to their early life beneficial effects. These alleles are described as exhibiting antagonistic pleiotropy where a gene with effects on multiple traits is seen to be antagonistically beneficial and detrimental [215, 218]. Numerous population based studies in the natural environment have tested the idea that when a population experiences a decline in extrinsic hazard levels, this allows for the force of natural selection to act later and later in life resulting in an extended lifespan. These experiments generally support the model. The third class of genes that affect lifespan are those that act to reduce the detrimental effects of genes in the former classes. Aging is associated with an accumulation of molecular damage and general loss of homeostasis. Class three genes counteract the accumulation of damage and hence may slow the rate of aging. For example, the over-expression of *old-1*, *hsp-70* or *hsp-16* in *C. elegans* extends lifespan. These genes have an influence on aging but are the product of selection in early life for an effective stress response [100] (Walker and Lithgow, in press). The existence of this class of genes is consistent with the descriptions of aging by Kirkwood, who proposed that a trade-off between maintenance and reproductive processes determines aging rate [219].

Does slowed aging result in lowered Darwinian fitness?
The genetic influence on aging is therefore not due to direct selection on aging characters. Rather, aging is a non-adaptive process in which the genes that influence the rate of aging either do not affect fitness or have been selected due to beneficial effects early in life. It follows that aging may be influenced by many genes and is caused by a range of distinct physiological and molecular processes. In keeping with this view, genetic variants of *Drosophila melanogaster (D. melanogaster)* or *C. elegans* that exhibit extended lifespan commonly exhibit phenotypic trade offs such as reduced or delayed fertility although the exceptions to this should prove highly informative.

Despite a well-developed theoretical framework for the understanding of the evolutionary origins of aging, very little experimentation has been undertaken that directly test these ideas. Breeding experiments with *D. melanogaster* have demonstrated that longevity is in part an outcome of selecting for late-life reproduction. Due to the self-fertilizing mode of reproduction, *C. elegans* populations are highly inbred and show no heterosis and are unsuitable for such breeding experiments [220]. However, the existence of long-lived mutants and recombinant inbred lines provide a great opportunity to test evolutionary theories. At first, the existence of Age mutations appear to be inconsistent with some of the predictions of evolution theories but it is becoming increasingly clear that this is not the case.

The difficulty in reconciling Age mutants with evolution theory stems from a series of observations that separate longevity from any detrimental effects on fertility or development. For example, many mutations in genes encoding components of the insulin-signaling pathway extend lifespan without any apparent cost to the organism. This suggests that longevity does not have any physiological cost in *C. elegans* and consequently that aging must be under some direct regulation and natural selection. There are three main reasons why this view is wrong.

First, the modification of an allele that normally has no effect in early life but has a detrimental effect in late life may extend lifespan. Such an allele could fall into the first class of genes that arise through mutation accumulation. The "wildtype" allele could simply have arisen without a fitness cost but acts to shorten the post-reproductive lifespan under laboratory conditions.

Second, it is important to understand that the observed actions of genes are highly dependant on the environment in which they are studied. The true effects of the specific mutation on its Darwinian fitness can only be assessed in the natural environment. All observations to date on the relative characteristics of the Age mutants have been made under standard laboratory conditions. Hence the observed absence of a cost of longevity under these conditions does not inform on the evolutionary origins of the gene in relation to aging.

Third, Age mutations occur in genes that are highly pleiotropic. For example, the insulin signaling genes have influence over growth, fertility, metabolism, stress resistance and development as described earlier in this chapter. Mutations in *daf-2* fall into phenotypic classes that share either few or many of these associated phenotypes. Whilst longevity has been separated to some extent from each of these characteristics, the influence of these genes on early life traits shows that the genes are very likely to be subject to strong selective pressure in early life.

One set of experiments was undertaken to reconcile the observations made on Age mutants with evolution theory [221]. In a direct test of the relative fitness of an Age mutant, replicate mixed populations of wildtype and long-lived *age-1(hx546)* mutants were cultured over multiple generations and the relative frequencies of the mutant allele determined at each generation. Under normal laboratory culturing conditions on agar plates with plenty of *E. coli* as a food source, no non-random change in the allele frequencies was observed. The long-lived mutant therefore exhibited no direct fitness cost under these conditions. A second set of culturing conditions was undertaken to more closely mimic that experienced by wild *C. elegans*. In this set of experiments, the mixed populations were subjected to periods of starvation for up to four days and the result was a rapid decline in the frequency of the longevity-associated allele consistent with a large fitness deficit. The cost itself is the result of a reduction in adult fertility following starvation (Jenkins and Lithgow, unpublished data). This result is consistent with antagonistic pleiotropy, in that the wildtype form of the *age-1* gene is required for normal early fertility but has an unspecified detrimental effect on late life.

The results observed by changing the environmental conditions are revealing. The observed cost of maintaining the *age-1* mutant allele is an important example of a single gene that exhibits antagonistic pleiotropy. But the lack of a fitness cost under normal laboratory conditions also reveals interesting features about the nature of aging genes. It suggests that under normal laboratory conditions there is no physiological cost to longevity. This is enlightening due to the common extrapolation of evolution theories to the idea of physiological trade off. It suggests that rather than an obligate cost on other life history traits, the trade off for longevity is environment specific and longevity is attainable without any dramatic alteration in other life history traits.

References

1. Brenner S (1988). In: Wood W, ed. The Nematode *Caenorhabditis elegans*. New York: CSHL Press, pp. ix–xiii.
2. Metzstein MM, Stanfield GM, Horvitz HR (1998). Genetics of programmed cell death in *C. elegans*: past, present and future. *Trends Genet*. 14: 410–16.
3. Johnson TE, Wood WB (1982). Genetic analysis of life-span in *Caenorhabditis elegans*. *Proc Natl Acad Sci USA* 79: 6603–7.
4. Klass MR (1983). A method for the isolation of longevity mutants in the nematode *Caenorhabditis elegans* and initial results. *Mech Ageing Dev*. 22: 279–86.

5. Friedman DB, Johnson TE (1988). A mutation in the *age-1* gene in *Caenorhabditis elegans* lengthens life and reduces hermaphrodite fertility. *Genetics.* 118: 75–86.
6. Friedman DB, Johnson TE (1988). Three mutants that extend both mean and maximum life span of the nematode, *Caenorhabditis elegans*, define the *age-1* gene. *J Gerontol.* 43: B102–9.
7. Johnson TE (1990). Increased life-span of *age-1* mutants in *Caenorhabditis elegans* and lower Gompertz rate of aging. *Science* 249: 908–12.
8. Johnson TE (2002). A personal retrospective on the genetics of aging. *Biogerontology* 3: 7–12.
9. Brenner S (1974). The genetics of *Caenorhabditis elegans. Genetics* 77: 71–94.
10. Anderson P (1995). Mutagenesis. *Methods Cell Biol.* 48: 31–58.
11. Fire A, Albertson D, Harrison SW, Moerman DG (1991). Production of antisense RNA leads to effective and specific inhibition of gene expression in *C. elegans* muscle. *Development* 113: 503–14.
12. Gems D, Riddle DL (2000). Genetic, behavioral and environmental determinants of male longevity in *Caenorhabditis elegans. Genetics* 154: 1597–610.
13. Klass M, Hirsh D (1976). Non-ageing developmental variant of *Caenorhabditis elegans. Nature* 260: 523–5.
14. Riddle DL, Albert PS (1997). Genetic and environmental regulation of dauer larva development. In: Riddle DL, Blumenthal T, Meyer BJ, Priess JR, eds. *C. elegans* II. New York: CSHP, pp. 739–68.
15. Anderson GL (1978). Responses of dauer larvae of *Caenorhabditis elegans* (Nematoda: Rhabditidae) to thermal stress and oxygen deprivation. *Canadian J Zoology* 56: 1786–91.
16. Klass MR (1977). Aging in the nematode *Caenorhabditis elegans*: major biological and environmental factors influencing life span. *Mech Ageing Dev.* 6: 413–29.
17. Lithgow GJ, White TM, Melov S, Johnson TE (1995). Thermotolerance and extended life span conferred by single-gene mutations and induced by thermal stress. *Proc Natl Acad Sci USA* 92: 7540–4.
18. Brooks A, Lithgow GJ, Johnson TE (1994). Mortality rates in a genetically heterogeneous population of *Caenorhabditis elegans. Science* 263: 668–71.
19. Vaupel JW, Johnson TE, Lithgow GJ (1994). Rates of mortality in populations of *Caenorhabditis elegans. Science* 266: 826.
20. Gems D (2000). An integrated theory of ageing in the nematode *Caenorhabditis elegans. J Anat.* 197 Pt 4: 521–8.
21. Michalski AI, Johnson TE, Cypser JR, Yashin AI (2001). Heating stress patterns in *Caenorhabditis elegans* longevity and survivorship. *Biogerontology* 2: 35–44.
22. Yashin AI, Cypser JW, Johnson TE, Michalski AI, Boyko SI, Novoseltsev VN (2002). Heat shock changes the heterogeneity distribution in populations of *Caenorhabditis elegans*: does it tell us anything about the biological mechanism of stress response? *J Gerontol A Biol Sci Med Sci.* 57: B83–92.
23. Carey JR, Liedo P, Orozco D, Vaupel JW (1992). Slowing of mortality rates at older ages in large medfly cohorts. *Science* 258: 457–61.
24. Fukui HH, Xiu L, Curtsinger JW (1993). Slowing of age-specific mortality rates in *Drosophila melanogaster. Exp Gerontol.* 28: 585–99.
25. Curtsinger JW, Fukui HH, Townsend DR, Vaupel JW (1992). Demography of genotypes: failure of the limited life-span paradigm in *Drosophila melanogaster. Science* 258: 461–3.

26. Vaupel JW, Carey JR (1993). Compositional interpretations of medfly mortality. *Science* 260: 1666–7.
27. Fabian T, Johnson TE (1994). Production of age-synchronous mass cultures of *Caenorhabditis elegans*. *J Gerontol Biol Sci*. 49: B145–56.
28. Bolanowski MA, Jacobson LA, Russell RL (1983). Quantitative measures of aging in the nematode *Caenorhabditis elegans*: II. Lysosomal hydrolases as markers of senescence. *Mech Ageing Dev*. 21: 295–319.
29. Bolanowski MA, Russell RL, Jacobson LA (1981). Quantitative measures of aging in the nematode *Caenorhabditis elegans*. I. Population and longitudinal studies of two behavioral parameters. *Mech Ageing Dev*. 15: 279–95.
30. Garigan D, Hsu AL, Fraser AG, Kamath RS, Ahringer J, Kenyon C (2002). Genetic analysis of tissue aging in *Caenorhabditis elegans*. A role for heat-shock factor and bacterial proliferation. *Genetics* 161: 1101–12.
31. Herndon LA, Schmeissner PJ, Dudaronek JM, *et al.* (2002). Stochastic, genetic factors influence tissue-specific decline in ageing *C. elegans*. *Nature* 419: 808–14.
32. Guarente L, Kenyon C (2000). Genetic pathways that regulate ageing in model organisms. *Nature* 408: 255–62.
33. Gems D, Partridge L (2001). Insulin/IGF signalling and ageing: seeing the bigger picture. *Curr Opin Genet Dev*. 11: 287–92.
34. Paradis S, Ailion M, Toker A, Thomas JH, Ruvkun G (1999). A PDK1 homolog is necessary and sufficient to transduce AGE-1 PI3 kinase signals that regulate diapause in *Caenorhabditis elegans*. *Genes Dev*. 13: 1438–52.
35. Dorman JB, Albinder B, Shroyer T, Kenyon C (1995). The *age-1* and *daf-2* genes function in a common pathway to control the lifespan of *Caenorhabditis elegans*. *Genetics* 141: 1399–406.
36. Barsyte D, Lovejoy DA, Lithgow GJ (2001). Longevity and heavy metal resistance in *daf-2* and *age-1* long-lived mutants of *Caenorhabditis elegans*. *FASEB J*. 15: 627–34.
37. Adachi H, Fujiwara Y, Ishii N (1998). Effects of oxygen on protein carbonyl and aging in *Caenorhabditis elegans* mutants with long (*age-1*) and short (*mev-1*) life spans. *J Gerontol A Biol Sci Med Sci*. 53A: B240–4.
38. Malone EA, Inoue T, Thomas JH (1996). Genetic analysis of the roles of *daf-28* and *age-1* in regulating *Caenorhabditis elegans* dauer formation. *Genetics* 143: 1193–205.
39. Fujiwara M, Ishihara T, Katsura I (1999). A novel WD40 protein, CHE-2, acts cell-autonomously in the formation of *C. elegans* sensory cilia. *Development* 126: 4839–48.
40. Apfeld J, Kenyon C (1999). Regulation of lifespan by sensory perception in *Caenorhabditis elegans*. *Nature* 402: 804–9.
41. Wicks SR, de Vries CJ, van Luenen HG, Plasterk RH (2000). CHE-3, a cytosolic dynein heavy chain, is required for sensory cilia structure and function in *Caenorhabditis elegans*. *Dev Biol*. 221: 295–307.
42. Vowels JJ, Thomas JH (1992). Genetic analysis of chemosensory control of dauer formation in *Caenorhabditis elegans*. *Genetics* 130: 105–23.
43. Starich TA, Herman RK, Kari CK, *et al.* (1995). Mutations affecting the chemosensory neurons of *Caenorhabditis elegans*. *Genetics* 139: 171–88.
44. Perkins LA, Hedgecock EM, Thomson JN, Culotti JG (1986). Mutant sensory cilia in the nematode *Caenorhabditis elegans*. *Dev Biol*. 117: 456–87.
45. Lakowski B, Hekimi S (1996). Determination of life-span in *Caenorhabditis elegans* by four clock genes. *Science* 272: 1010–13.
46. Wong A, Boutis P, Hekimi S (1995). Mutations in the *clk-1* gene of *Caenorhabditis elegans* affect developmental and behavioral timing. *Genetics* 139: 1247–59.

47. Felkai S, Ewbank JJ, Lemieux J, Labbe JC, Brown GG, Hekimi S (1999). CLK-1 controls respiration, behavior and aging in the nematode *Caenorhabditis elegans*. *EMBO J.* 18: 1783–92.
48. Jonassen T, Proft M, Randez-Gil F, *et al.* (1998). Yeast Clk-1 homologue (Coq7/Cat5) is a mitochondrial protein in coenzyme Q synthesis. *J Biol Chem.* 273: 3351–7.
49. Jonassen T, Larsen PL, Clarke CF (2001). A dietary source of coenzyme Q is essential for growth of long-lived *Caenorhabditis elegans clk-1* mutants. *Proc Natl Acad Sci USA* 98: 421–6.
50. Vajo Z, King LM, Jonassen T, *et al.* (1999). Conservation of the *Caenorhabditis elegans* timing gene *clk-1* from yeast to human: a gene required for ubiquinone biosynthesis with potential implications for aging. *Mamm Genome* 10: 1000–4.
51. Ewbank JJ, Barnes TM, Lakowski B, Lussier M, Bussey H, Hekimi S (1997). Structural and functional conservation of the *Caenorhabditis elegans* timing gene *clk-1*. *Science* 275: 980–3.
52. Branicky R, Shibata Y, Feng J, Hekimi S (2001). Phenotypic and suppressor analysis of defecation in *clk-1* mutants reveals that reaction to changes in temperature is an active process in *Caenorhabditis elegans*. *Genetics* 159: 997–1006.
53. Gorbunova V, Seluanov A (2002). CLK-1 protein has DNA binding activity specific to O(L) region of mitochondrial DNA. *FEBS Lett.* 516: 279–84.
54. Larsen PL, Clarke CF (2002). Extension of life-span in *Caenorhabditis elegans* by a diet lacking coenzyme Q. *Science* 295: 120–3.
55. Ahmed S, Alpi A, Hengartner MO, Gartner A (2001). *C. elegans* RAD-5/CLK-2 defines a new DNA damage checkpoint protein. *Curr Biol.* 11: 1934–44.
56. Benard C, McCright B, Zhang Y, Felkai S, Lakowski B, Hekimi S (2001). The *C. elegans* maternal-effect gene *clk-2* is essential for embryonic development, encodes a protein homologous to yeast Tel2p and affects telomere length. *Development* 128: 4045–55.
57. Lim CS, Mian IS, Dernburg AF, Campisi J (2001). *C. elegans clk-2*, a gene that limits life span, encodes a telomere length regulator similar to yeast telomere binding protein Tel2p. *Curr Biol.* 11: 1706–10.
58. Antebi A, Yeh WH, Tait D, Hedgecock EM, Riddle DL (2000). *daf-12* encodes a nuclear receptor that regulates the dauer diapause and developmental age in *C. elegans*. *Genes Dev.* 14: 1512–27.
59. Antebi A, Culotti JG, Hedgecock EM (1998). *daf-12* regulates developmental age and the dauer alternative in *Caenorhabditis elegans*. *Development* 125: 1191–205.
60. Snow MI, Larsen PL (2000). Structure and expression of *daf-12*: a nuclear hormone receptor with three isoforms that are involved in development and aging in *Caenorhabditis elegans*. *Biochim Biophys Acta* 1494: 104–16.
61. Hsin H, Kenyon C (1999). Signals from the reproductive system regulate the lifespan of *C. elegans*. *Nature* 399: 362–6.
62. Swoboda P, Adler HT, Thomas JH (2000). The RFX-type transcription factor DAF-19 regulates sensory neuron cilium formation in *C. elegans*. *Mol Cell* 5: 411–21.
63. Collet J, Spike CA, Lundquist EA, Shaw JE, Herman RK (1998). Analysis of *osm-6*, a gene that affects sensory cilium structure and sensory neuron function in *Caenorhabditis elegans*. *Genetics* 148: 187–200.
64. Kenyon C, Chang J, Gensch E, Rudner A, Tabtiang R (1993). A *C. elegans* mutant that lives twice as long as wild type. *Nature* 366: 461–4.
65. Gems D, Sutton AJ, Sundermeyer ML, *et al.* (1998). Two pleiotropic classes of *daf-2* mutation affect larval arrest, adult behavior, reproduction, longevity in *Caenorhabditis elegans*. *Genetics* 150: 129–55.

66. Apfeld J, Kenyon C (1998). Cell nonautonomy of *C. elegans daf-2* function in the regulation of diapause and life span. *Cell* 95: 199–210.
67. Kimura KD, Tissenbaum HA, Liu Y, Ruvkun G (1997). *daf-2*, an insulin receptor-like gene that regulates longevity and diapause in *Caenorhabditis elegans. Science* 277: 942–6.
68. Honda Y, Honda S (1999). The *daf-2* gene network for longevity regulates oxidative stress resistance and Mn-superoxide dismutase gene expression in *Caenorhabditis elegans. FASEB J.* 13: 1385–93.
69. Scott BA, Avidan MS, Crowder CM (2002). Regulation of hypoxic death in *C. elegans* by the insulin/IGF receptor homolog DAF-2. *Science* 296: 2388–91.
70. Lee RY, Hench J, Ruvkun G (2001). Regulation of *C. elegans* DAF-16 and its human ortholog FKHRL1 by the *daf-2* insulin-like signaling pathway. *Curr Biol.* 11: 1950–7.
71. Lakowski B, Hekimi S (1998). The genetics of caloric restriction in *Caenorhabditis elegans. Proc Natl Acad Sci USA* 95: 13091–6.
72. Avery L (1993). The genetics of feeding in *Caenorhabditis elegans. Genetics* 133: 897–917.
73. Arantes-Oliveira N, Apfeld J, Dillin A, Kenyon C (2002). Regulation of life-span by germ-line stem cells in *Caenorhabditis elegans. Science* 295: 502–5.
74. Kimble JE, White JG (1981). On the control of germ cell development in *Caenorhabditis elegans. Dev Biol.* 81: 208–19.
75. Henderson ST, Gao D, Lambie EJ, Kimble J (1994). *lag-2* may encode a signaling ligand for the GLP-1 and LIN-12 receptors of *C. elegans. Development* 120: 2913–24.
76. Tax FE, Thomas JH, Ferguson EL, Horvitz HR (1997). Identification and characterization of genes that interact with *lin-12* in *Caenorhabditis elegans. Genetics* 147: 1675–95.
77. Feng J, Bussiere F, Hekimi S (2001). Mitochondrial electron transport is a key determinant of life span in *Caenorhabditis elegans. Dev Cell* 1: 633–44.
78. Lundquist EA, Herman RK (1994). The *mec-8* gene of *Caenorhabditis elegans* affects muscle and sensory neuron function and interacts with three other genes: *unc-52*, *smu-1* and *smu-2*. *Genetics* 138: 83–101.
79. Lundquist EA, Herman RK, Rogalski TM, Mullen GP, Moerman DG, Shaw JE (1996). The *mec-8* gene of *C. elegans* encodes a protein with two RNA recognition motifs and regulates alternative splicing of *unc-52* transcripts. *Development* 122: 1601–10.
80. Berkowitz LA, Strome S (2000). MES-1, a protein required for unequal divisions of the germline in early *C. elegans* embryos, resembles receptor tyrosine kinases and is localized to the boundary between the germline and gut cells. *Development* 127: 4419–31.
81. Capowski EE, Martin P, Garvin C, Strome S (1991). Identification of grandchildless loci whose products are required for normal germ-line development in the nematode *Caenorhabditis elegans. Genetics* 129: 1061–72.
82. Tabish M, Siddiqui ZK, Nishikawa K, Siddiqui SS (1995). Exclusive expression of *C. elegans osm-3* kinesin gene in chemosensory neurons open to the external environment. *J Mol Biol.* 247: 377–89.
83. Bargmann CI, Hartwieg E, Horvitz HR (1993). Odorant-selective genes and neurons mediate olfaction in *C. elegans. Cell* 74: 515–27.
84. Thomas JH, Birnby DA, Vowels JJ (1993). Evidence for parallel processing of sensory information controlling dauer formation in *Caenorhabditis elegans. Genetics* 134: 1105–17.
85. Culotti JG, Russell RL (1978). Osmotic avoidance defective mutants of the nematode *Caenorhabditis elegans. Genetics* 90: 243–56.

86. Kawasaki I, Shim YH, Kirchner J, Kaminker J, Wood WB, Strome S (1998). PGL-1, a predicted RNA-binding component of germ granules, is essential for fertility in *C. elegans*. *Cell* 94: 635–45.
87. Murakami S, Johnson TE (1996). A genetic pathway conferring life extension and resistance to UV stress in *Caenorhabditis elegans*. *Genetics* 143: 1207–18.
88. Varkey JP, Muhlrad PJ, Minniti AN, Do B, Ward S (1995). The *Caenorhabditis elegans spe-26* gene is necessary to form spermatids and encodes a protein similar to the actin-associated proteins kelch and scruin. *Genes Dev.* 9: 1074–86.
89. Coburn CM, Mori I, Ohshima Y, Bargmann CI (1998). A cyclic nucleotide-gated channel inhibits sensory axon outgrowth in larval and adult *Caenorhabditis elegans*: a distinct pathway for maintenance of sensory axon structure. *Development* 125: 249–58.
90. Richmond JE, Davis WS, Jorgensen EM (1999). UNC-13 is required for synaptic vesicle fusion in *C. elegans*. *Nat Neurosci.* 2: 959–64.
91. Ailion M, Inoue T, Weaver CI, Holdcraft RW, Thomas JH (1999). Neurosecretory control of aging in *Caenorhabditis elegans*. *Proc Natl Acad Sci USA* 96: 7394–7.
92. Ailion M, Thomas JH (2000). Dauer formation induced by high temperatures in *Caenorhabditis elegans*. *Genetics* 156: 1047–67.
93. Ann K, Kowalchyk JA, Loyet KM, Martin TF (1997). Novel Ca^{2+}-binding protein (CAPS) related to UNC-31 required for Ca^{2+}-activated exocytosis. *J Biol Chem.* 272: 19637–40.
94. Avery L, Bargmann CI, Horvitz HR (1993). The *Caenorhabditis elegans unc-31* gene affects multiple nervous system-controlled functions. *Genetics* 134: 455–64.
95. Ogawa H, Harada S, Sassa T, Yamamoto H, Hosono R (1998). Functional properties of the *unc-64* gene encoding a *Caenorhabditis elegans* syntaxin. *J Biol Chem.* 273: 2192–8.
96. Saifee O, Wei L, Nonet ML (1998). The *Caenorhabditis elegans unc-64* locus encodes a syntaxin that interacts genetically with synaptobrevin. *Mol Biol Cell* 9: 1235–52.
97. Desai C, Garriga G, McIntire SL, Horvitz HR (1988). A genetic pathway for the development of the *Caenorhabditis elegans* HSN motor neurons. *Nature* 336: 638–46.
98. Kawano T, Ito Y, Ishiguro M, Takuwa K, Nakajima T, Kimura Y (2000). Molecular cloning and characterization of a new insulin/IGF-like peptide of the nematode *Caenorhabditis elegans*. *Biochem Biophys Res Commun.* 273: 431–6.
99. Gregoire FM, Chomiki N, Kachinskas D, Warden CH (1998). Cloning and developmental regulation of a novel member of the insulin-like gene family in *Caenorhabditis elegans*. *Biochem Biophys Res Commun.* 249: 385–90.
100. Murakami S, Johnson TE (2001). The OLD-1 positive regulator of longevity and stress resistance is under DAF-16 regulation in *Caenorhabditis elegans*. *Curr Biol.* 11: 1517–23.
101. Murakami S, Johnson TE (1998). Life extension and stress resistance in *Caenorhabditis elegans* modulated by the *tkr-1* gene. *Curr Biol.* 8: 1091–4.
102. Tissenbaum HA, Guarente L (2001). Increased dosage of a *sir-2* gene extends lifespan in *Caenorhabditis elegans*. *Nature* 410: 227–30.
103. Pierce SB, Costa M, Wisotzkey R, *et al.* (2001). Regulation of DAF-2 receptor signaling by human insulin and *ins-1*, a member of the unusually large and diverse *C. elegans* insulin gene family. *Genes Dev.* 15: 672–86.
104. Yokoyama K, Fukumoto K, Murakami T, *et al.* (2002). Extended longevity of *Caenorhabditis elegans* by knocking in extra copies of *hsp70F*, a homolog of *mot-2* (mortalin)/mthsp70/Grp75. *FEBS Lett.* 516: 53–7.

105. Morris JZ, Tissenbaum HA, Ruvkun G (1996). A phosphatidylinositol-3-OH kinase family member regulating longevity and diapause in *Caenorhabditis elegans*. *Nature* 382: 536–9.
106. Clancy DJ, Gems D, Harshman LG, *et al.* (2001). Extension of life-span by loss of CHICO, a *Drosophila* insulin receptor substrate protein. *Science* 292: 104–6.
107. Tatar M, Kopelman A, Epstein D, Tu MP, Yin CM, Garofalo RS (2001). A mutant *Drosophila* insulin receptor homolog that extends life-span and impairs neuroendocrine function. *Science* 292: 107–10.
108. Tatar M, Yin C (2001). Slow aging during insect reproductive diapause: why butterflies, grasshoppers and flies are like worms. *Exp Gerontol.* 36: 723–38.
109. Fabrizio P, Pozza F, Pletcher SD, Gendron CM, Longo VD (2001). Regulation of longevity and stress resistance by Sch9 in yeast. *Science* 292: 288–90.
110. Kapeller R, Cantley LC (1994). Phosphatidylinositol 3-kinase. *Bioessays* 16: 565–76.
111. Paradis S, Ruvkun G (1998). *Caenorhabditis elegans* Akt/PKB transduces insulin receptor-like signals from AGE-1 PI3 kinase to the DAF-16 transcription factor. *Genes Dev.* 12: 2488–98.
112. Lin K, Dorman JB, Rodan A, Kenyon C (1997). *daf-16*: An HNF-3/forkhead family member that can function to double the life-span of *Caenorhabditis elegans*. *Science* 278: 1319–22.
113. Ogg S, Paradis S, Gottlieb S, *et al.* (1997). The Fork head transcription factor DAF-16 transduces insulin-like metabolic and longevity signals in *C. elegans*. *Nature* 389: 994–9.
114. Ogg S, Ruvkun G (1998). The *C. elegans* PTEN homolog, DAF-18, acts in the insulin receptor-like metabolic signaling pathway. *Mol Cell* 2: 887–93.
115. Ren P, Lim CS, Johnsen R, Albert PS, Pilgrim D, Riddle DL (1996). Control of *C. elegans* larval development by neuronal expression of a TGF-beta homolog. *Science* 274: 1389–91.
116. Wolkow CA, Kimura KD, Lee MS and Ruvkun G (2000). Regulation of *C. elegans* life-span by insulin like signaling in the nervous system. *Science* 290: 147–50.
117. Dillin A, Crawford DK, Kenyon C (2002). Timing requirements for insulin/IGF-1 signaling in *C. elegans*. *Science* 298: 830–4.
118. Babar P, Adamson C, Walker GA, Walker DW, Lithgow GJ (1999). P13-kinase inhibition induces dauer formation, thermotolerance and longevity in *C. elegans*. *Neurobiol Aging* 20: 513–19.
119. Gerisch B, Weitzel C, Kober-Eisermann C, Rottiers V, Antebi A (2001). A hormonal signaling pathway influencing *C. elegans* metabolism, reproductive development, and life span. *Dev Cell* 1: 841–51.
120. Jia K, Albert PS, Riddle DL (2002). DAF-9, a cytochrome P450 regulating *C. elegans* larval development and adult longevity. *Development* 129: 221–31.
121. Troemel ER (1999). Chemosensory signaling in *C. elegans*. Bioessays, 21: 1011–20.
122. Ahmed S, Hodgkin J (2000). MRT-2 checkpoint protein is required for germline immortality and telomere replication in *C. elegans*. *Nature* 403: 159–64.
123. Sohal RS, Weindruch R (1996). Oxidative stress, caloric restriction, and aging. *Science* 273: 59–63.
124. Tavernarakis N, Driscoll M (2002). Caloric restriction and lifespan: a role for protein turnover? *Mech Ageing Dev.* 123: 215–29.
125. Turturro A, Hass BS, Hart RW (2000). Does caloric restriction induce hormesis? *Hum Exp Toxicol.* 19: 320–9.
126. Lithgow GJ, Walker GA (2002). Stress resistance as a determinate of *C. elegans* lifespan. *Mech Ageing Dev.* 123: 765–71.

127. Larsen PL (1993). Aging and resistance to oxidative damage in *Caenorhabditis elegans. Proc Natl Acad Sci USA* 90: 8905–9.
128. Vanfleteren JR (1993). Oxidative stress and ageing in *Caenorhabditis elegans. Biochem J.* 292: 605–8.
129. Lithgow GJ, White TM, Hinerfeld DA, Johnson TE (1994). Thermotolerance of a long-lived mutant of *Caenorhabditis elegans. J Gerontol Biol Sci.* 49: B270–6.
130. Walker GA, Walker DW, Lithgow GJ (1998). A relationship between thermotolerance and longevity in *Caenorhabditis elegans. J Investig Dermatol Symp Proc.* 3: 6–10.
131. Walker GA, Walker DW, Lithgow GJ (1998). Genes that determine both thermotolerance and rate of aging in *Caenorhabditis elegans. Ann NY Acad Sci.* 851: 444–9.
132. Parsell DA, Lindquist S (1994). Heat shock proteins and stress tolerance In: Morimoto RI, Tissieres A, Georgopoulos C, eds. *The Biology of Heat Shock Proteins and Molecular Chaperones.* New York: CSHL Press, pp. 457–94.
133. Walker GA, White TM, McColl G, *et al.* (2001). Heat shock protein accumulation is upregulated in a long-lived mutant of *Caenorhabditis elegans. J Gerontol A Biol Sci Med Sci.* 56: B281–7.
134. Halliwell B, Gutteridge JMC (1999). *Free Radicals in Biology and Medicine*, 3rd edn. Oxford: Oxford University Press.
135. Henderson ST, Johnson TE (2001). *daf-16* integrates developmental and environmental inputs to mediate aging in the nematode *Caenorhabditis elegans. Curr Biol.* 11: 1975–80.
136. Lin K, Hsin H, Libina N, Kenyon C (2001). Regulation of the *Caenorhabditis elegans* longevity protein DAF-16 by insulin/IGF-1 and germline signaling. *Nature Genetics* 28: 139–45.
137. Rattan SI (2000). Ageing, gerontogenes, and hormesis. *Indian J Exp Biol.* 38: 1–5.
138. Shama S, Lai CY, Antoniazzi JM, Jiang JC, Jazwinski SM (1998). Heat stress-induced life span extension in yeast. *Exp Cell Res.* 245: 379–88.
139. Caratero A, Courtade M, Bonnet L, Planel H, Caratero C (1998). Effect of a continuous gamma irradiation at a very low dose on the life span of mice. *Gerontology* 44: 272–6.
140. Maynard Smith J (1958). The effects of temperature and of egg-laying on the longevity of *Drosophila subobscura. J Exp Biol.* 35: 832–43.
141. Cypser JR, Johnson TE (2002). Multiple stressors in *Caenorhabditis elegans* induce stress hormesis and extended longevity. *J Gerontol A Biol Sci Med Sci.* 57: B109–14.
142. Johnson TE, Hartman PS (1988). Radiation effects on life span in *Caenorhabditis elegans. J Gerontol.* 43: B137–41.
143. Lithgow GJ (2001). Hormesis–a new hope for ageing studies or a poor second to genetics? *Hum Exp Toxicol.* 20: 301–3.
144. Harman D (1956). Aging: a theory based on free radical and radiation chemistry. *J Gerontol Biol Sci.* 11: 298–300.
145. Sohal R (2002). Oxidative stress hypothesis of aging (1,2). *Free Radic Biol Med.* 33: 573.
146. Martin GM, Austad SN, Johnson TE (1996). Genetic analysis of ageing: role of oxidative damage and environmental stresses. *Nature Genetics* 13: 25–34.
147. Beckman KB, Ames BN (1998). The free radical theory of aging matures. *Physiol Rev.* 78: 547–81.
148. Ishii N, Goto S, Hartman P (2002). Protein oxidation during aging of the nematode *Caenorhabditis elegans. Free Radic Biol Med.* 33: 1021.
149. Melov S, Ravenscroft J, Malik S, *et al.* (2000). Extension of life-span with superoxide dismutase/catalase mimetics. *Science* 289: 1567–9.
150. Boveris A, Chance B (1973). The mitochondrial generation of hydrogen peroxide. General properties and effect of hyperbaric oxygen. *Biochem J.* 134: 707–16.

151. Beckman KB, Ames BN (1997). Oxidants, antioxidants and aging In: Scandalios JG, ed. *Oxidative Stress and the Molecular Biology of Antioxidant Defences.* New York: CSHL Press, pp. 201–46.
152. Battino M, Ferri E, Gorini A, *et al.* (1990). Natural distribution and occurrence of coenzyme Q homologues. *Membr Biochem.* 9: 179–90.
153. Stadtman E (2002). Importance of individuality in oxidative stress and aging (1,2). *Free Radic Biol Med.* 33: 597.
154. Richter C, Schweizer M (1997). Oxidative stress in mitochondria In: Scandalios JG, ed. *Oxidative Stress and the Molecular Biology of Antioxidant Defences.* New York: CSHL Press, pp. 169–200.
155. Levine RL, Stadtman ER (2001). Oxidative modification of proteins during aging. *Exp Gerontol.* 36: 1495–502.
156. Brunk U, Terman A (2002). Lipofuscin: mechanisms of age-related accumulation and influence on cell function(1)(2). *Free Radic Biol Med.* 33: 611.
157. Yan LJ, Levine RL, Sohal RS (1997). Oxidative damage during aging targets mitochondrial aconitase. *Proc Natl Acad Sci USA* 94: 11168–72.
158. Yan LJ, Sohal RS (1998). Mitochondrial adenine nucleotide translocase is modified oxidatively during aging. *Proc Natl Acad Sci USA* 95: 12896–901.
159. Nakamura A, Yasuda K, Adachi H, Sakurai Y, Ishii N, Goto S (1999). Vitellogenin-6 is a major carbonylated protein in aged nematode, *Caenorhabditis elegans. Biochem Biophys Res Commun.* 264: 580–3.
160. Beckman KB, Ames BN (1998). Mitochondrial aging: open questions. *Ann NY Acad Sci.* 854: 118–27.
161. Melov S, Hertz GZ, Stormo GD, Johnson TE (1994). Detection of deletions in the mitochondrial genome of *Caenorhabditis elegans. Nucleic Acids Res.* 22: 1075–8.
162. Melov S, Lithgow GJ, Fischer DR, Tedesco PM, Johnson TE (1995). Increased frequency of deletions in the mitochondrial genome with age of *Caenorhabditis elegans. Nucleic Acids Res.* 23: 1419–25.
163. Wei YH, Lee HC (2002). Oxidative stress, mitochondrial DNA mutation, and impairment of antioxidant enzymes in aging. *Exp Biol Med (Maywood)* 227: 671–82.
164. Fridovich I (1995). Superoxide radical, superoxide dismutases. *Annu Rev Biochem.* 64: 97–112.
165. Suzuki N, Inokuma K, Yasuda K, Ishii N (1996). Cloning, sequencing and mapping of a manganese superoxide dismutase gene of the nematode *Caenorhabditis elegans. DNA Res.* 3: 171–4.
166. Hunter T, Bannister WH, Hunter GJ (1997). Cloning, expression, and characterization of two manganese superoxide dismutases from *Caenorhabditis elegans. J Biol Chem.* 272: 28652–9.
167. Fujii M, Ishii N, Joguchi A, Yasuda K, Ayusawa D (1998). A novel superoxide dismutase gene encoding membrane-bound, extracellular isoforms by alternative splicing in *Caenorhabditis elegans. DNA Res.* 5: 25–30.
168. Taub J, Lau JF, Ma C, *et al.* (1999). A cytosolic catalase is needed to extend adult lifespan in *C. elegans* daf- and *clk-1* mutants. *Nature* 399: 162–6.
169. Blum J, Fridovich I (1983). Superoxide, hydrogen peroxide, and oxygen toxicity in two free-living nematode species. *Arch Biochem Biophys.* 222: 35–43.
170. Darr D, Fridovich I (1995). Adaptation to oxidative stress in young, but not in mature or old, *Caenorhabditis elegans. Free Radic Biol Med.* 18: 195–201.
171. Di Mascio P, Murphy ME, Sies H (1991). Antioxidant defense systems: the role of carotenoids, tocopherols, and thiols. *Am J Clin Nutr.* 53: 194S–200S.

172. Viarengo A, Burlando B, Ceratto N, Panfoli I (2000). Antioxidant role of metallothioneins: a comparative overview. *Cell Mol Biol.* 46: 407–17.
173. Sohal RS (2002). Role of oxidative stress and protein oxidation in the aging process(1,2). *Free Radic Biol Med.* 33: 37–44.
174. Vanfleteren JR, Braeckman BP (1999). Mechanisms of life span determination in *Caenorhabditis elegans. Neurobiol Aging* 20: 487–502.
175. Johnson TE, Cypser J, de Castro E, *et al.* (2000). Gerontogenes mediate health and longevity in nematodes through increasing resistance to environmental toxins and stressors. *Exp Gerontol.* 35: 687–94.
176. Johnson TE, Henderson S, Murakami S, *et al.* (2002). Longevity genes in the nematode *Caenorhabditis elegans* also mediate increased resistance to stress and prevent disease. *J Inherit Metab Dis.* 25: 197–206.
177. Houthoofd K, Braeckman B, Lenaerts I, *et al.* (2002). Ageing is reversed, and metabolism is reset to young levels in recovering dauer larvae of *C. elegans. Exp Gerontol.* 37: 1015.
178. Stenmark P, Grunler J, Mattsson J, Sindelar PJ, Nordlund P, Berthold DA (2001). A new member of the family of di-iron carboxylate proteins. Coq7 (*clk-1*), a membrane-bound hydroxylase involved in ubiquinone biosynthesis. *J Biol Chem.* 276: 33297–300.
179. Miyadera H, Kano K, Miyoshi H, Ishii N, Hekimi S, Kita K (2002). Quinones in long-lived *clk-1* mutants of *Caenorhabditis elegans. FEBS Lett.* 512: 33–7.
180. Ishii N, Takahashi K, Tomita S, *et al.* (1990). A methyl viologen-sensitive mutant of the nematode *Caenorhabditis elegans. Mutat Res.* 237: 165–71.
181. Ishiguro H, Yasuda K, Ishii N, *et al.* (2001). Enhancement of oxidative damage to cultured cells and *Caenorhabditis elegans* by mitochondrial electron transport inhibitors. *IUBMB Life* 51: 263–8.
182. Senoo-Matsuda N, Yasuda K, Tsuda M, *et al.* (2001). A defect in the cytochrome b large subunit in complex II causes both superoxide anion overproduction and abnormal energy metabolism in *Caenorhabditis elegans. J Biol Chem.* 276: 41553–8.
183. Kayser EB, Morgan PG, Sedensky MM (1999). GAS-1: a mitochondrial protein controls sensitivity to volatile anesthetics in the nematode *Caenorhabditis elegans. Anesthesiology* 90: 545–54.
184. Kayser EB, Morgan PG, Hoppel CL, Sedensky MM (2001). Mitochondrial expression and function of GAS-1 in *Caenorhabditis elegans. J Biol Chem.* 276: 20551–8.
185. Scandalios JG (1997). Molecular genetics of superoxide dismutases in plants. In: Scandalios JG, ed. *Oxidative Stress and the Molecular Biology of Antioxidant Defenses.* New York: CSHL Press, pp. 527–68.
186. Harrington LA, Harley CB (1988). Effect of vitamin E on lifespan and reproduction in *Caenorhabditis elegans. Mech Ageing Dev.* 43: 71–8.
187. Adachi H, Ishii N (2000). Effects of tocotrienols on life span and protein carbonylation in *Caenorhabditis elegans. J Gerontol A Biol Sci Med Sci.* 55: B280–5.
188. Wu Z, Smith JV, Paramasivam V, Butko P, Khan I, Cypser JR, *et al.* (2002). Ginkgo biloba extract EGb 761 increases stress resistance and extends life span of *Caenorhabditis elegans. Cell Mol Biol.* 48: 725–31.
189. Melov S (2002). Therapeutics against mitochondrial oxidative stress in animal models of aging. *Ann NY Acad Sci.* 959: 330–40.
190. Lander HM (1997). An essential role for free radicals and derived species in signal transduction. *FASEB J.*11: 118–24.
191. Nemoto S, Takeda K, Yu ZX, Ferrans VJ, Finkel T (2000). Role for mitochondrial oxidants as regulators of cellular metabolism. *Mol Cell Biol.* 20: 7311–18.

192. Finkel T, Holbrook NJ (2000). Oxidants, oxidative stress and the biology of ageing. *Nature* 408: 239–47.
193. Martindale JL, Holbrook NJ (2002). Cellular response to oxidative stress: signaling for suicide and survival. *J Cell Physiol.* 192: 1–15.
194. Patterson GI, Padgett RW (2000). TGF beta-related pathways. Roles in *Caenorhabditis elegans* development. *Trends Genet.* 16: 27–33.
195. Savage-Dunn C (2001). Targets of TGF beta-related signaling in *Caenorhabditis elegans. Cytokine Growth Factor Rev.* 12: 305–12.
196. Yu H, Larsen PL (2001). DAF-16-dependent and independent expression targets of DAF-2 insulin receptor-like pathway in *Caenorhabditis elegans* include FKBPs. *J Mol Biol.* 314: 1017–28.
197. Cherkasova V, Ayyadevara S, Egilmez N, Shmookler RR (2000). Diverse *Caenorhabditis elegans* genes that are upregulated in dauer larvae also show elevated transcript levels in long-lived, aged, or starved adults. *J Mol Biol.* 300: 433–48.
198. Kenyon C (2001). A conserved regulatory system for aging. *Cell* 105: 165–8.
199. Montgomery MK, Fire A (1998). Double-stranded RNA as a mediator in sequence-specific genetic silencing and co-suppression. *Trends Genet.* 14: 255–8.
200. Lund J, Tedesco P, Duke K, Wang J, Kim S, Johnson T (2002). Transcriptional profile of aging in *C. elegans. Curr Biol.* 12: 1566.
201. Hill AA, Hunter CP, Tsung BT, Tucker-Kellogg G, Brown EL (2000). Genomic analysis of gene expression in *C. elegans. Science* 290: 809–12.
202. Link CD (1995). Expression of human beta-amyloid peptide in transgenic *Caenorhabditis elegans. Proc Natl Acad Sci USA* 92: 9368–72.
203. Fay DS, Fluet A, Johnson CJ, Link CD (1998). *In vivo* aggregation of beta-amyloid peptide variants. *J Neurochem.* 71: 1616–25.
204. Link CD, Johnson CJ (2002). Reporter transgenes for study of oxidant stress in *Caenorhabditis elegans. Methods Enzymol.* 353: 497–505.
205. Fonte V, Kapulkin V, Taft A, Fluet A, Friedman D, Link CD (2002). Interaction of intracellular beta amyloid peptide with chaperone proteins. *Proc Natl Acad Sci USA* 99: 9439–44.
206. Faber PW, Alter JR, MacDonald ME, Hart AC (1999). Polyglutamine-mediated dysfunction and apoptotic death of a *Caenorhabditis elegans* sensory neuron. *Proc Natl Acad Sci USA* 96: 179–84.
207. Satyal SH, Schmidt E, Kitagawa K, Sondheimer N, Lindquist S, Kramer JM, *et al.* (2000). Polyglutamine aggregates alter protein folding homeostasis in *Caenorhabditis elegans. Proc Natl Acad Sci USA* 97: 5750–5.
208. Krobitsch S, Lindquist S (2000). Aggregation of huntingtin in yeast varies with the length of the polyglutamine expansion and the expression of chaperone proteins. *Proc Natl Acad Sci USA* 97: 1589–94.
209. Morley JF, Brignull HR, Weyers JJ, Morimoto RI (2002). The threshold for polyglutamine-expansion protein aggregation and cellular toxicity is dynamic and influenced by aging in *Caenorhabditis elegans. Proc Natl Acad Sci USA* 99: 10417–22.
210. Parker JA, Connolly JB, Wellington C, Hayden M, Dausset J, Neri C (2001). Expanded polyglutamines in *Caenorhabditis elegans* cause axonal abnormalities and severe dysfunction of PLM mechanosensory neurons without cell death. *Proc Natl Acad Sci USA* 98: 13318–23.
211. Nass R, Hall DH, Miller DM III, Blakely RD (2002). Neurotoxin-induced degeneration of dopamine neurons in *Caenorhabditis elegans. Proc Natl Acad Sci USA* 99: 3264–9.
212. Rose MR (1991). *Evolutionary Biology of Aging*. New York: Oxford University Press.

213. Partridge L (2001). Evolutionary theories of ageing applied to long-lived organisms. *Exp Gerontol.* 36: 641–50.
214. Charlesworth B (1980). *Evolution in Age-Structured Populations.* Cambridge: Cambridge University Press.
215. Medawar PB (1952). *An Unsolved Problem of Biology.* London: HK Lewis.
216. Hamilton WD (1966). The moulding of senescence by natural selection. *J Theor Biol.* 12: 12–45.
217. Edney EB, Gill RW (1968). Evolution of senescence and specific longevity. *Nature* 220: 281–2.
218. Williams GC (1957). Pleiotropy, natural selection, and the evolution of senescence. *Evolution* 11: 398–411.
219. Kirkwood TB (2002). Evolution of ageing. *Mech Ageing Dev.* 123: 737–45.
220. Johnson TE, Hutchinson EW (1993). Absence of strong heterosis for life span and other life history traits in *Caenorhabditis elegans. Genetics* 134: 465–74.
221. Walker DW, McColl G, Jenkins NL, Harris J, Lithgow GJ (2000). Evolution of lifespan in *C. elegans. Nature* 405: 296–7.

Aging in Birds

Donna J. Holmes
Department of Biological Sciences, University of Idaho, Moscow ID 83844-3051, USA

Introduction

The class Aves is a diverse group consisting of approximately 9600 species of birds in 29 orders [1–3] which, on the whole, are remarkable for their longevity. Body weights of birds vary from smallest to largest by a factor of about 40 000-fold [4, 5], and the range of variation in avian metabolic rates, life histories, life spans and aging patterns is equally striking [6–9]. Birds can live three to four times longer than mammals of equal body mass (Table 1). A 20 g laboratory mouse, for example, rarely lives over four years, but documented maximum life spans of small songbirds often exceed 10 years. The most paradoxical examples of avian longevity by far are the hummingbirds. They are the smallest of all birds, with the highest metabolic rates, yet individuals of some species have survived up to 14 years in the wild (Calder, personal communication) [10, 11].

Since lifetime energy expenditures of birds – and the corresponding potential for damage from reactive oxygen species – can exceed 15 times those for similar-sized mammals, comparative gerontologists have suggested that long-lived birds must have very effective adaptations for preventing or repairing aging-related cellular and molecular damage [7–9, 12–15]. A new and growing body of research supports this premise, but the exact biochemical mechanisms responsible for birds' longevity and resistance to aging-related deterioration still are poorly understood.

Although birds have been featured in the literature more often as models for aging studies, both in the laboratory and the field, since the early 1990s, the quality of these studies is variable. The most intensive studies of aging still tend to be limited to domestic poultry species (most notably, chickens and quail), which are unusually short-lived for birds of their size. There is still far too little dialogue or collaboration between field ornithologists and biogerontologists, or between researchers investigating basic mechanisms of aging and avian ecology and evolution.

H.D. Osiewacz (ed.), Aging of Organisms, 201–219.

Table 1. Life spans and approximate lifetime energy expenditures in mice, humans and some typical birds

Species	Maximum documented longevity (years)	Body mass (g)	Estimated lifetime energy expenditure (kcal/g)
Mouse (*Mus domesticus*)	6 (c)	20	250
Human (*Homo sapiens*)	122 (c)	50 000	800
Japanese Quail (*Coturnix japonica*)	6	100	460
House Wren (*Troglodytes aëdon*)	7	10	1002
American Robin (*Turdus migratorius*)	14	77	1336
Budgerigar (*Melopsittacus undulatus*)	19+ (c)	40	1440
Raven (*Corvus corax*)	69 (c)	1200	1400
Cardinal (*Cardinalis cardinalis*)	16	45	1535
Starling (*Sturnus vulgaris*)	21	71	2100
Canary (*Serinus canaria*)	24 (c)	22	3200
Eurasian Goldfinch (*Carduelis carduelis*)	27 (c)	16	4000
Broad-tailed Hummingbird (*Selasphorus platycercus*)	5	14	8834

Data from refs 19–22. Records for most wild birds are based on banding records from the field, and probably underestimate potential life span in captivity. Most longevities were validated using similar values from more than one large sample for a given species. Lifetime energy expenditures were calculated using published estimates of basal or resting metabolic rates.

(c) = record from captivity

In this chapter I begin by describing patterns of avian aging in broad terms, and discussing possible evolutionary causes of the long life spans and correlated life-history traits characterizing many birds. Next I describe current progress in the comparative study of avian aging, including evidence for birds' apparent resistance to damage from reactive oxygen molecules, and particular aspects of reproductive aging and costs which merit further study. A final section is devoted to recent studies of avian telomeres, and their potential relevance to the field of aging. Throughout the chapter, I emphasize the use and potential utility of birds as animal models for the study of basic aging processes, and the exceptional anti-aging adaptations suggested to be operating in longer-lived bird species.

Avian aging: evolution and general themes

Finch [6] provided a useful system for classifying animal aging patterns using three general categories: (1) rapid senescence culminating in sudden death; (2) gradual senescence with finite life span; and (3) negligible (or as yet undetectable) senescence. Aging patterns in birds generally fall into the second category. There are no known semelparous birds. Compared with other vertebrates, including ectotherms, most birds age moderately slowly and exhibit gradual senescence overall. Some bird species, however – most notably seabirds (including shorebirds and albatrosses in orders Charadriiformes and Procellariiformes) – show extremely slow, possibly even negligible, reproductive aging in the wild. Hence this last group are of considerable interest as potential models of slow aging and delayed fertility loss.

Evolutionary theory predicts that organisms with relatively high mortality rates will evolve rapid maturation and high reproductive rates early in life. In this scenario, there is weak selection for the long-term somatic maintenance that prevents aging. On the other hand, organisms protected against predation, disease, or accidental injury are free from selection for early, intense reproductive effort. These organisms, in turn, are predicted to evolve long life spans and effective mechanisms for long-term somatic maintenance, including prevention of the biochemical insults thought to be responsible for aging [16–18].

Mortality rates for birds are generally high until they can fly competently. Once they have entered the breeding population, however, mortality rates even for small songbirds are very low relative to those for small mammals. Annual adult survival rates for small land birds (mostly passerines) have been estimated at over 65%; for raptors (hawks, eagles, etc.) and some seabirds, adult survival averages over 90% [3, 6, 23–25]. In contrast, less skillful fliers, including galliform poultry species, have survival rates as low as 20%. The long life spans and slow aging rates typical of most birds therefore are consistent with theoretical predictions.

Aging patterns documented to date in the class Aves can be subdivided into three themes (Table 2). As I will use them here, each "theme" is intended as a *composite sketch* of aging in the species it characterizes, with some details omitted for the sake of comparison. As data become available for a more diverse cross-section of wild birds aging in captivity, this scheme may need substantial revision.

Short-lived birds with moderately rapid aging

The first avian aging theme is based primarily on extensive information available for domestic chickens (*Gallus domesticus*) and Japanese quail (*Coturnix japonica*) (Table 2; Figure 1). Well-documented examples of this theme are thus far limited to the weakly flying order Galliformes, in which young are precocial (develop rapidly) and both sexes mature quickly and produce relatively large broods. It is characterized by moderately rapid aging (as evidenced by steady increases in mortality rate, as well as increasing incidence of pathologies of aging in captive populations) with relatively precipitous reproductive declines far in advance of the deterioration of other systems [26–35]. Fertility in both sexes declines abruptly in these birds after the first breeding year and all but ceases thereafter. Healthy post-reproductive life spans of at least 25% of the total life span are routinely seen in both females and males; males often outlive females. The maximum documented longevity for domestic quail is about six years and, for chickens, around 15 years.

This pattern is not limited to domestic poultry bred for meat and egg production, but also is seen in wild Red Junglefowl (the nearest relative of the domestic chicken) and wild pheasants and quail in captivity. Longevities for wild European quail, *Coturnix coturnix*, moreover, are comparable to those of Japanese quail [36]. This

Table 2. Avian aging themes: relative timing of development and aging in representative species

	Theme 1 Japanese Quail[a] (*Coturnix japonica*)	Theme 2 Great Tit[b] (*Parus major*)	Theme 3 California Gull[b] (*Larus californicus*)
Body weight	100 grams	50 grams	600 g
Age at maturity	4 weeks	1+ year	3 years
Age at peak reproduction	1 year	4 years	Steady increases for 20+ years
Age of significant reproductive decline	15 months	5 years	No documented decline 20+ years
Postreproductive life span (of total life span)	25%	Undocumented	Undocumented

Data from [a]domesticated birds in captivity; [b]wild birds in nature.
Refs. 27, 32, 35, 47, 48

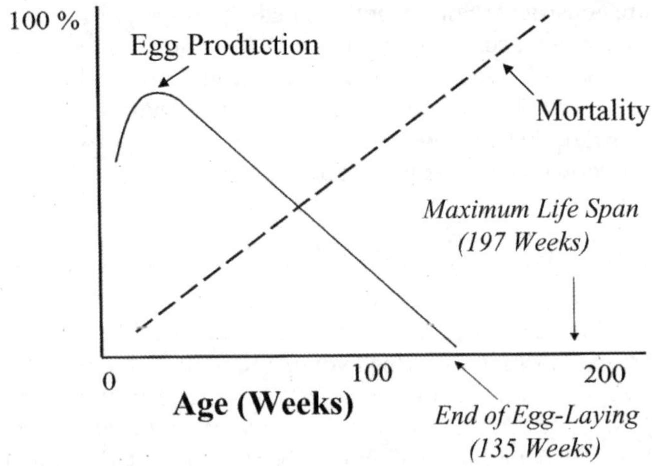

Figure 1. Age-related changes in mortality and egg production in domestic Japanese Quail. [Data from ref. 27].

theme of relatively fast aging overall, with an early reproductive decline, is remarkably similar in its outlines and timing to that seen in laboratory rats and mice. But aging in most small mammals, when scaled for body size, is faster than that in galliform birds, despite the fact that lifetime energy expenditures of these short-lived birds are considerably higher than those of similar-sized mammals (Table 1).

Table 3. *Patterns of age-related decline in two ptarmigan species*

	White-tailed Ptarmigan (*Lagopus leucurus*)	Willow Ptarmigan (*Lagopus lagopus*)
Habitat	Alpine tundra	Subarctic tundra
Life Span	5–6 years	5–6 years
Female Mortality (>3 years)	Significant increases	No significant increases
Fecundity (>3 years)	Significant declines	No significant declines
Fledging Success (>3 years)	Significant increases	No significant increases

Data from ref. 37.

There is at least one known, and very interesting, exception to this pattern of aging in galliform birds, particularly with respect to reproductive aging. Karen Wiebe and Kathy Martin combined longitudinal and cross-sectional data to compare age-related changes in mortality and reproduction for wild populations of White-tailed and Willow Ptarmigans, short-lived (5 to 6 years), grouse-like birds [37]. Both live in harsh tundra environments, but White-tails and Willows specialize in alpine and subarctic areas, respectively. Wiebe and Martin reported a significant mortality-rate increase in White-tailed females three years and older, but not in older Willow females (Table 3). Their most surprising finding, however, was that, while the oldest White-tailed females showed clear declines in egg-laying date and fecundity, they had the highest fledging success overall, apparently compensating for reduced fecundity with better parenting. These age-related effects were less pronounced for Willow Ptarmigan females.

This pattern of increased reproductive success with age in White-tailed Ptarmigan (albeit superimposed on a much shorter life span) resembles that of many seabirds with much lower adult mortality rates (see section on "Exceptionally long-lived birds with slow aging and undocumented reproductive declines"). It is much more difficult to explain from an evolutionary standpoint, however. Wiebe and Martin suggest that younger ptarmigan are adaptively constrained from peak investment in reproduction until they have maximized their foraging experience and other parenting skills. This rationale, however, fails to explain the apparent lack of corresponding increases in actual fecundity in older ptarmigans.

Moderately long-lived, slowly-aging birds

The second avian aging theme is one of slower maturation than seen in the first theme, coupled with moderate longevity (over 75 years for some larger parrot species in captivity) and relatively slow aging (Table 2; Figures 2 and 3). This pattern

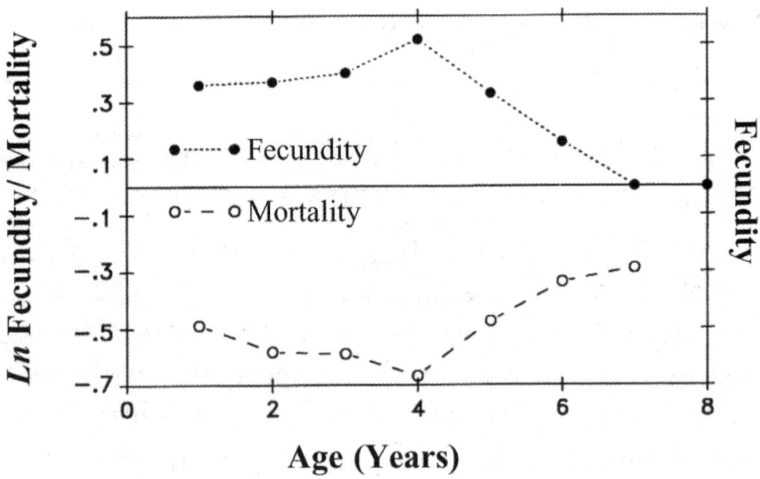

Figure 2. Age-related changes in mortality and fecundity in the Great Tit, a small European passerine, based on banding records in the wild from over 3000 breeding males and females combined. Maximum recorded life span for this species is 9 years. [Data from ref. 38; reprinted with permission from Holmes and Austad 1995, Copyright Clearance Center.]

probably characterizes the majority of birds, including songbirds (order Passeriformes), which comprise over half of all avian species. While aging in these birds is similar in its general outlines to the "quail-chicken" theme, it is slower by an order of magnitude or more.

Passerine songbirds, for example, usually take a couple of years to reach maximum breeding proficiency; this peak is followed by the increasing mortality with age indicative of senescence, with correlated declines in fertility [38–41]. Even small (under 50 g) songbirds have average life spans of over several years in the wild, and their reproductive aging occurs relatively slowly – more than twice as slowly as that of similar-sized captive rodents. Parrots and pigeons, which have long been domesticated, may also fit into this category, but fertility declines in parrots have not been clearly characterized. Reliable longevity records for the budgerigar parakeet (average weight 35–40 g) exceed 15 years [7, 8].

Unlike the case for domestic poultry, much of the evidence consistent with this pattern of aging is based on mark-recapture records for wild birds. The quality of the data for many field studies is excellent, and there are now clear indications of aging in a number of wild bird populations [7–9]. Demographic data from nature, however, usually do not give a clear picture of the potential postreproductive life span in these birds. Senescent individuals are so vulnerable to disease, predation and other natural sources of mortality that they usually disappear quickly from study populations. As

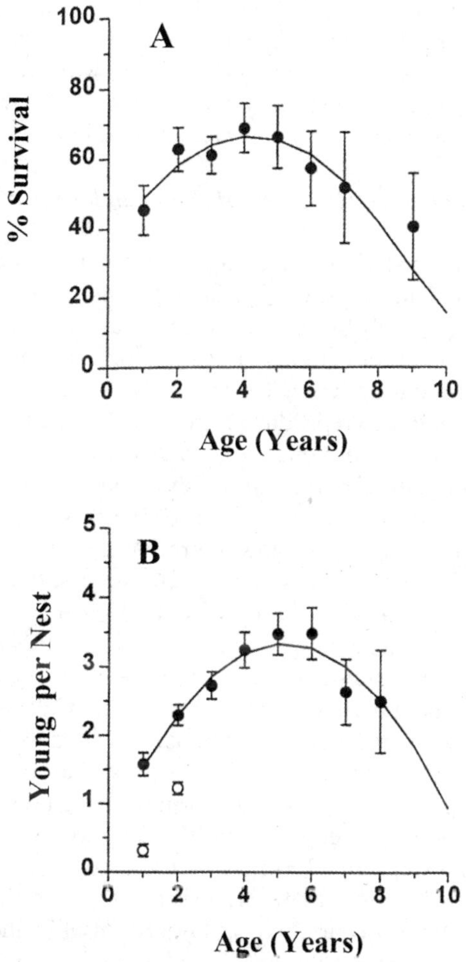

Figure 3. *Age-related changes in (A) survival and (B) young fledged per nest in the European Sparrowhawk, a small falcon. Reprinted with permission from Newton and Rothery 1997, Ecological Society of America.*

more data from zoos and captive breeding colonies of wild birds accumulate (e.g., R. Ricklefs, unpublished analysis), a clearer picture will emerge of patterns of reproductive and postreproductive aging in passerines, raptors, parrots and other birds that appear to fit this profile.

In American Kestrels (small predatory birds; order Falconiformes) maintained in a captive breeding colony, reproduction tends to wane after eight to 10 years, with a subsequent postreproductive period of up to several more years (J. French, personal communication). This pattern is also seen in the larger Peregrine Falcon, which can live over 15 years in captivity: female fecundity begins to decline around seven years of age [42]. I expect that most captive female birds will eventually show midlife reproductive declines and corresponding postreproductive life spans of 20 to 30% of their total life span, as do women and other female mammals in captivity [34, 43].

Exceptionally long-lived birds with slow aging and undocumented reproductive declines

Many pelagic seabirds (order Procellariiformes, which includes albatrosses and shearwaters, and Charadriiformes, e.g., gulls and terns) spend extraordinarily long periods in the air, and generally breed in very remote areas exposed to few predators. Their life histories and reproductive styles, therefore, are presumably shaped by mortality rates that are unusually low even for birds. They mature late (sometimes over 10 years after fledging) and reproduce extremely slowly, often laying only a single egg and successfully rearing under one young per year [3, 6, 44, 45]. One of the best studied seabirds, the songbird-sized Common Tern, often lives over 20 years in nature; some larger albatrosses have recorded life spans of 50 years or more [6, 25, 46, 47].

Long life spans, slow aging and slow reproduction are not the only features distinguishing the life histories of these birds from the first two avian aging themes. Data from long-term demographic studies of several seabird species in nature show convincingly that reproduction changes with age in these birds in a remarkable way. Instead of exhibiting the steady age-related reproductive declines typical of most birds, some seabirds, including Common Terns and California Gulls, actually *increase* the number of offspring successfully fledged even as they are entering the oldest age classes in the population. This means that they often reach their reproductive peak at ages when mortality risk is increasing most quickly, presumably as a result of deterioration of other, non-reproductive aspects of fitness. In a study of 24 very old (18+ years) Common Terns, Nisbet and co-workers [46–48] found that these birds were better breeders than any other age group. Senescent declines in reproduction in terns, if they occur at all, seem to be offset by increased reproductive effort, better parenting, or generally high fitness of old individuals. (Terns that did not reproduce, however, showed aging-related endocrine deficits.) Pugesek [49–50] has reached similar conclusions for aging California Gulls of all but the very oldest (20+ years) parental age classes (Figure 4).

Some larger, exceptionally long-lived, slowly-reproducing raptors, like condors, may also fit the theme of very slow, if not negligible, aging. Many of these species are rare or endangered, though, and little information is available on changes with fertility or reproductive success with age.

Avian aging research: current progress and prospectus

Bird species are still relatively unexploited as comparative animal models for delayed aging and long life span. This is true despite the fact that small, relatively long-lived pet birds like finches and parakeets are easily maintained in captivity, and are widely used in neurobiological research. Small birds hold a great deal of promise as

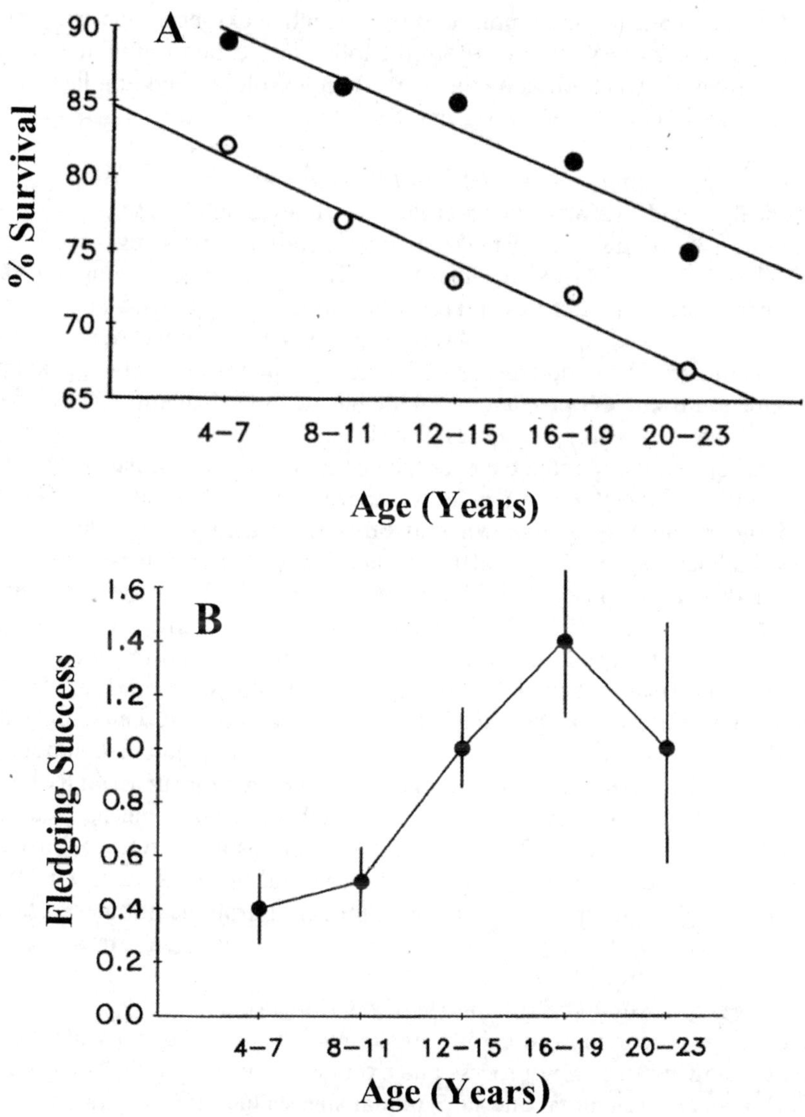

***Figure 4**. Age-related changes in (A) survival and (B) fledging success in a breeding colony of California Gulls. Reprinted with permission from Pugesek and Diem 1983, Ecological Society of America.*

laboratory models for the study of cellular adaptations for the prevention, defense, and repair of damage to macromolecules from reactive oxygen species and products of non-enzymatic glycosylation [7–9]. In the following section I also suggest several additional areas in which birds might play a unique role in shedding light on basic aging processes in homeothermic vertebrates.

Birds and the prevention of oxygen free-radical damage

A great deal of evidence now supports the idea that reactive oxygen species (ROS), produced as a by-product of normal oxidative metabolic processes, are responsible for the cell damage and fitness declines typically associated with aging. Given their high lifetime energy expenditures and potential susceptibility to ROS damage, birds – particularly long-lived species – should exhibit very effective defenses against oxidative damage. These defenses could take several forms, including adaptively lower mitochondrial ROS production, inducible defenses, like antioxidant enzymes, and constitutive or structural defenses [9, 51].

On the whole, studies conducted over the last decade or so comparing endogenous ROS generation and cellular responses to experimentally imposed oxidative stresses in birds and mammals have shown that birds from a range of orders (including canaries, budgies, pigeons and starlings) indeed have better defenses against ROS damage, and that they probably utilize a combination of cellular protective mechanisms [15, 52–63, but see 14]. Most of the accumulated evidence now supports the idea that long-lived birds have a much more efficient mitochondrial metabolism, producing as much as 10 times fewer ROS per oxygen molecule used [15, 51, 52, 54–56, 58]. Superior defenses against ROS damage have been documented convincingly, as well. Cultured avian cells appear to be better than their mammalian counterparts at resisting oxidative stresses of various kinds, and they may incur less DNA damage [57, 59, 62, 63]. At least some of these avian cellular defense mechanisms require active DNA transcription [67]. Moreover, several distantly related bird species (pigeon, budgie and canary) have apparent structural defenses against ROS in the form of low levels of saturated fatty acids in mitochondrial membranes of heart and liver [53, 60, 61].

Avian energy utilization and the caloric restriction model

The experimental reduction of caloric intake by 25% or more is a widely accepted method for slowing the aging process and extending life span in laboratory rodents, as well as many other animal and non-animal species [64, 65]. In rats and mice, the effects of long-term caloric restriction include reduced circulating levels of sex steroids, increases in glucocorticoid stress hormones, and alterations in carbohydrate metabolism, including lower insulin levels and improved insulin resistance. Fertility is impaired in both sexes and fat stores drop significantly. The exact mechanism responsible for the anti-aging effects of caloric restriction, however, is still not understood. The effects of short-term food restriction are a frequent subject of study in poultry species, but to my knowledge caloric restriction has never been continued through the end of the natural life span for any bird, or interpreted in a standard biogerontological context.

Birds generally maintain levels of glucose and lipids in blood significantly higher than those of mammals, presumably to satisfy the metabolic demands of flight [3, 7]. Lipids in avian blood tend to be less saturated (and hence possibly less subject to oxidative processes) than those in mammalian blood [16, 19].

Glucagon, rather than insulin, is the dominant pancreatic hormone in many birds (especially seed-eating species), and the ratio of glucagon to insulin produced by avian pancreatic islets is generally substantially higher than in mammals [66, 67]. This difference in the ratio of insulin to glucagon may have to do with birds' need to keep glucose levels high, even during fasting.

These aspects of avian blood chemistry and carbohydrate physiology are intriguing in light of their potential relevance to the slow rate of avian aging, yet they are remarkably understudied. It would be interesting to explore the effects of long-term caloric restriction in experimental populations of small, domestic granivorous birds (both long- and short-lived), and to compare its effects on carbohydrate and lipid metabolism in birds and mammals. While maximum life span in these birds can exceed 15 years, fertility drops off in females within a few years of maturity. The anti-aging effects of caloric restriction, if any, would probably become evident within six to eight years.

Avian reproductive aging and lifetime costs of reproduction

The physiology of reproductive aging in birds – with the exception of a few short-lived, rapidly aging poultry species – is poorly understood. While the underlying physiological events remain unexplored, patterns of age-related reproductive changes in females of various wild bird populations are, however, much better documented now than they were ten or fifteen years ago, as are the possible costs of reproduction in terms of lifelong fitness and aging [9, 37–42, 44–50, 68–81]. As I have already noted, postreproductive life spans are probably typical of a variety of captive birds enjoying optimal care, as they are for mammals and humans.

The poultry industry has long provided an impetus for the study of age-related changes in fertility in chickens, turkey, quail and ducks bred for meat or egg-laying. The best studied model for reproductive and neuroendocrine aging in a male bird is the Japanese quail; for female birds, it is the domestic laying hen. Since the large, meaty breasts of male domestic turkeys render them incompetent at copulation, there is also a great deal of research currently directed toward improving techniques for artificial insemination and sperm storage in these birds [82].

Reproductive aging: female birds

The ovarian changes and neuroendocrinological events associated with reproductive aging have been studied in considerable detail for strains of female chickens bred for prodigious egg-laying [30, 31, 33, 34, 83–86]. Females are mature by about 6 months of age, and typically have a production period of about 1 year. By about two years of age, egg production by hens typically has dropped by as much as 90%. This decline in fertility is correlated with decreasing levels of progesterone, follicle-stimulating hormone (FSH), and follicular responsiveness to luteinizing hormone (LH), as well as declining numbers of viable oocytes (note that declines in estrogen are not included in this scenario, in contrast to the typical mammalian aging pattern). Specific ovarian changes include decreases in available yolky preovulatory follicles and increased numbers of atretic follicles, as well as deterioration of cellular communication between follicular granulosa and oocytes.

Conventional wisdom holds that female birds, like female mammals, form all of their primary oocytes during embryonic development or shortly after birth. The little comparative evidence available supports this view [26, 87, 88], but species from only a few avian orders have been examined. The role of actual oocyte depletion in avian reproductive aging remains remarkably understudied.

Follicular apoptosis (the programmed cell death that is a normal part of ovarian development and preparation of the follicular hierarchy for ovulation, as well as a function of aging) is a very active research area, as is the role of the germinal disc of the developing egg in preventing follicular atresia and apoptosis [89, 100]. A.L. Johnson and his colleagues are engaged in ongoing and intensive investigation of key molecular pathways regulating signaling between aging chicken oocytes and supporting follicular cells [90–95]. No comparable studies have been published for any long-lived wild or domestic bird.

There has been little research, moreover, on endocrinological correlates of reproductive aging in wild birds. One notable exception is a study I discussed in an earlier section by Ian Nisbet, Mary Ann Ottinger and colleagues [46–48], in which blood from Common Terns during the breeding season was assayed for levels of reproductive steroids. Their results suggest that female terns maintain remarkably stable endocrine status as they age. While there was some suggestion of a decline in estrogen in the oldest (16–21 years) breeding females, this was not statistically reliable. LH increased little in birds up to 16 years of age. Obviously the delayed reproductive aging characteristic of seabirds like terns is very interesting to comparative gerontologists, and merits a great deal more investigation.

One of the greatest challenges now facing the biomedical research community is human infertility and menopause. Currently, the most intensively used models for human ovarian aging are short-lived rodents and chickens. With moderate adjustments for differences in their breeding biology and behavior, longer-lived, small cage birds should lend themselves well to studies of reproductive aging, and provide an opportunity for studying relatively slow ovarian aging and biochemical adaptations for delayed fertility loss.

Reproductive aging: male birds

The domestic Japanese Quail is by far the best understood male bird model of reproductive and neuroendocrine aging, primarily as a result of extensive work by Mary Ann Ottinger and collaborators [32, 97–105]. Male quail are considerably longer-lived than females; this difference may be due, in part, to the demands of egg production by females. Males exhibit exhibit gradual declines in sperm production with age, as roosters do [96]. Male quail also show decreases in testis weight and declines in spermatogenesis by about 3 years. Subsequently, plasma androgen levels fall, and levels of LH and FSH show a decrease and an increase, respectively. Testosterone-dependent courtship behaviors decline with age; Ottinger has linked these deficits to declines in gonadotropin-releasing hormone (GnRH) by the hypothalamus and a resulting failure of gonadotropin secretion. In contrast to the scenario in quails, the study of wild terns discussed above revealed no clear age-associated declines in males in plasma levels of testosterone, progesterone, or LH [48].

Avian costs of reproduction

A central tenet of evolutionary life-history theory is that sexual reproduction is energetically costly, and investment in reproduction reduces the energy available for parental somatic maintenance and repair. Hence natural selection is predicted to force organisms to make adaptive trade-offs between present and future reproductive investment, as well as between reproduction and their own fitness and longevity [106, 107]. Birds lend themselves especially well to long-term studies of reproduction and its possible costs for several reasons. First, individuals in wild populations can be captured, banded and subsequently observed in substantial numbers and through repeated breeding episodes. Nests and young also can be monitored, and reproductive investment effectively manipulated simply by adding or subtracting eggs in nests. Over the past 20 years or so, avian ecologists have been using their study populations to test theoretical predictions about costs of reproduction in terms of subsequent survival, reproduction and aging patterns.

With some exceptions [see, for example, refs 68 and 71], the majority of published studies (primarily using small songbirds) are consistent with the idea of reproductive costs and trade-offs. Results of these studies that are consistent with the presence of costs have taken the form of (1) reduced survivorship of breeders vs. non-breeders; (2) correlations between early reproductive investment and reduced survivorship later in life; (3) correlations between later breeding onset and bigger clutches or broods; (4) experimental brood enlargement resulting in smaller clutches later on; (5) brood enlargement resulting in earlier onset of reproductive senescence; or some combination of these effects [69, 70, 72–81]. In some species, such as the European collared flycatcher [74], costs of reproduction may be detected only in individuals in relatively poor condition, and are not seen as readily in more fit breeders. Predation rates on young, as well as adult mortality rates, have also been shown to be associated with timing and costs of reproduction in some birds [69], emphasizing the potential complexity of adaptive trade-offs and their measurement in natural systems.

Even more recently, ornithologists have tried to quantify the exact costs of production of eggs and young, as well as foraging costs. They are now beginning attempts to identify the proximate physiological mechanisms underlying putative reproductive costs (both metabolic and in terms of subsequent reproduction). They have used a combination of field and laboratory approaches [see, for example, refs 2, 75, 76, 78–81]. Domestic finches experimentally forced to work harder immediately preceding a breeding episode have been shown in at least one study to exhibit reduced reproductive success later on [76]. There is some evidence, moreover, that reproduction (either in terms of physiological readiness or actual investment) has a negative impact on immune status, particularly for adults in poor condition or with compromised nutritional status [see, for example, refs 72, 74, 80]. This suggests that reproductive costs may be incurred in the form of increased susceptibility to disease or parasitism. Although a few field studies have shown correlations between high reproductive investment early in life and more rapid parental senescence [see, for example, refs 70, 72, 74, 80], the exact mechanisms whereby reproduction may hasten the aging process are not at all clear.

Avian telomeres and aging

Telomeres are sections of highly conserved nucleotide repeats located on the ends of linear chromosomes that assist in DNA replication. Since telomeres have been shown to shorten with age in human somatic cells both *in vivo* and *in vitro*, telomore length has been suggested to function as a molecular time-keeper for cellular senescence [108]. Age-related changes in telomere length, however, as well as in the activity of the telomere-length-maintaining enzyme, telomerase, vary a great deal among species and are not clearly correlated with life span. The mechanisms underlying teleomere shortening also may vary substantially between mammalian orders. Comparative studies relating telomere dynamics to senescence rates in more vertebrate taxa are still needed.

A recent study by Mary Delaney and colleagues [109] included the first published comparisons of telomere DNA arrays within and between members of 18 bird species in several different orders, including domestic and wild poultry, raptors, and parrot, finch and crane species. Telomere sequences from erythrocytes in most of these birds (with the exception of the American Bald Eagle and Northern Goshawk) were 5- to 10-fold longer than those in mammals, and longer than for any vertebrates examined thus far. This is particularly interesting in light of the fact that avian genomes are generally considerably smaller and more conserved than those of mammals. In addition, comparison among chickens of different ages showed shortening of telomeres with age in erythrocytes and sperm *in vivo*; this finding correlates with earlier reports of *in-vivo* chicken telomere shortening. A great deal more data are needed, however, before patterns of changing telomere length with age can be fully addressed for birds or any other group of vertebrates.

Acknowledgments

I would like to thank D. Kristan and B. Parks for their editorial comments on the manuscript. I was supported during the writing of this chapter by Biomedical Research Infrastructure (National Institutes of Health) grant P20 RR16454-01 to the University of Idaho.

References

1. Sibley CG, Monroe BL Jr (1990). *Distribution and Taxonomy of Birds of the World.* New Haven, Connecticut: Yale University Press.
2. Sibley CG, Monroe BL Jr (1993). *A Supplement to Distribution and Taxonomy of Birds of the World.* New Haven, Connecticut: Yale University Press.
3. Gill FB (1995). *Ornithology*, 2nd edn. New York: WH Freeman and Co.

4. Bennett PM, Owens IPF (2002). *Evolutionary Ecology of Birds. Life Histories, Mating Systems and Extinction.* Oxford: Oxford University Press.
5. Dunning JB Jr (1992). *CRC Handbook of Avian Body Masses.* Boca Raton, Florida: CRC Press.
6. Finch CE (1990). *Longevity, Senescence, and the Genome.* Chicago: University of Chicago Press.
7. Holmes DJ, Austad SN (1995). Birds as animal models for the comparative biology of aging: a prospectus. *J Gerontol Biol Sci.* 50A: B59–66.
8. Holmes DJ, Austad SN (1995). The evolution of avian senescence patterns: implications for understanding primary aging processes. *Am Zool.* 35: 307–17.
9. Holmes DJ, Flückiger R, Austad S (2001). Comparative biology of aging in birds: an update. *Exp Gerontol.* 36: 869–83.
10. Calder WA (1985). The comparative biology of longevity and lifetime energetics. *Exp Gerontol.* 20: 161–70.
11. Calder WA (1990). Avian longevity and aging. In: Harrison D, ed. *Genetic Effects on Aging II.* Caldwell, New Jersey: elford Press, pp. 185–204.
12. Monnier VM (1990). Nonenzymatic glycosylation: the Maillard reaction and the aging process. *J Gerontol Biol Sci.* 45: B105–11.
13. Monnier VM, Sell DR, Ramanakoppa HN, Miyata S (1991). Mechanisms of protection against damage mediated by the Maillard reaction in aging. *Gerontology* 37, 152–65.
14. Ku H, Sohal RS (1993). Comparison of mitochondrial prooxidant generation and antioxidant defenses between rat and pigeon: possible basis of variation in longevity and metabolic potential. *Mech Ageing Dev.* 72: 67–76.
15. Barja G, Cadenas S, Roja C, Pérez-Campo R, López-Torres M (1994). Low mitochondrial free radical production per unit O_2 consumption can explain the simultaneous presence of high longevity and high aerobic metabolic rate in birds. *Free Rad Res.* 21: 317–28.
16. Williams GC (1957). Pleiotropy, natural selection and the evolution of senescence. *Evolution* 11: 398–411.
17. Edney EB, Gill RW (1967). Evolution of senescence and specific longevity. *Nature* 220: 281–2.
18. Partridge L, Barton NH (1993). Optimality, mutation and the evolution of ageing. *Nature* 362: 305–11
19. Lasiewski RC, Dawson WR (1967). A re-examination of the relation between standard metabolic rate and body weight in birds. *Condor* 69: 13–23.
20. Altman PL, Dittmer DS (1972). *Biology Data Book*, Vol. 1. Bethesda, Maryland: Federation of the American Sciences for Experimental Biology.
21. Gavrilov VM, Dolnik VR (1982). Basal metabolic rate, thermoregulation and existence energy in birds: world data. In: Ilyichev VD, Gavrilov VM, eds. *Acta XVIII. Congress Internationalis Ornithologici.* Moscow: Academy of Sciences of the USSR, pp. 421–66.
22. Carey JR, Judge DS (2000). *Longevity Records: Life Spans of Mammals, Birds, Amphibians, Reptiles, and Fish.* Odense Monographs on Population Aging 8. Odense, Denmark: Odense University Press.
23. Ricklefs RE (1973). Fecundity, mortality, and avian demography. In: Farner DS, ed. *Breeding Biology of Birds.* Washington, DC: National Academy of Science, pp. 366–435.
24. Welty JC (1982). *The Life of Birds*, 3rd edn. New York: Saunders.
25. Perrins CR, Birkhead TR (1983). *Avian Ecology.* New York: Chapman and Hall.

26. Franchi LL, Mandl AM, Zuckerman S (1962). The development of the ovary and the process of oogenesis. In: Zuckerman S, ed. *The Ovary*, New York: Academic Press, pp. 1–88.
27. Woodard AE, Abplanalp H (1971). Longevity and reproduction in Japanese quail maintained under stimulatory lighting. *Poultry Science* 50: 688–92.
28. Tokarz RR (1978). Oogonial proliferation, oogenesis, and folliculogenesis in nonmammalian vertebrates. In: Jones RE, ed. *The Vertebrate Ovary: Comparative Biology and Evolution*. New York: Plenum Press, New York, pp. 145–79.
29. Williams JB, Sharp PJ (1978). Ovarian morphology and ovarian follicular development in laying broiler hens and commercial egg producing hens. *Br Poultr Sci.* 19: 387–95.
30. Williams JB, Sharp PJ (1987). Age-dependent changes in the hypothalamo-pituitary-ovarian axis of the laying hen. *J Reprod.* 53: 141–6.
31. Bahr JM, Palmer SS (1989). The influence of aging on ovarian function. *CRC Crit Rev Poultry Biol.* 2: 103–10.
32. Ottinger MA (1991). Neuroendocrine and behavioral determinants of reproductive aging. *Crit Rev Poultry Biol.* 3: 131–42.
33. Palmer SS, Bahr JM (1992). Follicle stimulating hormone increases serum oestradiol-17β concentrations, number of growing follicles and yolk deposition in aging hens (*Gallus gallus domesticus*). with increased egg production. *Br Poultry Sci.* 33: 403–14.
34. vom Saal FS, Finch CE, Nelson JF (1994). Natural history and mechanisms of reproductive aging in humans, laboratory rodents, and other selected vertebrates. In: Knobil E, Neill JD, eds. *The Physiology of Reproduction*. New York: Raven Press, pp. 1213–314.
35. Ottinger MA (2001). Quail and other short-lived birds. *Exp Gerontol.* 36: 859–68.
36. Puigcerver M, Gallego S, Rodrigues-Teijeiro JD, Senar JC (1992). Survival and mean life span of the quail *Coturnix c. coturnix*. *Bird Study* 39: 120–3.
37. Wiebe KL, Martin K (1998). Age-specific patterns of reproduction in White-tailed and Willow Ptarmigan *Lagopus leucurus* and *L. lagopus*. *Ibis* 140: 14–24.
38. McCleery RH, Perrins CM (1988). Lifetime reproductive success in the Great Tit. In: Clutton-Brock TH, ed. *Reproductive Success: Studies of Individual Variation in Contrasting Breeding Systems*. Chicago: University of Chicago Press, pp. 136–53.
39. Newton I (1989). *Lifetime Reproduction in Birds*. London: Academic Press.
40. Gustafsson L, Pärt T (1991). Acceleration of senescence in the collared flycatcher *Ficedula albicollis* by reproductive costs. *Nature* 347: 279–81.
41. Newton I, Rothery P (1997). Senescence and reproductive value in sparrowhawks. *Ecology* 78: 1000–8.
42. Clum NJ (1995). Effects of aging and mate retention on reproductive success of captive female peregrine falcons. *Am Zool.* 35: 329–39.
43. Holmes DJ (2002). Is postmenopausal life-span a gift of modern health or a product of natural selection? SAGE KE, Science Online, February.
44. Ollasen JC, Dunnett GM (1988). Variations in breeding success in fulmars. In: Clutton-Brock TH, ed. *Reproductive Success: Studies of Individual Variation in Contrasting Breeding Systems*. Chicago: University of Chicago Press, pp. 263–78.
45. Ricklefs RE (1990). Seabird life histories and the marine environment: some speculations. *Colonial Waterbirds* 13: 1–6.
46. Nisbet IC, Finch CE, Thompson N, Russek-Cohen E, Proudman JA, Ottinger, MA (1999). Endocrine patterns during aging in the common tern (*Sterna hirundo*). *Gen Comp Endocrinol.* 114: 279–86.

47. Nisbet ICT, Apanius V, Friar MS (2002). Breeding performance of very old Common Terns. *J Field Ornithol.* 73: 117–24.
48. Ottinger MA, Nisbet ICT, Finch CE (1995). Aging and reproduction: comparative endocrinology of the common tern and Japanese quail. *Am Zool.* 35: 299–306.
49. Pugesek BH (1981). Increased reproductive effort with age in the California gull (*Larus californicus*). *Science* 212: 822–3.
50. Pugesek BH, Diem KL (1983). A multivariate study of the relationship of parental age to reproductive success in California gulls. *Ecology* 64: 829–39.
51. Barja G (1998). Mitochondrial free radical production and aging in mammals and birds. *Ann N Acad Sci.* 20: 224–38.
52. Barja G, Cadenas S, Rojas C, López-Torres M, Pérez-Campo R (1994). A decrease in free radical production near critical targets as a cause of maximum longevity in animals. *Comp Biochem Physiol Biochem Mol Biol.* 108: 501–12.
53. Pamplona R, Prat J, Cadenas S, *et al.* (1996). Low fatty acid unsaturation protects against lipid peroxidation in liver mitochondria from long-lived species: the pigeon and human case. *Mech Ageing Dev.* 86: 53–66.
54. Herrero A, Barja G (1997). Sites and mechanisms responsible for the low rate of free radical production of heart mitochondria in the long-lived pigeon. *Mech Ageing Dev.* 98: 95–111.
55. Barja G, Herrero A (1998). Localization at complex I and mechanism of the higher free radical production of brain nonsynaptic mitochondria in the short-lived rat than in the longevous pigeon. *J Bioenerg Biomembr.* 30: 235–43.
56. Herrero A, Barja G (1998). H_2O_2 production of heart mitochondria and aging rate are slower in canaries and parakeets than in mice: sites of free radical generation and mechanisms involved. *Mech Ageing Dev.* 103: 133–46.
57. Ogburn CE, Austad SN, Holmes DJ, et al. (1998). Cultured renal epithelial cells from birds and mice: enhanced resistance of avian cells to oxidative stress and DNA damage. *J Gerontol Biol Sci.* 53A: B287–92.
58. Perez-Campo R, López-Torres M, Cadenas S, Rojas C, Barja G (1998). The rate of free radical production as a determinant of the rate of aging: evidence from the comparative approach. *J Comp Physiol B* 168: 149–58.
59. Herrero A, Barja G (1999). 8-oxo-deoxyguanosine levels in heart and brain mitochondrial and nuclear DNA of two mammals and three birds in relation to their different rates of aging. *Aging (Milano)*, 11: 294–300.
60. Pamplona R, Portero-Otín M, Requena JR, Thorpe SR, Herrero A, Barja G (1999). A low degree of fatty acid unsaturation leads to lower lipid peroxidation and lipoxidation-derived protein modification in heart mitochondria of the longevous pigeon than in the short-lived rat. *Mech Ageing Dev.* 106: 283–96.
61. Pamplona R, Portero-Otín M, Riba D, *et al.* (1999). Heart fatty acid unsaturation and lipid peroxidation, and aging rate, are lower in the canary and the parakeet than in the mouse. *Aging (Milano)*, 11: 44–9.
62. Jaensch S, Cullen L, Morton L, Raidal SR (2001). Normobaric hyperoxic stress in budgerigars: non-enzymic antioxidants. *Comp Biochem Physiol Part C* 128: 181–7.
63. Ogburn CE, Carlberg K, Ottinger MA, Holmes DJ, Martin GM, Austad SN (2001). Exceptional resistance to oxidative damage in long-lived birds requires active gene expression. *J Gerontol Biol Sci.* 56: B468–74.
64. Yu BP, ed. (1999). *Methods in Aging Research*. Boca Raton, Florida: CRC Press.
65. Masoro EJ, Austad SN, eds. (2001). *Handbook of the Biology of Aging*, 5th edn. New York: Academic Press.

66. Whittow GC, ed. (2000). *Sturkie's Avian Physiology*. New York: Academic Press.
67. Pollock, C (2002). Carbohydrate regulation in avian species. *Semin Avian Exotic Pet Med.* II: 55–6.
68. Smith JM (1981). Does high fecundity reduce survival in song sparrows? *Evolution* 35: 1142–8.
69. Ekman J, Askenmo C (1986). Reproductive cost, age-specific survival and a comparison of the reproductive strategy in two European tits (Genus *Parus*). *Evolution* 40: 159–68.
70. Gustafsson L, Pärt T (1990). Acceleration of senescence in the collared flycatcher *Ficedula albicollis* by reproductive costs. *Nature* 347: 279–81.
71. Wheelwright NT, Leary J, Fitzgerald C (1991). The costs of reproduction in tree swallows (*Tachycineta bicolor*). *Can J Zool.* 69: 2540–7.
72. Gustafsson L, Nordling D, Andersson MS, Sheldon BC, Qvarnström A (1994). Infectious diseases, reproductive effort and the cost of reproduction in birds. *Phil Trans R Soc Lond B Biol Sci.* 346: 323–31.
73. Viallefon, A, Cooke F, Lebreton J-D (1995). Age-specific costs of first-time breeding. *Auk* 112: 67–6.
74. Cichoñ M, Olejniczak P, Gustafsson L (1997). The effect of body condition on the cost of reproduction in female collared flycatchers *Ficedula albicollis*. *Ibis*, 140: 128–30.
75. Bentley GE, Demas GE, Nelson RJ, Ball GF (1998). Melatonin, immunity and the cost of reproductive state in male European starlings. *Proc R Soc Lond B Biol Sci.* 265: 1191–5.
76. Deerenberg C, Overkamp GJF (1999). Hard work impinges on fitness: an experimental study with zebra finches. *Anim Behav.* 58: 173–9.
77. Ghalambor CK, Martin TE (2001). Fecundity-survival trade-offs and parental risk-taking in birds. *Science* 292: 494–7.
78. Visser ME, Lessells CM (2001). The costs of egg production and incubation in great tits (*Parus major*). *Proc Roy Soc Lond B* 268: 1271–7.
79. Orell MU, Belda EJ (2002). Delayed cost of reproduction and senescence in the willow tit *Parus montana*. *J Anim Ecol.* 71: 55–64.
80. Saino N, Ambrosini R, Martinelli R, Møller AP (2002). Mate fidelity, senescence in breeding performance and reproductive trade-offs in the barn swallow. *J Anim Ecol.* 71: 309–19.
81. Vezina F, Williams TD (2002). Metabolic costs of egg production in the European Starling (*Sturnus vulgaris*). *Physiol Biochem.* 75: 377–85.
82. Bakst MR (1998). Structure of the avian oviduct with emphasis on sperm storage in poultry. *J Exp Zool.* 282: 618–26.
83. Waddington D, Perry MM, Gilbert AB, Hardie MA (1985). Follicular growth and atresia in the ovaries of hens (*Gallus domesticus*) with diminished egg production rates. *J Reprod Fertil.* 74: 399–405.
84. Johnson PA, Dickerman RW, Bahr JM (1986). Decreased granulosa cell luteinizing hormone sensitivity and altered thecal estradiol concentration in the aged hen (*Gallus domesticus*). *Biol Reprod.* 35: 641–6.
85. Joyner CJ, Peddie MJ, Taylor TG (1987). The effects of age on egg production in the domestic hen. *Gen Comp Endocrinol.* 65: 331–6.
86. Chun S-Y, McGee EA, Hsu SY, *et al.* (1999). Restricted expression of WT1 messenger ribonucleic acid in immature ovarian follicles: uniformity in mammalian and avian species and maintenance during reproductive senescence. *Biol Reprod.* 60: 365–73.
87. Guraya SS (1989). *Ovarian Follicles in Reptiles and Birds.* Heidelberg: Springer-Verlag.
88. Lombardi J (1998). *Comparative Vertebrate Reproduction.* Boston: Kluwer.

89. Tischkau SA, Bahr JM (1996). Avian germinal disc region secretes factors that stimulate proliferation and inhibit progesterone production by granulosa cells. *Biol Reprod.* 54: 865–80.
90. Yao HH, Volentine KK, Bahr JM (1998). Destruction of the germinal disc region of an immature preovulatory chicken follicle induces atresia and apoptosis. *Biol Reprod.* 59: 516–21.
91. Johnson AL (2000). Granulosa cell apoptosis: conservation of cell signaling in an avian ovarian model system. *Biol Signals Recept.* 9: 96–101.
92. Johnson AL, Tilly JL (1988). Effects of vasoactive intestinal peptide on steroid secretion and plasminogen activator activity in granulosa cells of the hen. *Biol Reprod.* 296–303.
93. Johnson AL, Bridgham JT, Wagner B (1996). Characterization of a chicken luteinizing hormone receptor (cLH-R) complementary deoxyribonucleic acid, and expression of cLH-R messenger ribonucleic acid in the ovary. *Biol Reprod.* 55: 304–9.
94. Bridgham JT, Johnson AL (2001). Expression and regulation of Fas antigen and tumor necrosis factor receptor type I in hen granulosa cells. *Biol Reprod.* 65: 733–9.
95. Johnson AL, Langer JS, Bridgham JT (2002). Survivin as a cell cycle-related and antiapoptotic protein in granulosa cells. *Endocrinology* 143: 3405–13.
96. Payne F (1952). Cytological changes in the pituitary, thyroids, adrenals, and sex glands of the aging fowl. In: Lansing A, ed. *Cowdry's Problems of Ageing*. New York: Plenum Press, pp. 381–402.
97. Ottinger MA (1983). Sexual behavior and endocrine changes during reproductive maturation and aging in the avian male. In: Balthazart J, Pröve E, Gilles R, eds. *Hormones and Behavior in Higher Vertebrates.* Berlin: Springer-Verlag, pp. 350–67.
98. Ottinger MA, Duchala CS, Masson, M (1983). Age-related reproductive decline in the male Japanese Quail. *Horm Behav.* 17: 197–207.
99. Balthazart J, Turek R, Ottinger MA (1984). Altered brain metabolism of testosterone is correlated with reproductive decline in aging quail. *Horm Behav.* 18: 330–45.
100. Gorham SL, Ottinger MA (1986). Sertoli cell tumors in Japanese Quail. *Avian Dis.* 30: 337–9.
101. Ottinger MA, Balthazart J (1986). Altered endocrine and behavioral responses with reproductive aging in the male Japanese Quail. *Horm Behav.* 20: 83–94.
102. Ottinger MA (1996). Aging in the avian brain: neuroendocrine considerations. *Sem Avian Exotic Pet Med.* 5: 172–7.
103. Panzica GC, García-Ojeda E, Viglietti-Panzica C, Thompson NE, Ottinger MA (1996). Testosterone effects on vasotocinergic innervation of sexually dimorphic medial preoptic nucleus and lateral septum during aging in male quail. *Brain Res.* 712: 190–8.
104. Ottinger MA, Thompson N, Viglietti-Panzica C, Panzica GC (1997). Neuroendocrine of GnRH and behavior during aging in birds. *Brain Res Bull.* 44: 471–7.
105. Ottinger MA (1998). Male reproduction: testosterone, gonadotropins, and aging. In: Mobbs CV, ed. *Functional Endocrinology of Aging*, vol. 29. Basel: Karger, pp. 105–26.
106. Williams GC (1966). Natural selection, the cost of reproduction, and a refinement of Lack's principle. *Am Nat.* 100: 687–90.
107. Stearns SC (1976). Life-history tactics: a review of the ideas. *Q Rev Biol Syst.* 51: 3–47.
108. Harley CB (1995). Telomeres and aging. In: Blackburn EH, Greider CW, eds. *Telomeres.* Cold Spring Harbor: Cold Spring Harbor Laboratory Press, pp. 247–63.
109. Delaney ME, Krupkin AB, Miller MM (2000). Organization of telomere sequences in birds: evidence for arrays of extreme length and for *in vivo* shortening. *Cytogenet Cell Genet.* 90: 139–45.

Exploring the Mechanism of Aging Using Rodent Models

Yuji Ikeno and Holly Van Remmen
Department of Physiology, University of Texas Health Science Center at San Antonio, and South Texas Veterans Health Care System, Audie L. Murphy Division, San Antonio, Texas 78284-7756, USA

Using caloric restriction to probe the mechanism of aging

Life-extending effect of caloric restriction

Interventions of the aging processes by various experimental manipulations provide gerontologists with the opportunity to examine the basic mechanisms underlying aging processes. More than 60 years ago, McCay *et al.* first reported that restriction of food intake of rats soon after weaning increases the life span [1]. Since this seminal report, caloric restriction (CR) has represented the most effective and well-known intervention by which to explore aging in experimental gerontology [2, 3]. CR increases both median and maximum life spans, delays the appearance, frequency, and severity of many age-related diseases, and attenuates much of the physiological decline associated with aging in rodents [2–5].

Since McCay's original finding that CR increases longevity, gerontologists have tried to characterize and understand the potential mechanisms of aging by determining the mechanism of CR using various rat and mouse (various strains and both genders) and nonmammalian species. In aging studies, population survival curves provide essential information on maximum and median life spans as well as the age-specific mortality of the population. Although the maximum life span of a population is commonly considered the more reliable index for aging processes, the survival curve (i.e., the mortality of the population) may be a better biomarker for age-specific morbidity. CR extends both the maximum and median life spans compared to *ad libitum* fed counterparts when implemented in young adult life [6, 7]. Another index of aging that is widely used is Gompertzian analyses of mortality characteristics which exploits the relationship between mortality rate and the life span of human population. Several investigators have attempted to apply Gompertz's analysis to assess the effect of CR [8–10]. Holehan and Merry [8] showed that the mortality rate

H.D. Osiewacz (ed.), *Aging of Organisms,* 221–246.

doubling time was slowed by 40% in CR compared to *ad libitum* rats, strong evidence of the anti-aging effect of CR.

Putative underlying mechanisms

The life-extending effects of CR are associated with changes in a variety of physiological parameters, i.e., most physiological processes remain in a youthful state at advanced ages in CR rodents. Based on a variety of metabolic characteristics observed in CR animals, several factors have been implicated as underlying mechanisms of the life-extending effect of CR. Early on, McCay hypothesized that the rate of growth is a potential key modulator of aging, i.e., longevity and the growth rate are inversely related [1]. He and his colleagues' observations using various feeding paradigms were consistent with the initial hypothesis of an inverse relationship between growth rate and longevity. Later, McCay and colleagues measured the basal metabolic rate of CR animals. Since CR animals have a reduced intake of calories and a slower growth rate, it is a reasonable assumption that the CR animals may also have a reduced metabolic rate. This idea led to the "rate-of-living" theory of aging, which proposes that life span is inversely proportional to metabolic rate [11]. Interestingly, McCay *et al.* found that CR animals actually had higher cal/kg/h metabolisms than *ad libitum* fed animals. McCarter *et al.* [12] found that the metabolic rate in rats per 24 h per unit lean body mass (metabolic mass) was not decreased by CR after the initial transient decline which lasted less than 6 weeks. In addition, CR has been associated with elevated physical activity (and presumably, an elevated metabolic rate). Yu *et al.* [7] noted that Fischer 344 rats subjected to CR (40% fewer calories than *ad libitum* fed controls) maintained high levels of spontaneous activity throughout their life span in contrast to the *ad libitum* fed rats. Similarly, Holloszy and Schechtman [13] found that a decrease in calorie intake by 30% in male, Long-Evans rats elicited a much higher voluntary running activity over the life span than that observed in *ad libitum* controls. Therefore, the evidence suggests that the anti-aging effect of CR is not due to reduced metabolic rate in these animals.

Several other mechanisms have been proposed as the basis for the aging process and the action of CR on aging. These include: a retardation in growth rate [1], a decrease in energy availability for cell proliferation [3, 14], reduced DNA damage and accumulation of mutations, a reduction plasma glucose levels and non-enzymatic glycation, a decrease in GH-IGF-1 axis and, finally, a reduction in oxidative stress [15–18]. Each of the hypotheses mentioned above, as they apply to CR, are discussed individually in the sections that follow.

Alterations in growth rate and cell turnover

The retardation of growth rate is one of the most obvious responses to CR. Because the effect is so dramatic, McCay *et al.* [1] proposed that slowing the growth rate is the basis for the anti-aging action of CR. However, several lines of evidence indicate that the relationship between the retardation of growth rate and longevity may be merely correlative and not causal. Yu *et al.* [7] showed that CR initiated at 6 months of age (after the rapid growth phase) is as effective as CR begun at 6 weeks of age (prior to the rapid growth phase) in extending life span, delaying age-associated physiological decline, and retarding the progression and delaying the occurrence of age-associated diseases. Therefore, based on this consideration, the idea that the action of CR resides in the ability to retard growth rate seems unlikely. However, this issue needs to be addressed carefully because of the recent observations that the small sized animals live longer [19].

Change in cell proliferation activity and the rate of cell death could play important roles in various patho-physiological changes of aging, and many studies have demonstrated that CR can modulate these potentially important factors. Mitotic activity is reduced by CR in a number of tissue including skin, bone marrow derived stromal and epithelial cells, leukocytes, intestinal and mammary epithelium, liver and kidney [20–23]. These results suggest that a number of tissues of CR animals accumulate fewer replicative events than do those of *ad libitum* fed animals. Fewer replicative events, possibly via reduced mutational events, could play a role in the reduced incidence of cancer in the CR animals.

Elimination of damaged cells could be beneficial for the tissue to maintain the physiological functions. CR enhances apoptosis in the liver of aging mice [24], and in the small intestine and colon of aging rats [25], which may be an important component of the anti-aging effect of CR. Apoptosis of preneoplastic cells is enhanced preferentially by CR in rats. Therefore, the enhanced apoptosis by CR could protect the animals from carcinogenesis [26]. Enhanced apoptosis in liver by CR is also associated with the decreased incidence in tumors in liver [27]. Thus, changes in cell proliferation and cell death by CR could be important mechanisms of anti-aging and anti-tumor effects of CR.

Energy availability

Weindruch and Walford proposed that CR retards the aging processes by decreasing the energy available for cell proliferation, thereby shifting the physiological state from one of growth to one of maintenance and repair [3, 14]. The strength of the proposal lies in the well-known anti-tumorigenic effect of CR. Moreover, a further reduction of energy availability induced by exercise in the 40% CR group elicited additional decreases in the incidence of total neoplastic disease, compared to the sedentary CR group [28]. Several phenomena argue against this proposal. First, although exercise is associated with a reduced incidence of neoplastic disease, it has no effect on the incidence of age-related, non-neoplastic disease, and, in fact, actually increases the occurrence of fatal cardiomyopathy (Ikeno and McCarter, unpublished data). Second, exercised rats with mild CR (10%) exhibit the same incidence of neoplastic disease (although with a slightly delayed occurrence) as the sedentary *ad libitum* fed rats, even though the former treatment increases median and maximum life spans. Finally, the energy availability hypothesis, which invokes cell proliferation mechanisms, fails to explain the effects of CR on post-mitotic organs. The suppression of age-related changes by CR was not confined only to mitotic tissues; it was observed in post-mitotic tissues (brain, heart, and skeletal muscles) as well. Brain development occurs prior to the initiation of CR, thereby effectively removing any beneficial effects of CR via changes in cell proliferation. Although the brain weights of *ad libitum* fed and CR rats were not different, CR suppressed a number of age-related physiological changes in the organ [29]. Thus, although reduction of energy availability may contribute to the anti-tumor action of CR, it is unlikely that a decrease in energy availability is the essential underlying mechanism responsible for the effect of CR on the aging processes.

DNA damage, repair, and mutation frequency

It is likely that damage to DNA plays an important role in aging. In addition, DNA repair capacity appears to become less efficient with age, and the resulting accumulation of damaged DNA could have serious patho-physiological consequences. DNA repair capacity decreases with age in mouse splenocytes [30] and in rat liver and kidney cells [31], and CR retards the age-related decline in DNA repair capacity [32, 33]. However, it appears that the effect of CR on DNA repair depends on the type of DNA damage [34], possibly because the ability of CR to increase DNA repair is not universal; activities of DNA polymerases were found to be increased by CR in some, but not all, brain regions of the rat [35].

Mutation frequency also increases with age. Dollé *et al.* [36] demonstrated that mutant frequencies at the lacZ transgene increase with age in the liver, from about 3×10^{-5} in the young adults to about 10×10^{-5} in old animals (about 30 months), while such an increase was virtually absent in the brain. More recently, the pattern of organ-specificity in age-related mutation accumulation was expanded with the observation of age-related increases in mutation frequencies in spleen, heart and small intestine, but not in testes [37]. These age-related accumulation of mutations are retarded by CR [38].

Reduced plasma glucose and non-enzymatic glycation

One very interesting effect of CR is that it causes profound decreases in plasma glucose and insulin levels [39, 40]. CR lowered fasting plasma glucose levels compared to the AL counterpart in mice [41]. Masoro and colleagues examined the plasma glucose and insulin levels in AL and CR rats. These studies showed that the CR rats maintained lower plasma glucose levels through most of the day, and mean 24 h plasma glucose levels in CR rats was approximately 15 mg/dl below those of AL rats though the life span. Plasma insulin levels were also markedly lower in CR than AL rats [40]. In spite of the lower levels of plasma glucose and insulin, CR rats utilize glucose at the same rate as AL rats as fuel per gram lean body mass [39], which indicates that CR rats use glucose and insulin more efficiently. The lowering effect of CR on plasma glucose (about a 15% decrease in average diurnal values of CR rats in comparison with *ad libitum* fed rats) has been confirmed in another rat strain [42], in mice [43], and in non-human primates subjected to long-term restriction of calorie intake [44].

Cerami [45] proposed that glucose may be a mediator of aging because the non-enzymatic reactions of glucose with proteins and nucleic acids yield advanced glycosylation end products. These non-enzymatic reactions correlate with free radical generation and are associated with functional deterioration [46]. Therefore, a decrease in plasma glucose levels could lead to a decrease in advanced glycation end products and the associated age-related damage [47]. The study by Cefalu [42] showed that CR decreases the age-dependent accumulation of glycation/glycoxidation products in rat skin collagen. The non-enzymatic glycation could play an important role in aging because a synergistic interaction among free radicals, glycation, and advanced Maillard reactions could exerts deleterious effects in tissues during aging.

A decrease in GH-IGF-1 axis

CR has been shown to result in a reduction in the levels of several hormones/growth factors in plasma, e.g., GH [48], TH [49] and IGF-1 [50]. Reduced levels of growth hormone (GH) and the resulting suppression of peripheral IGF-1 levels are assumed to play key roles in anti-aging effects of CR. IGF-1 is mitogenic, and mitotic activity is reduced by CR in a number of tissues [20–23]. The reduction of plasma IGF-1 and the consequent reduced mitogenic actions of IGF-1, may be related to the delayed and/or suppressed occurrence of fatal neoplastic disease, because substantial evidence suggests that GH and IGF-1 levels play important roles in tumor development through their potent mitogenic and anti-apoptotic effects [51]. Some non-neoplastic diseases also seem to be affected by changes in GH and IGF-1 levels, such as a decrease in the severity of chronic nephropathy. GH transgenic mice, which have pathologically high levels of GH, exhibit early death that appears to be related primarily to pathological changes in their kidneys (glomerulonephritis and glomerulosclerosis) [52]. Therefore, reduced levels of GH may play an important role in the delayed progression of these lesions CR rodents. The possibility that changes in the endocrine system, especially in GH and IGF1 levels, and related subsequent physiological changes play an important role in anti-aging effect of CR has been examined in various animals models [53–55].

Oxidative stress

The "free radical theory of aging" as originally proposed by Harman [56] has thus far weathered the test of time, and is currently the only theory of aging that has maintained wide support. The basic premise of this theory is that the rate of oxygen consumption correlates with the rate of formation of reactive oxygen species (ROS). Because these radicals are highly reactive, they may cause extensive DNA, protein, and lipid damage [57], which, in turn, could lead to tissue injury, carcinogenesis [58], and aging. It is becoming increasingly clear, however, that the production of ROS represents only one facet of the cellular oxidant/anti-oxidant system. Accordingly, a shift in the balance between oxidative stress and the protective capacity of cellular antioxidant defense systems may lie at the root of the aging process and could explain the appearance of age-related diseases.

One consistent finding of CR is that it ameliorates the damage caused by oxidative stress [15, 16, 18, 59]. For example CR has been shown to reduce the level of oxidative damage in tissues as measured by a decrease in lipofuscin [60, 61], lipid peroxidation [16, 62–66], protein oxidation [67, 68], and DNA oxidation [69–72]. Even with the same metabolic rate as *ad libitum* fed rats, CR rats have a lower ROS production and higher superoxide dismutase activity within 6 weeks of CR initiation

[17, 73]. Lopez-Torres *et al.* [74] recently found that long-term CR significantly decreased H_2O_2 production of rat liver mitochondria (47% reduction) and significantly reduced oxidative damage to mtDNA (46% reduction). CR also reduces the peroxidizability of membrane lipids and maintains the integrity of membrane structure and fluidity during aging [16]. In heart mitochondria, Kim *et al.* [59] showed that CR lowers malondialdehyde production and elevates concentrations of antioxidant enzymes (e.g., superoxide dismutase, selenium-dependent glutathione peroxidase, and glutathione-S-transferase). In addition, CR modulates both the amount of lipid and fatty acid composition [75]. Thus, the anti-aging action of dietary restriction has been consistently correlated to a decrease in oxidative damage to a variety of molecules.

Pathological changes

Many functional and structural changes take place with age in all organs. These changes are generally categorized as physiological aging or pathological aging, depending on the degree of functional and/or morphological deterioration. Because all animals develop diseases associated with aging, a tight link may exist between the physiological and pathological aging processes and the pathological analyses of the disease pattern may offer insight into the mechanisms underlying the aging processes. Similarly, any intervention of age-related pathologies, such as occurs with CR, may also offer insight into the mechanisms of aging.

McCay *et al.* [1] found that CR also suppressed the occurrence of diseases. Since then, many studies have shown that CR delays and/or suppresses the occurrence of various age-related diseases in rodents, including nephropathy, cardiomyopathy, cataracts and autoimmune disease [4, 76–78]. CR also reduces and/or retards the occurrence of neoplastic lesions, which are the major age-related lesions in rodents [79]. A study showed the direct correlation between the amount of calorie intake and incidence of neoplastic lesions [80]. In studies conducted by our aging group, 40% CR suppressed and/or delayed the occurrence of total neoplastic and non-neoplastic diseases in Fisher 344 rats [4]. CR retards the occurrence/progression of chronic nephropathy. Similar beneficial effects on chronic nephropathy was observed by restriction of protein intake in the absence of calorie restriction or by using soy protein as the protein source [77]. However, the beneficial effects of these manipulations were less than the effects of CR, and elicited little effect on aging *per se*.

A recent study by our group in C57BL/6 mice (Ikeno and Nelson, unpublished data) also demonstrates that CR suppresses the incidence of several age-related histological changes. For example, we observed that the incidence of lymphoma increased with age, and the total number of histological changes increase dramatically with age in C57BL/6 mice. And these histological changes were suppressed by CR in both young and old mice as would be expected (no CR mice were examined at 18 months). These data are in agreement with similar data published by Bronson and Lipman [81], in which the authors claim the total number of histological changes could be a useful marker of the rate of aging.

Some of other interesting effects of CR on age-related lesions are: CR slows the development of lens cataracts in the cataract-prone Emory mouse strain [82]; CR also retards the age-related decrease in the proliferative capacity of mouse lens epithelial cells, although it is not still clear this action plays a role in the retardation of cataract formation by CR; and CR also decrease blood pressure in spontaneously hypertensive, stroke-prone (SHR-SP) rats [83].

Mouse models to study the role of oxidative damage in aging

Oxidative stress hypothesis of aging

Aging is a multifactorial process involving a complex interaction of a variety of metabolic processes and tissues. Because of this complexity, the mechanism behind the aging process has remained elusive. One of the most widely accepted theories of aging is the *free radical* or *oxidative stress theory of aging*. The basis of this theory is that free radicals and other reactive oxygen species (ROS), such as peroxides, cause random progressive and irreversible accumulation of oxidative damage that accumulates over time and contribute to the process of aging and various age-associated diseases [56, 84]. Although this theory is currently one of the most popular explanations for how aging occurs at the biochemical level, most of the evidence in support of this theory still remains correlative. For example, there is a considerable body of evidence in support of the oxidative stress theory of aging that comes from a series of recent studies with invertebrates. The *age-1* mutants of *C. elegans*, which were selected for increased longevity, are more resistant to hydrogen peroxide and paraquat and have increased activity of superoxide dismutase and catalase [85, 86]. Another *C. elegans* mutant that has an extended life span, *daf-2*, is resistant to oxidative stress and has increased levels of Mn superoxide dismutase [87]. Using synthetic mimetics with superoxide dismutase and catalase activity, Melov *et al.* [88] were able to increase the mean life span of wild type *C. elegans* by 44%. Perhaps the most direct evidence for the oxidative stress theory of aging has come from transgenic *Drosophila*, in which the antioxidant enzymes Cu/Zn-superoxide dismutase (CuZnSOD) or Cu/ZnSOD and catalase have been overexpressed. These studies show a correlation between increased expression of these two antioxidant enzymes and increased oxidative stress resistance in long lived strains [89–91]. Conversely, loss of catalase and/or CuZnSOD activity by mutations decreases the resistance of *Drosophila* to oxidative stress and dramatically reduces life span [92, 93]. A study by Tower's laboratory showed that overexpression of CuZnSOD alone resulted in an increase in life span of up to 48%. In addition, CuZnSOD overexpression targeted selectively to motorneurons has also been shown to result in a dramatic increase in life span (40%) and also an increase in resistance to paraquat and g-irradiation [94].

Although the studies in invertebrates strongly suggest that protection against oxidative damage is critically important in aging, it is critical that these experiments be replicated in mammals. This is not a straightforward task, however, as investigating the relationship between oxidative stress and aging in mammalian models is much more complex than in invertebrates. The feasibility of such investigations has been increased by important technologic advances that have occurred in the last few decades which allow investigators to utilize genetic engineering techniques to alter a specific gene or process in mice. These manipulations make it possible to in order to study the effect of physiologic stressors such as oxidative stress/damage on the aging process. Some of the mouse models currently available to study the effect of altered antioxidant defense on the aging process are described in the sections below.

Transgenic mice with enhanced antioxidant defense

CuZnSOD transgenic mice

The first transgenic mouse to be produced with increased expression of an antioxidant enzyme was generated using a 15-kb fragment containing the human 10-kb *Sod1* gene [95]. The overexpression of Cu/Zn-SOD in these transgenic mice various among different tissues and is greatest in the brain [95]. The mice in which Cu/Zn-SOD expression is increased are resistant several types of oxidative stress including focal cerebral ischemia [96], pulmonary oxygen toxicity [97] cold-induced injury [98], and neurotoxicity of glutamate [99] and *N*-methyl-4-phenyl-1,2,3,6-tetrahydropyridine (MPTP) [100]. However, increased activity of CuZnSOD in these transgenic mice has also been associated with a number of detrimental effects. For example, pathological changes have been reported in neuromuscular junctions in tongue and hindlimb [101, 102]; a reduction in serotonin uptake has been found in platelets [103] and a decrease in prostaglandin synthesis in the kidney and cerebellum of transgenic mice has been measured [104]. In addition, these mice have been reported to undergo premature thymic involution starting at 3–4 months of age compared to 10–12 months in control mice [105, 106].

To study the effect of increased CuZnSOD activity on aging, a life span study was conducted using CuZnSOD transgenic mice. The mice in this study had 1.5- to 5-fold higher activities of Cu/ZnSOD in various tissues compared to the non transgenic control mice [107]. A total of 198 nontransgenic mice, 200 hemizygous mice, and 98 homozygous mice were housed under barrier conditions, and no significant difference in life span was found between the nontransgenic, hemizygous and homozygous transgenic mice. Thus, although increased expression of superoxide dismutase appears to alter life span in *Drosophila*, overexpression of Cu/ZnSOD does not appear to enhance the life span of mice.

Catalase transgenic mice

There have been no reports to date on the effect of alterations in catalase expression on life span of mammals. Several mouse models that overexpress catalase have been generated; however, in contrast to the CuZn transgenic mice which were generated using a fragment of human DNA containing the endogenous *Sod1* promoter, the first catalase transgenic mouse models were generated using tissue specific promoters that express catalase in specific tissues. For example, overexpression of catalase (up to 600-fold) in heart using mouse catalase cDNA driven by a heart-specific promoter has been shown to suppress oxidative injury induced by both short term and long term doxorubicin treatment, ischemia-reperfusion, and hypoxia-reoxygenation in the heart [108, 109]. Overexpression of catalase specifically in liver (3 to 4-fold) inhibits the effect of peroxisome proliferators on DNA synthesis in the transgenic mice [110]. A 50-fold overexpression of catalase in pancreatic islets of transgenic mice produced using a human catalase cDNA under the control of an insulin promoter provided a marked protection of islets against hydrogen peroxide and significantly reduced the diabetogenic effect of streptozocin *in vivo* [111]. Using a catalase cDNA driven by a CMV enhancer element and chick-β-actin promoter, Schriner *et al.* [112] generated transgenic mice with increased expression of peroxisomally located catalase in several tissues. A second transgenic model was generated by this group using a nuclear localization sequence in the cDNA construct to target catalase expression to the nucleus (catalase is normally found compartmentalized to the peroxisomes and is not present in the nucleus). The increase in either peroxisomal or nuclear catalase had no effect on levels of oxidative damage to DNA or mutation frequency of a reporter transgene in either transgenic mouse model, suggesting that hydrogen peroxide may not play a critical role in generation of DNA oxidative damage *in vivo*.

Although an increase in catalase activity in some of the models described above resulted in protection against tissue specific acute insults, transgenic mice with ubiquitous transgene expression would be potentially much more valuable to effectively study the complex interaction of metabolic processes involved in the aging process. Using an alternative approach in which large fragments of human genomic DNA containing introns and exons of the gene of interest as well as the 5′- and 3′-flanking regions containing the regulatory sequences required for proper expression of the transgene, our laboratory has produced and characterized transgenic lines [Tg(CAT) and Tg(SOD1)] that overexpress catalase and CuZnSOD, respectively [113]. We included the introns and the flanking regions of the genes in the transgene construct to increase the likelihood that the regulatory elements controlling expression of the gene would remain intact in the transgene and transmit global increased expression of the transgene. The Tg(CAT) transgenic mice were generated using an 80 kb human genomic DNA containing the 33 kb human CAT gene as well as 41 kb of the 5′ and 6 kb of the 3′ flanking regions obtained by screening a human P1 library. The catalase mRNA levels were 2- to 5-fold higher, and catalase activity levels were 2- to 4-fold higher in the various tissues of the hemizygous Tg(CAT) mice compared to wild type mice. Tg(SOD1) mice were generated using a 64 kb human genomic DNA containing the 10 kb human SOD1 gene and the 27 kb of both 5′ and 3′ flanking regions. The mRNA levels for CuZnSOD were 2- to 12- fold higher and the CuZnSOD activity levels were 2- to 5- fold higher in the hemizygous Tg(SOD1) mice compared to wild type mice. As predicted, the expression pattern of the human CAT and SOD1 transgenes was found to be similar to that of the endogenous genes. For example, in the Tg(CAT) mice, the catalase transgene expression is the highest in liver and kidney, where the endogenous catalase activity levels are the highest, and low in brain and muscle, where the endogenous catalase levels are low. We hypothesize that these transgenic mouse models, Tg(CAT) and Tg(Sod1) will be effective models to test the effect of ubiquitous overexpression of antioxidant enzymes on oxidative stress/damage in aging and age-related pathologies.

The antioxidant defense system is complex and involves many different components; therefore, a balanced expression of antioxidant enzymes may be essential for survival. It is possible that increasing the activity of a single antioxidant enzyme could result in an imbalance in the system, e.g., increased activity of only CuZnSOD activity without a concurrent increase in either of the hydrogen peroxide-detoxifying enzymes, catalase or glutathione peroxidase could potentially result in accumulation of harmful hydrogen peroxide acts [114]. Because of this, manipulations on the expression of one antioxidant enzyme may not be sufficient to have a beneficial effect on life span. By crossing the Tg(CAT) mice with Tg(SOD1) mice described above, we hope to be able dissect the interactions and the effects of catalase and CuZnSOD on aging. Thus, aging colonies of the Tg(CAT), Tg(SOD1) and Tg(CATxSOD1) mice are currently under study by our laboratory.

Manganese superoxide dismutase transgenic mice

Manganese superoxide dismutase (MnSOD) is located in the mitochondrial matrix, where it plays a critical role in protecting the mitochondria from oxidant stress by enzymatically scavenging superoxide anions produced as a by-product of the respiratory chain. Because the mitochondria are a major site of production of reactive oxygen species, it has been suggested that oxidative damage within the mitochondria can exert a compounding effect with age accelerating oxidative damage to mitochondrial, cytosolic and nuclear proteins, lipids and DNA, and leading eventually to dysfunctional or defective mitochondria [115]. Thus, transgenic mice with an increased expression of MnSOD would be a potentially valuable model for studying the relationship between mitochondrial oxidative damage and aging.

Several lines of transgenic mice that overexpress MnSOD have been produced. The first MnSOD transgenic mouse model was produced using the human MnSOD gene under the control of the human SP-C (surfactant protein C) promoter. These mice have a 60% increase in the expression of MnSOD in the mitochondria of distal respiratory epithelial lung cells and show an increased resistance to oxidative stress when exposed to 95% oxygen [116]. Another MnSOD transgenic mouse was produced using the human MnSOD gene under the control of the human B-actin promoter. The human MnSOD transgene expressed at a relatively high level in muscle, heart, lung, and tongue, at a lower level in brain, kidney, spleen, and eye and at a very low level in the liver. Although total lung MnSOD activity is increased 180% in these transgenic mice, they did not show an increased resistance to exposure to >99% oxygen [117]. Increased expression of MnSOD in these mice showed increased protection against doxorubicin [118, 119] in heart ($\sim$2 to 3-fold) and increased resistance to cardiac ischemia reperfusion injury [120]. In brain of these transgenic mice, the increase in MnSOD results in a resistance to neuronal apoptosis and reduced injury following ischemic insult, MPTP and methamphetamine (MA) toxicity [121–123]. More recently, a transgenic mouse model has been developed using the endogenous mouse gene TgN(*Sod2*)274Cje<tg274>. The construct used to produce the mice contained a 13 kb genomic *Sod2* clone isolated from C57BL/6J mice, containing 2 kb of the native *Sod2* promoter. Overexpression is approximately 2-to 3-fold in several tissues of these mice including kidney heart and brain [124]. Interestingly high expression of MnSOD (up to 6 to 10- fold overexpression in some tissues) in another transgenic line described in this study B6<tg11>, is characterized by abnormalities in growth and fertility. Both male and femaleB6<tg11> mice are smaller than control littermates before weaning and on into adulthood. Increased MnSOD activity in the testis was associated with abnormal mitochondria in Leydig cells leading to arrested sperm development. In females, increased MnSOD activity is associated with decreased litter sizes. We are currently establishing aging colonies of the TgN(*Sod2*)274Cje<tg274> mice with 2-fold overexpression to investigate the effect of increased MnSOD activity on life span and pathology.

GPx1 transgenic mice

Transgenic mice that overexpress human GPX under the direction of the human metallothionein IIA promoter have been produced [125]. These mice exhibit an increase in the expression of GPX in several regions of the brain and show no signs of abnormal development or behavior. Transgenic mice overexpressing the glutathione peroxidase (GPx1) show significant protection against I/R damage in brain and kidney [126, 127].

Thioredoxin transgenic mice

Thioredoxin 1 (Trx1) is the major cellular protein disulfide reducing molecule and a hydrogen donor for a variety of molecules and is thus an important player in defense against oxidative stress. Trx plays an important role as a hydrogen donor for reductive enzymes such as the peroxiredoxins and methionine sulfoxide reductase, which is involved in redox regulation of proteins by maintaining methionine residues in proteins in a reduced state [128, 129]. Trx is also directly involved in the reduction of disulfides in proteins in all cells, through thiol-disulfide exchange reactions, which are rapid and readily reversible. Thioredoxin transgenic mice [Tg(*hTrx1*)$^{+/0}$] have been generated using the *hTrx1* cDNA fused to the β-actin promoter [130]. The Tg(*hTrx1*)$^{+/0}$ mice show increased resistance to oxidative stress as evidenced by reduced lipid peroxidation in liver, increased resistance to bleomycin-induced lung fibrosis, increased resistance to ischemic injury in brain and increased resistance to adriamycin mediated cardiotoxicity [130–132]. Preliminary data from our lab suggests that the levels of protein thiols are higher and the protein carbonyl levels are lower in the Tg(*hTrx1*)$^{+/0}$ mice compared to wild-type controls (Ikeno, unpublished data). A study by Dr. Yodoi's laboratory has shown that the Trx transgenic mice have an increased life span compared to the wild-type control mice, maintained under conventional housing conditions [131, 133]. Recently, Moskovitz *et al.* [134] have produced knockout mice lacking the *MsrA* gene (*MsrA*$^{-/-}$) which encodes methionine sulfoxide reductase. These mice show an enhanced sensitivity to oxidative stress and have shorter life span under both normal and hyperoxic conditions compared to the wild-type mice. This exciting result may indicate that the cellular redox systems could play important roles in oxidative stress in tissues and aging. One of the major cellular redox systems is Trx system, and one of fundamental functions of Trx is to provide electrons for reduction of methionine sulfoxide reductase. Therefore, the possible life-extension by overexpression of Trx is currently being explored in our laboratory.

Reduced expression of antioxidant genes in knockout mice

Knockout mice in which a portion of the gene is deleted and the expression of the gene reduced are another important tool that can be used to investigate the effect of a reduction in a particular gene or pathway on the aging process. Mice that are heterozygous for a mutated gene are the most physiologically relevant for aging research studies. A decrease in activity of approximately 50% found in the heterozygous knockout models is comparable to the magnitude of the decrease seen in most enzymes with age (50% or less) [135–138]. In addition, changes in gene expression that occur with caloric restriction are also generally in the range of 30 to 50% [139, 140]. Thus, the changes in gene expression that occur with age or are altered by caloric restriction are in the same range as the changes in expression found in transgenic mice heterozygous for a null mutation.

Knockout mice for most of the genes coding for the key components of antioxidant defense have been generated, e.g., *Sod1*, *Sod2*, *Sod3*, *Gpx1* and *Gpx4*. Interestingly, the homozygous, the null mutation has been shown to be lethal only for the MnSOD (*Sod2*) [141] and phospholipid hydroperoxide glutathione peroxidase (*Gpx4*) genes [142]. The $Gpx4^{-/-}$ fetuses die very early during embryonic development (less than 8 days) or in the case of the MnSOD mutants, the mice survive for only a short time after birth. Mice lacking MnSOD may also die embryonically, or they may survive 1 to 18 days after birth depending on the genetic background [141, 143, 144]. In contrast, homozygous mutations in *Sod1*, *Sod3*, and *GPx1* are not lethal, and mice with these null mutations appear phenotypically normal. Interestingly, mice null for catalase have not been generated.

MnSOD knockout mice

There are currently two knockout mouse models in which the MnSOD has been disrupted. In the first model, exon 3, which codes for the active site of the gene, has been replaced with a portion of the neomycin gene [141], while in the second MnSOD knockout mouse was produced by a disruption of exons 1 and 2 [143]. In both knockout models, the homozygous ($Sod2^{-/-}$) mutants exhibited no detectable MnSOD activity in any of the tissues studied, and the $Sod2^{-/-}$ phenotype is lethal. Mice lacking MnSOD die within 1 to 18 days from dilated cardiomyopathy or neurodegeneration depending on the genetic background [144]. The $Sod2^{-/-}$ mutants produced by Epstein's laboratory [141] are small, pale, and hypothermic state compared to wildtype mice or the heterozygous mice ($Sod2^{+/-}$). After 4 to 5 days, 37% of the $Sod2^{-/-}$ mice died with nearly all the animals dead by day 10. At death, the $Sod2^{-/-}$ mice had enlarged hearts with a dilated left ventricular cavity and reduced left ventricular wall thickness, which suggested cardiomyopathy. In contrast, the $Sod2^{-/-}$ mice produced by Lebovitz *et al.* [143] could not be distinguished from their littermate controls at birth. However, they exhibited a diminished growth rate and usually died within 18 days after birth. Although no skeletal abnormalities were observed, they did exhibit reductions in both adipose tissue and skeletal muscle. Analysis of the brain and spinal cord by electron microscopy revealed that the $Sod2^{-/-}$ mice demonstrated degenerative injury to large CNS neurons, particularly in the basal ganglia and brainstem, characterized by extensive mitochondrial damage, loss of polysomes, and clearing of the cytoplasm. In addition, progressive motor disturbances were characterized by weakness, rapid fatigue, and circling behavior. Interestingly, only 10% of these $Sod2^{-/-}$ mice died of cardiomyopathy.

In contrast, mice that are heterozygous for the MnSOD gene appear normal and exhibit normal breeding behavior. We have studied the relationship between oxidative damage and aging in *Sod2*$^{+/-}$ knockout mice, which have approximately 50% of the Mn-Superoxide dismutase (MnSOD) activity of wild-type mice in all tissues [145]. *Sod2*$^{+/-}$ mice exhibit alterations in mitochondrial function and increased mitochondrial oxidative damage as early as 2 to 4 months of age [146]. In young mice, the activities of aconitase and NADH-oxidoreductase, mitochondrial enzymes that are sensitive to inactivation by oxidative stress, are decreased from 30 to 60% in mitochondria from liver, heart or skeletal muscle while cytosolic aconitase and glutamine synthetase activities are unchanged. In liver from young *Sod2*$^{+/-}$ mice, oxidative damage to nuclear and mitochondrial DNA is increased approximately 1.7- to 1.4-fold, respectively [146]. Mitochondrial respiration is also altered in the young *Sod2*$^{+/-}$ mice as shown by a significant decrease in the respiratory control ratio for substrates metabolized by complexes I, II, and III for liver mitochondria and complex I for heart mitochondria isolated from *Sod2*$^{+/-}$ mice compared to wild type mice [147]. Overall, young *Sod2*$^{+/-}$ mice have increased levels of oxidative damage to mitochondrial proteins, greater levels of oxidative damage to both nuclear and mitochondrial DNA, alterations in mitochondrial function and increased sensitivity to oxidative stress induced apoptosis. The accumulation of oxidative damage with age has been suggested as a primary factor in the age-related decline of physiological function and in the pathogenesis of age-related diseases. Our laboratory is currently investigating whether the extent of free radical damage with age is altered in the knockout mice and whether the decrease in the activity of Mn-SOD has any effect on life span or the incidence of age-related pathology in these animals.

CuZnSOD knockout mice

Mice deficient in CuZnSOD have also been generated [148–150]. The *Sod2*$^{-/-}$ mice show no detectable Cu/ZnSOD activity while the *Sod2*$^{+/-}$ mice show approximately 50% of the Cu/ZnSOD activity found in wild-type control mice [148]. *Sod2*$^{-/-}$ mice were reported to show a normal phenotype up to 16 months of age except for a marked reduction in fertility in the female *Sod2*$^{-/-}$ mice [149]. McFadden *et al.* [151] found that the age-related loss of cochear cells, which is associated with hearing loss in humans and rodents, is accelerated in the *Sod2*$^{-/-}$ mice. More recently, motor neuron pathology has been observed in the *Sod2*$^{-/-}$ mice. For example, Shefner *et al.* [152] found that motor unit numbers in the hindlimbs of *Sod1*$^{-/-}$ mice were reduced early in life and declined slowly with age, suggesting a possible impairment in reinnervation of denervated muscle fibers in the Cu/ZnSOD mutant mice. Flood *et al.* [153] observed mild denervation of hindlimb of the *Sod2*$^{-/-}$ mice, along with behavioral and physiological motor deficits at 5-months of age.

Transgenic mice deficient in p66^{shc}

The protein p66^{shc}, appears to play a role in response of cells to oxidative stress, becoming phosphorylated when cells are exposed to stress. A study by Migliaccio *et al.* [154] reported that homozygous knockout mice null for p66^{shc} showed increased resistance to oxidative stress and an increase in life span. Cells from the mice homozygous for the null mutation in p66shc showed increased resistance to apoptosis induced by either UV irradiation or hydrogen peroxide. The $p66^{shc}$ null mice are resistant to paraquat toxicity. In a survival study, the mice lacking $p66^{shc}$ lived approximately 30% longer than the control, wild type littermates. The survival data need to be considered with some caution however, as the study was done with only a small number of mice (14 wild-type and 15 homozygous $p66^{shc}$ knockout mice) and the mean survival of the mice was relatively short, suggesting that the quality of the animal housing environment may have been suboptimal.

MsrA knockout mice

Recently, Moskovitz *et al.* [134] produced mice lacking the *MsrA* gene ($MsrA^{-/-}$), which encodes methionine sulfoxide reductase (MsrA). MsrA catalyzes the thioredoxin dependent reduction of free and protein-bound methionine sulfoxide to methionine. MsrA may also play an important role in antioxidant defense, acting as an antioxidant enzyme, scavenging ROS by facilitating conversion of protein methionine residues between oxidized and reduced forms [129]. Alternatively, MsrA may act as a repair enzyme maintaining critical methionine residues in a reduced state. The MsrA protein is highly expressed in liver, kidney, pigment epithelial cells of the retina, macrophages, cerebellum, and brain neurons. $MsrA^{-/-}$ mice show increased sensitivity to oxidative stress (100% oxygen) and have a shorter life span under both normal and hyperoxic conditions compared to wild-type mice. This result may indicate that the cellular redox systems could play important roles in oxidative stress in tissues and aging.

Mouse models to study IGF-1/insulin signaling

Substantial evidence suggests an important role for genetic control in the aging process. Most of the supportive evidence comes from studies in yeast, *Caenorhabditis elegans*, and the fruitfly, *Drosophila melanogaster* [155–158]. The Ames dwarf mice are the first mouse line to show a delayed aging process and a remarkable life span extension from a single gene mutation [159]. Soon after this exciting discovery, another dwarf mouse model (Snell dwarf mice) was observed to have a longer lifespan compared to its siblings [55]. The single mutation in these mice causes developmental arrest of anterior pituitary, resulting in mice that are deficient in three major hormones, i.e., growth hormone (GH), prolactin (PRL), and thyroid stimulating hormone (TSH) [160–162]. Interestingly these long-lived dwarf mice showed a number of similar physiological characteristics: (1) small body size; (2) reduced core body temperature; (3) delayed sexual maturation; (4) reduced plasma glucose levels and increased insulin sensitivity; and (5) reduced or absent levels of various hormones and growth factors, i.e., GH, TSH, insulin, and insulin-like growth factor-I (IGF-1) [160, 163]. These characteristics are similar to the effects of CR as described in the previous section. The extended life span of Ames and Snell dwarf mice is of great interest and has led to an extensive characterization of the mice to try to identify the underlying mechanism(s) of the anti-aging action [163, 164]. Possible links between oxidative stress, aging, and age-related pathology in Ames and Snell dwarf mice are discussed below.

Ames dwarf mice

Ames dwarf mice are homozygous for a recessive mutation at a gene named Prophet of Pit-1 (*Prop-1*), which causes developmental arrest in the pituitary gland of these mice [162]. As noted above these mice are deficient in GH, PRL, and TSH. In 1996, Brown-Borg *et al.* [159] reported that a single gene mutation delayed aging and extended the lifespan of Ames dwarf mice. The extension of life span is approximately 50% in males and over 60% in females. Ames dwarf mice are very intriguing animals for aging research not only because of their remarkable life span extension, but also because of the similar physiological characteristics that are similar to the DR mice [164]. Endocrine changes in Ames dwarf mice are found to lead to the various physiological changes [163]. At birth, Ames dwarf mice are approximately normal in size, however, their growth rate is less than that of their normal siblings. The body weight of adult Ames dwarf mice is approximately one third that of normal siblings. Deficiency of growth hormone (GH) can result in suppression of plasma IGF-1 levels, and may play an important role in their small body size. GH is also well known for its action to decrease insulin sensitivity. Therefore, deficiency of GH could also play important roles in reduced plasma glucose and insulin. Plasma thyroxin levels are extremely low due to the reduced levels of TSH. The reduced levels of thyroid hormone could presumably lower the metabolic rate and reduce the core body temperature. Prolactin (PRL) deficiency is associated with delayed sexual maturation and clearly affects the fertility of Ames dwarf mice.

Ames dwarf mice have been shown to have higher activities of antioxidant enzymes, i.e., Cu/Zn superoxide dismutase and catalase, in various organs compared to the normal siblings [165–167]. Upregulation of these antioxidant enzymes could contribute to higher resistance of Ames dwarf mice to various oxidative stress. Ames dwarf mice show significantly reduced oxidative damage to DNA and protein [165], and they are more resistant to paraquat administration [168]. Therefore, it seems that the increased antioxidant enzyme activities and the consequent protection from oxidative stress could play an important role in their life-extension and delayed occurrence of various age-related diseases.

Pathological analyses of Ames dwarf mice demonstrated that the percentage of tumor-bearing mice (the percent of mice having neoplastic lesions in the group) and the tumor burden (the number of different tumors found in a mouse) were similar in dwarf and normal control mice. However, Ames dwarf mice show a retardation in the occurrence of fatal neoplastic diseases, and the severity of some neoplastic lesions was significantly less compared with their normal siblings. The severity of glomerulonephritis, which is a major age-related non-neoplastic lesion in these mice, was also significantly less in Ames dwarf mice compared with their normal siblings. Furthermore, Ames dwarf mice showed significantly less disease burden and slower age-related accumulation of various pathological changes compared with their normal siblings [169].

Deficiency of growth hormone and the resulting suppression of peripheral IGF-1 levels are assumed to play key roles in the delayed occurrence of diseases in Ames dwarf mice. The reduction of plasma IGF-1 and the consequent reduced mitogenic actions of IGF-1, may be related to the delayed occurrence of fatal neoplastic disease, because substantial evidence suggests that GH and IGF-1 levels play important roles in tumor development through their potent mitogenic and anti-apoptotic effects [51]. Some non-neoplastic diseases also seem to be affected by changes in GH and IGF-1 levels, as was shown by the lessened severity of glomerulonephritis in Ames dwarf mice. Studies of GH transgenic mice, which have pathologically high levels of GH, showed that the early death of these animals appears to be related primarily to pathological changes in their kidneys [glomerulonephritis and glomerulosclerosis] [52]. Although the evidence suggest that the reduced levels of GH may play an important role in the delayed progression of these lesions in Ames dwarf mice, the exact underlying mechanism that acts as a mediator between hormonal changes and patho-physiological changes remain to be examined.

Snell dwarf mice

Snell dwarf mice are homozygous for a recessive mutation at the *Pit-1* gene [170]. Since the *Pit-1* gene regulates the differentiation of anterior pituitary cells, the mutation in this gene causes the deficient in GH, PRL, and TSH. These hormonal profiles are identical to that of Ames dwarf mice [164]. Snell dwarf mice also showed the extension of their lifespan (around 42%) as expected based on the data of Ames dwarf mice [55, 171]. Other physiological characteristics are also very similar to the ones of Ames dwarf mice [164].

Growth hormone receptor knockout mice

Since the Ames and Snell dwarf mice showed the deficiency of three major hormones, it is important to have animal models to examine the role of each hormone (GH, TSH, or PRL) in their anti-aging effects. In 1997, Zhou *et al.* [172] successfully generate the knock out mice for GH receptor/GH binding protein (GHR/GHBP). This animal model allowed gerontologists to dissect out the role of GH-IGF-1 axis in patho-physiology of aging. These mice showed some different physiological characteristics compare to the Ames and Snell dwarf mice as expected. Because of absence of GHR, biological actions of GH are completely blocked although these mice showed elevated plasma GH levels. Like dwarf mice, deficiency of biological actions of GH results in suppression of plasma IGF-1 levels, which could presumably retard their postnatal growth. Although these mice have small body size, the adult GHR KO mice body weight is slightly less than half of that of normal siblings. Endocrine changes in GHR KO mice showed some similarities and difference compared to the Ames and Snell dwarf mice. Plasma PRL levels are elevated [173]. In spite of the elevated plasma PRL levels, GHR KO mice showed the delayed sexual maturation and reduced their fertility [164, 172, 173]. Plasma levels of thyroid hormones (both thyroxine (T4) and triiodo-thyronnine (T3)) are slightly lower compared to normal siblings, which may play a role in slight decrease in core body temperature of these mice [164]. Lack of biological actions of GH and suppression of plasma IGF-1 levels seem to have similar effects on plasma insulin and glucose levels, and insulin sensitivity compared to the Ames and Snell dwarf mice, i.e., plasma glucose levels are normal or reduced; plasma insulin levels are reduced; and insulin sensitivity is enhanced [164]. These mice also showed a significant life extension compared to their normal siblings [174], which is a strong evidence that GH-IGF-1 axis plays important role in longevity.

Conclusion

Rodent models are an invaluable resource for studying the mechanism of mammalian aging. The observation by McCay in the 1930's that restriction of food intake resulted in a significant increase in life span in mice was a fundamental advance for aging research and has since provided many insights into the potential mechanisms of aging. Because of its anti- aging effects, caloric restriction has been one of the most active research area in experimental gerontology In recent years, the availability of transgenic and knockout mouse models has facilitated the study of potential mechanisms of aging. Several mouse models with altered antioxidant defense systems are available to test the role of oxidative stress/damage in the aging process. Dwarf mice and GH receptor knockout mice are currently providing important clues regarding the role of IGF-1 and insulin signaling in the aging process.

Acknowledgments

This work was supported by a Merit Review grant from the Department of Veteran Affairs (HVR), AICR Grant (#01A069) (YI), a grant from the American Cancer Society (HVR).

References

1. McCay CM, Crowell MF, Maynard LA (1935). The effect of retarded growth upon the length of life span and upon the ultimate body size. *J Nutr.* 10: 63–79.
2. Masoro EJ, Yu BP, Bertrand HA (1982). Action of food restriction in delaying the aging process. *Proc Natl Acad Sci USA* 79: 4239–41.
3. Weindruch R, Walford RL (1988). *The Retardation of Aging and Disease by Dietary Restriction.* Illinois: Thomas, Springfield.
4. Maeda H, Gleiser CA, Masoro EJ, Murata I, McMahan CA, Yu BP (1985). Nutritional influences on aging of Fischer 344 rats: II. Pathology. *J Gerontol.* 40: 671–88.
5. Yu BP (1994). How diet influences the aging process of the rat. *Proc Soc Exp Biol Med.* 205: 97–105.
6. Yu BP, Masoro EJ, Murata I, Bertrand HA, Lynd FT (1982). Life span study of SPF Fischer 344 male rats fed ad libitum or restricted diets: longevity, growth, lean body mass and disease. *J Gerontol.* 37: 130–41.
7. Yu BP, Masoro EJ, McMahan CA (1985). Nutritional influences on aging of Fischer 344 rats: I. Physical, metabolic, and longevity characteristics. *J Gerontol.* 40: 657–70.
8. Holehan AM, Merry BJ (1986). The experimental manipulation of ageing by diet. *Biol Rev.* 61: 329–68.
9. Sacher GA (1977). Life table modification and life prolongation. In: Finch CE, Hayflick L, eds. *Handbook of the Biology of Aging*. New York: Van Nostrand Reinhold, New York, pp. 582–638.

10. Neafsey PJ, Boxenbaum H, Ciraulo DA, Fournier DJ (1989). A Gompertz age-specific mortality-rate model of aging – modification by dietary restriction in rats. *Drug Metabol Rev.* 21: 351–65.
11. Pearl, R (1928). *The Rate of Living*. New York: Alfred Knopf
12. McCarter RJ, McGee JR (1989). Transient reduction of metabolic rate by food restriction. *Am J Physiol.* 257: E175–9.
13. Holloszy JO, Schechtman KB (1991). Interaction between exercise and food restriction: effects on longevity of male rats. *J Appl Physiol.* 70: 1529–35.
14. Walford RL, Harris SB, Weindruch R (1987). Dietary restriction and aging: historical phases, mechanisms and current directions. *J Nutr.* 117: 1650–4.
15. Kim JD, McCarter RJ, Yu BP (1996). Influence of age, exercise, and dietary restriction on oxidative stress in rats. *Aging (Milano)* 8: 123–9.
16. Laganiere S, Yu BP (1987). Anti-lipoperoxidation action of food restriction. *Biochem Biophys Res Commun.* 145: 1185–91.
17. Lee DW, Yu BP (1990). Modulation of free radicals and superoxide dismutase by age and dietary restriction. *Aging Clin Exp Res.* 2: 357–62.
18. Koizumi A, Weindruch R, Walford RL (1987). Influences of dietary restriction and age on liver enzyme activities and lipid peroxidation in mice. *J Nutr.* 117: 361–7.
19. Bartke A, Brown-Borg H, Mattison J, Kinney B, Hauck S, Wright C (2001). Prolonged longevity of hypopituitary dwarf mice. *Exp Gerontol.* 36: 21–8.
20. Pashko LL, Schwartz AG (1992). Reversal of food restriction-induced inhibition of mouse skin tumor promotion by adrenalectomy. *Carcinogenesis* 13: 1925–8.
21. Koizumi A, Tsukada M, Wada Y, Masuda H, Weindruch R (1992). Mitotic activity in mice is suppressed by energy restriction-induced torpor. *J Nutr.* 122: 1446–53.
22. Wolf NS, Penn PE, Jiang D, Fei RG, Pendergrass WR (1995). Caloric restriction: conservation of *in vivo* cellular replicative capacity accompanies life-span extension in mice. *Exp Cell Res.* 217: 317–23.
23. Lok E, Scott FW, Mongeau R, Nera EA, Malcolm S, Clayson DB (1990). Calorie restriction and cellular proliferation in various tissues of the female Swiss Webster mouse. *Cancer Lett.* 51: 67–73.
24. Muskhelishvili L, Hart RW, Turturro A, James SJ (1995). Age-related changes in the intrinsic rate of apoptosis in livers of diet-restricted and ad libitum-fed B6C3F1 mice. *Am J Pathol.* 147: 20–4.
25. Holt PR, Moss SF, Heydari AR, Richardson A (1998). Diet restriction increases apoptosis in the gut of aging rats. *J Gerontol.* 53: B168–72.
26. Grasl-Kraupp B, Bursch W, Ruttkay-Nedecky B, Wagner A, Lauer B, Schulte-Hermann R (1994). Food restriction eliminates preneoplastic cells through apoptosis and antagonizes carcinogenesis in rat liver. *Proc Natl Acad Sci USA* 91: 9995–9.
27. James SJ, Muskhelishvili L, Gaylor DW, Turturro A, Hart R (1998). Upregulation of apoptosis with dietary restriction: implications for carcinogenesis and aging. *Environ Health Perspect.* 106(Suppl. 1): 307–12.
28. McCarter RJ, Shimokawa I, Ikeno Y, Higami Y, Hubbard GB, Yu BP (1997). Physical activity as a factor in the action of dietary restriction on aging: effects in Fischer 344 rats. *Aging (Milano)* 9: 73–9.
29. Choi J-H, Yu BP (1995). Brain synaptosomal aging: free radicals and membrane fluidity. *Free Radic Biol Med.* 18: 133–9.
30. Licastro F, Weindruch R, Davis LJ, Walford RL (1988). Effect of dietary restriction upon the age-associated decline of lymphocyte DNA repair activity in mice. *Age* 11: 48–52.

31. Weraarchakul N, Strong R, Wood WG, Richardson A (1989). The effect of aging and dietary restriction on DNA repair. *Exp Cell Res.* 181: 197–204.
32. Guo ZM, Heydari AR, Richardson A (1998). Nucleotide excision repair of actively transcribed versus nontranscribed DNA in rat hepatocytes: effect of age and dietary restriction. *Exp Cell Res.* 245: 228–38.
33. Guo ZM, Heydari AR, Wu W-T, Yang H, Sabia MR, Richardson A (1998). The characterization of gene specific DNA repair by primary cultures of rat hepatocytes. *J Cell Physiol.* 176(2): 314–22.
34. Haley-Zitlin V, Richardson A (1993). Effect of dietary restriction on DNA repair and DNA damage. *Mutation Res.* 295: 237–45.
35. Prapurna DR, Rao KS (1996). Long-term effects of caloric restriction initiated at different ages on DNA polymerases in rat brain. *Mech Ageing Dev.* 92: 133–42.
36. Dolle ME, Giese H, Hopkins CL, Martus HJ, Hausdorff JM, Vijg J (1997). Rapid accumulation of genome rearrangements in liver but not in brain of old mice. *Nature Genetics* 17: 431–4.
37. Dolle ME, Snyder WK, Gossen JA, Lohman PH, Vijg J (2000). Distinct spectra of somatic mutations accumulated with age in mouse heart and small intestine. *Proc Natl Acad Sci USA* 97: 8403–8.
38. Dempsey JL, Pfeiffer M, Morley AA (1993). Effect of dietary restriction on *in vivo* somatic mutation in mice. *Mutation Res.* 291: 141–5.
39. Masoro EJ, McCarter RJ, Katz MS, McMahan CA (1992). Dietary restriction alters characteristics of glucose fuel use. *J Gerontol.* 47: B202–8.
40. Masoro EJ, Katz MS, McMahan CA (1989). Evidence for the glycation hypothesis of aging from the food-restricted rodent mode. *J Gerontol.* 44: B20–2.
41. Koizumi A, Masuda H, Wada Y, *et al.* (1989). Caloric restriction perturbs the pituitary-ovarian axis and inhibits mouse mammary tumor virus production in a high-spontaneous-mammary-tumor-incidence mouse strain (C3H/SHN). *Mech Ageing Dev.* 49: 93–104.
42. Cefalu WT, Bell-Farrow AD, Wang ZQ, *et al.* (1995). Caloric restriction decreases age-dependent accumulation of the glycoxidation products, N epsilon-(carboxymethyl)lysine and pentosidine, in rat skin collagen. *J Gerontol A.* 50: B337–41.
43. Harris SB, Gunion MW, Rosenthal MJ, Walford RL (1994). Serum glucose, glucose tolerance, corticosterone and free fatty acids during aging in energy restricted mice. *Mech Ageing Dev.* 73: 209–21.
44. Kemnitz JW, Roecker EB, Weindruch R, Elson DF, Baum ST, Bergman RN (1994). Dietary restriction increases insulin sensitivity and lowers blood glucose in rhesus monkeys. *Am J Physiol.* 266: E540–7.
45. Cerami A (1985). Hypothesis: glucose as a mediator of aging. *J Am Geriatr Soc.* 33: 626–34.
46. Mullarkey CJ, Edelstein D, Brownlee M (1990). Free radical generation by early glycation products: a mechanism for accelerated atherogenesis in diabetes. *Biochem Biophys Res Commun.* 173: 932–9.
47. Kristal BS, Yu BP (1992). An emerging hypothesis: synergistic induction of aging by free radicals and maillard reactions. *J Gerontol.* 47: B107–14.
48. Armario A, Montero JL, Jolin T (1987). Chronic food restriction and the circadian rhythms of pituitary-adrenal hormones, growth hormone and thyroid-stimulating hormone. *Ann Nutr Metab.* 31: 81–7.
49. Herlihy JT, Stacy C, Bertrand HA (1990). Long-term food restriction depresses serum thyroid hormone concentrations in the rat. *Mech Ageing Dev.* 53: 9–16.

50. Breese CR, Ingram RL, Sonntag WE (1991). Influence of age and long-term dietary restriction on plasma insulin-like growth factor-1 (IGF-1), IGF-1 gene expression, and IGF-1 binding proteins. *J Gerontol.* 46: B180–7.
51. Bustin SA, Jenkins PJ (2001). The growth hormone-insulin-like growth factor-I axis and colorectal cancer. *Trends Mol Med.* 7: 447–54.
52. Wanke R, Wolf E, Hermanns W, Folger S, Buchmuller T, Brem G (1992). The GH-transgenic mouse as an experimental model for growth research: clinical and pathological studies. *Horm Res.* 37(Suppl 3): 74–87.
53. Bartke A, Turyn D (2001). Mechanisms of prolonged longevity: mutants, knock-outs, and caloric restriction. *J Anti-Aging Med.* 4: 197–203.
54. Coschigano KT, Clemmons D, Bellush LL, Kopchick JJ (2000). Assessment of growth parameters and life span of GHR/BP gene-disrupted mice. *Endocrinology* 141: 2608–13.
55. Flurkey K, Papaconstantinou J, Miller RA, Harrison DE (2001). Lifespan extension and delayed immune and collagen aging in mutant mice with defects in growth hormone production. *Proc Natl Acad Sci USA* 98: 6736–41.
56. Harman D (1956). Aging: a theory based on free radical and radiation chemistry. *J Gerontol.* 11: 298–300.
57. Ames BN, Shigenaga MK, Hagen TM (1993). Oxidants, antioxidants, and the degenerative diseases of aging. *Proc Natl Acad Sci USA* 90: 7915–22.
58. Sun Y (1990). Free radicals, antioxidant enzymes, and carcinogenesis. *Free Rad Biol Med.* 8: 583–99.
59. Kim JD, Yu BP, McCarter RJ, Lee SY, Herlihy JT (1996). Exercise and diet modulate cardiac lipid peroxidation and antioxidant defenses. *Free Rad Biol Med.* 20: 83–8.
60. De AK, Chipalkatti S, Aiyar AS (1983). Some biochemical parameters of ageing in relation to dietary protein. *Mech Ageing Dev.* 21: 37–48.
61. Enesco HE, Kruk P (1981). Dietary restriction reduces fluorescent age pigment accumulation in mice. *Exp Gerontol.* 16: 357–61.
62. Chipalkatti S, De AK, Aiyar AS (1983). Effect of diet restriction on some biochemical parameters related to aging in mice. *J Nutr.* 113: 944–50.
63. Davis LJ, Tadolini B, Biagi PL, Walford RL, Licastro F (1993). Effect of age and extent of dietary restriction on hepatic microsomal lipid peroxidation potential in mice. *Mech Ageing Dev.* 72: 155–63.
64. Levin G, Cogan U, Mokady S (1992). Food restriction and membrane fluidity. *Mech Ageing Dev.* 62: 137–41.
65. Pieri C, Falasca M, Marcheselli F, *et al.* (1992). Food restriction in female Wistar rats: V. Lipid peroxidation and antioxidant enzymes in liver. *Arch Gerontol Geriatr.* 14: 93–9.
66. Yu BP, Lee D-W, Choi J-H (1991). Prevention of free radical damage by food restriction In: Fishbein L, ed. *Biological Effects of Dietary Restriction.* New York: Springer-Verlag, pp. 191–7.
67. Youngman LD, Park JK, Ames BN (1992). Protein oxidation associated with aging is reduced by dietary restriction of protein or calories. *Proc Natl Acad Sci USA* 89: 9112–16.
68. Sohal RS, Ku H-H, Agarwal S, Forster MJ, Lal H (1994). Oxidative damage, mitochondrial oxidant generation and antioxidant defenses during aging and in response to food restriction in the mouse. *Mech Ageing Dev.* 74: 121–33.
69. Chen LH, Snyder DL (1992). Effects of age, dietary restriction and germ-free environment on glutathione-related enzymes in Loubund-Wistar rats. *Arch Gerontol Geriatr.* 14: 17–26.

70. Youngman LD (1993). Protein restriction (PR) and caloric restriction (CR) compared: Effects on DNA damage, carcinogenesis, and oxidative damage. *Mutation Res.* 295: 165–79.
71. Sohal RS, Agarwal S, Candas M, Forster MJ, Lal H (1994). Effect of age and caloric restriction on DNA oxidative damage in different tissues of C57BL/6 mice. *Mech Ageing Dev.* 76: 215–24.
72. Djuric Z, Lewis SM, Lu MH, *et al.* (2002). Effect of varying caloric restriction levels on female rat growth and 5-hydroxymethyl-2′-deoxyuridine in DNA. *Toxicol Sci.* 66: 125–30.
73. Yu BP, Yang R (1996). Critical evaluation of the free radical theory of aging. A proposal for the oxidative stress hypothesis. *Ann NY Acad Sci.* 786: 1–11.
74. Lopez-Torres M, Gredilla R, Sanz A, Barja G (2002). Influence of aging and long-term caloric restriction on oxygen radical generation and oxidative DNA damage in rat liver mitochondria. *Free Rad Biol Med.* 32: 882–9.
75. Leon TI, Lim BO, Yu BP, Lim Y, Jeon EJ, Park DK (2001). Effect of dietary restriction on age-related increase of liver susceptibility to peroxidation in rats. *Lipid* 36: 589–93.
76. Cheney KE, Liu RK, Smith GS, Leung RE, Mickey MR, Walford RL (1980). Survival and disease patterns in C57BL/6J mice subjected to undernutrition *Exp Gerontol.* 15: 237–58.
77. Iwasaki K, Gleiser CA, Masoro EJ, McMahan CA, Seo EJ, Yu BP (1988). The influence of dietary protein source in longevity and age-related disease processes of Fischer rats. *J Gerontol.* 43: B5–12.
78. Pollard M, Luckert PH (1985). Tumorigenic effects of direct- and indirect-acting chemical carcinogens in rats on a restricted diet. *J Natl Cancer Inst.* 74: 1347–9.
79. Smith GS, Walford RL, Mickey MR (1973). Lifespan and incidence of cancer and other diseases in selected long-lived inbred mice and their F1 hybrids. *J Natl Cancer Inst.* 50: 1195–213.
80. Ross MH, Lustbader ED, Bras G (1982). Dietary practices of early life and spontaneous tumors of the rat. *Nutr Cancer* 3: 150–67.
81. Bronson RT, Lipman RD (1991). Reduction in rate of occurrence of age related lesions in dietary restricted laboratory mice. *Growth Dev Aging* 55: 169–84.
82. Taylor A, Jacques PF, Epstein EM (1995). Relations among aging, antioxidant status, and cataract. *Am J Clin Nutr.* 62: 1439S–47S.
83. Stevens H, Knollema S, De Jong G, Korf J, Luiten PG (1988). Long-term food restriction, deprenyl, and nimodipine treatment on life expectancy and blood pressure of stroke-prone rats. *Neurobiol Aging* 19: 273–6.
84. Sohal R, Weindruch R (1996). Oxidative stress, caloric restriction, and aging. *Science* 273: 59–67.
85. Vanfleteren JR (1993). Oxidative stress and ageing in *Caenorhabditis elegans*. *Biochem J.* 292: 605–8.
86. Larsen PL (1993). Aging and resistance to oxidative damage in *Caenorhabditis elegans*. *Proc Natl Acad Sci USA* 90: 8905–9.
87. Honda Y, Honda S (1999). The *daf-2* gene network for longevity regulates oxidative stress resistance and Mn-superoxide dismutase gene expression in *Caenorhabditis elegans*. *FASEB J.* 13: 1385–93.
88. Melov S, Ravenscroft J, Malik S, *et al.* (2000). Extension of life-span with superoxide dismutase/catalase mimetics. *Science* 289: 1567–9.

89. Arking R, Buck S, Berrios A, Dwyer S, Baker GT, III (1991). Elevated paraquat resistance can be used as a bioassay for longevity in a genetically based long-lived strain of *Drosophila. Dev Genet.* 12: 362–70.
90. Dudas SP, Arking R (1995). A coordinate upregulation of antioxidant gene activities is associated with the delayed onset of senescence in a long-lived strain of *Drosophila. J Gerontol.* 50: B117–27.
91. Tyler RH, Brar H, Singh M, *et al.* (1993). The effect of superoxide dismutase alleles on aging in *Drosophila. Genetica* 91: 143–9.
92. Mackay WJ, Bewley GC (1989). The genetics of catalase in *Drosophila melanogaster*: isolation and characterization of acatalasemic mutants. *Genetics* 122: 643–52.
93. Phillips JP, Campbell SD, Michaud D, Charbonneau M, Hilliker AJ (1989). Null mutation of copper/zinc superoxide dismutase in *Drosophila* confers hypersensitivity to paraquat and reduced longevity. *Proc Natl Acad Sci USA* 86: 2761–5.
94. Parkes TL, Elia AJ, Dickinson D, Hilliker AJ, Phillips JP, Boulianne G (1998). Extension of *Drosophila* lifespan by overexpression of human SOD1 in motorneurons. *Nature Genetics* 19: 171–4.
95. Epstein CJ, Avraham KB, Lovett M, *et al.* (1987). Transgenic mice with increased Cu/Zn-superoxide dismutase activity: animal model of dosage effects in Down syndrome. *Proc Natl Acad Sci USA* 84: 8044–8.
96. Kinouchi H, Epstein CJ, Mizui T, Carlson E, Chen SF, Chan PH (1991). Attenuation of focal cerebral ischemic injury in transgenic mice overexpressing CuZn superoxide dismutase. *Proc Natl Acad Sci USA* 88: 11158–62.
97. White CW, Avraham KB, Shanley PF, Groner Y (1991). Transgenic mice with expression of elevated levels of copper-zinc superoxide dismutase in the lungs are resistant to pulmonary oxygen toxicity. *J Clin Invest.* 1991;87: 2162–8.
98. Chan PH, Yang GY, Chen SF, Carlson E, Epstein CJ (1991). Cold-induced brain edema and infarction are reduced in transgenic mice overexpressing CuZn-superoxide dismutase. *Ann Neurol.* 29: 482–6.
99. Chan PH, Chu L, Chen SF, Carlson EJ, Epstein CJ (1990). Reduced neurotoxicity in transgenic mice overexpressing human copper-zinc-superoxide dismutase. *Stroke* 21: 80–2.
100. Przedborski S, Kostic V, Jackson-Lewis V, *et al.* (1992). Transgenic mice with increased Cu/Zn-superoxide dismutase activity are resistant to *N*-methyl-4-phenyl-1,2,3,6-tetrahydropyridine- induced neurotoxicity. *J Neurosci.* 12: 1658–67.
101. Avraham KB, Schickler M, Sapoznikov D, Yarom R, Groner Y (1988). Down's syndrome: abnormal neuromuscular junction in tongue of transgenic mice with elevated levels of human Cu/Zn-superoxide dismutase. *Cell* 54: 823–9.
102. Avraham KB, Sugarman H, Rotshenker S, Groner Y (1991). Down's syndrome: morphological remodelling and increased complexity in the neuromuscular junction of transgenic CuZn-superoxide dismutase mice. *J Neurocytol.* 20: 208–15.
103. Schickler M, Knobler H, Avraham KB, Elroy-Stein O, Groner Y (1989). Diminished serotonin uptake in platelets of transgenic mice with increased Cu/Zn-superoxide dismutase activity. *EMBO J.* 8: 1385–92.
104. Minc-Golomb D, Knobler H, Groner Y (1991). Gene dosage of CuZnSOD and Down's syndrome: diminished prostaglandin synthesis in human trisomy 21, transfected cells and transgenic mice. *EMBO J.* 10: 2119–24.
105. Nabarra B, Casanova M, Paris D, *et al.* (1997). Premature thymic involution, observed at the ultrastructural level, in two lineages of human-SOD-1 transgenic mice. *Mech Ageing Dev.* 96: 59–73.

106. Nabarra B, Casanova M, Paris D, *et al.* (1996). Transgenic mice overexpressing the human CuZn-SOD gene: ultrastructural studies of a premature thymic involution model of Down's Syndrome (Trisomy 21). *Lab Invest*. 74: 617–26.
107. Huang TT, Carlson EJ, Gillespie AM, Shi Y, Epstein CJ (2000). Ubiquitous overexpression of CuZn superoxide dismutase does not extend life span in mice. *J Gerontol.*55: B5–9.
108. Kang YJ, Chen Y, Epstein PN (1996). Suppression of doxorubicin cardiotoxicity by overexpression of catalase in the heart of transgenic mice. *J Biol Chem*. 271: 12610–16.
109. Kang YJ, Sun X, Chen Y, Zhou Z (2002). Inhibition of Doxorubicin chronic toxicity in catalase-overexpressing transgenic mouse hearts. *Chem Res Toxicol.* 15: 1–6.
110. Nilakantan V, Li Y, Spear BT, Glauert HP (1996). Increased liver-specific expression of catalase in transgenic mice. *Ann NY Acad Sci*. 804: 542–53.
111. Xu B, Moritz JT, Epstein PN (1999). Overexpression of catalase provides partial protection to transgenic mouse beta cells. *Free Rad Biol Med.* 27: 830–7.
112. Schriner SE, Ogburn CE, Smith AC, *et al.* (2000). Levels of DNA damage are unaltered in mice overexpressing human catalase in nuclei. *Free Rad Biol Med.* 29: 664–73.
113. Chen X, Mele J, Giese H, *et al.* (2002). A strategy for the uniquitous overexpression of human catalase and CuZn superoxide dismutase genes in transgenic mice. *Mech Ageing Dev.* 124: 219–27.
114. Prasher VP, Cheung M (1993). Down's syndrome, dementia, and superoxide dismutase. *Br J Psychiatr.* 163: 552.
115. Harman D (1972). The biologic clock: the mitochondria? *J Am Geriat Soc.* 20: 145–7.
116. Wispe JR, Warner BB, Clark JC, *et al.* (1992). Human Mn-superoxide dismutase in pulmonary epithelial cells of transgenic mice confers protection from oxygen toxicity. *J Biol Chem*. 267: 23937–41.
117. Ho YS, Vincent R, Dey MS, Slot JW, Crapo JD (1998). Transgenic models for the study of lung antioxidant defense: enhanced manganese-containing superoxide dismutase activity gives partial protection to B6C3 hybrid mice exposed to hyperoxia. *Am J Respir Cell Mol Biol.* 18: 538–47.
118. Yen HC, Oberley TD, Vichitbandha S, Ho YS, St Clair DK (1996). The protective role of manganese superoxide dismutase against adriamycin-induced acute cardiac toxicity in transgenic mice. *J Clin Invest*. 98: 1253–60.
119. Yen HC, Oberley TD, Gairola CG, Szweda LI, St Clair DK (1999). Manganese superoxide dismutase protects mitochondrial complex I against adriamycin-induced cardiomyopathy in transgenic mice. *Arch Biochem Biophys*. 362: 59–66.
120. Chen Z, Siu B, Ho YS, *et al.* (1998). Overexpression of MnSOD protects against myocardial ischemia/reperfusion injury in transgenic mice. *J Mol Cell Cardiol.* 30: 2281–9.
121. Maragos WF, Jakel R, Chesnut D, *et al.* (2000). Methamphetamine toxicity is attenuated in mice that overexpress human manganese superoxide dismutase. *Brain Res*. 878: 218–22.
122. Keller JN, Kindy MS, Holtsberg FW, *et al.* (1998). Mitochondrial manganese superoxide dismutase prevents neural apoptosis and reduces ischemic brain injury: suppression of peroxynitrite production, lipid peroxidation, and mitochondrial dysfunction. *J Neurosci.* 18: 687–97.
123. Klivenyi P, St Clair D, Wermer M, *et al.* (1998). Manganese superoxide dismutase overexpression attenuates MPTP toxicity. *Neurobiol Dis.* 5: 253–8.

124. Raineri I, Carlson EJ, Gacayan R, *et al.* (2001). Strain-dependent high-level expression of a transgene for manganese superoxide dismutase is associated with growth retardation and decreased fertility. *Free Rad Biol Med.* 31: 1018–30.
125. Mirault M, Tremblay A, Furling D, *et al.* (1994). Transgenic glutathione peroxidase mouse models for neuroprotection studies. *Ann NY Acad Sci.* 738: 104–15.
126. Ishibashi N, Weisbrot-Lefkowitz M, Reuhl K, Inouye M, Mirochnitchenko O (1999). Modulation of chemokine expression during ischemia/reperfusion in transgenic mice overproducing human glutathione peroxidases. *J Immunol.* 163: 5666–77.
127. Weisbrot-Lefkowitz M, Reuhl K, Perry B, Chan PH, Inouye M, Mirochnitchenko O (1998). Overexpression of human glutathione peroxidase protects transgenic mice against focal cerebral ischemia/reperfusion damage. *Brain Res.* 53: 333–8.
128. Levine RL, Berlett BS, Moskovitz J, Mosoni L, Stadtman ER (1999). Methionine residues may protect proteins from critical oxidative damage. *Mech Ageing Dev.* 107: 323–32.
129. Moskovitz J, Berlett BS, Poston JM, Stadtman ER (1997). The yeast peptide-methionine sulfoxide reductase functions as an antioxidant *in vivo*. *Proc Natl Acad Sci USA* 94: 9585–9.
130. Takagi Y, Mitsui A, Nishiyama A, Nozaki K, Sono H, Gon Y, Hashimoto N, Yodoi J (1999). Overexpression of thioredoxin in transgenic mice attenuates focal ischemic brain damage. *Proc Natl Acad Sci USA* 96: 4131–6.
131. Nakamura H, Mitsui A, Yodoi J (2002). Thioredoxin overexpression in transgenic mice. *Methods Enzymol.* 347: 436–40.
132. Shioji K, Kishimoto C, Nakamura H, *et al.* (2002). Overexpression of thioredoxin-1 in transgenic mice attenuates adriamycin-induced cardiotoxicity. Circulation 106: 1403–9.
133. Mitsui A, Hamuro J, Nakamura H, *et al.* (2002). Overexpression of human thioredoxin in transgenic mice controls oxidative stress and life span. *Antioxid Redox Signal.* 4: 693–6.
134. Moskovitz J, Bar-Noy S, Williams WM, Requena J, Berlett BS, Stadtman ER (2001). Methionine sulfoxide reductase (MsrA) is a regulator of antioxidant defense and lifespan in mammals. *Proc Natl Acad Sci USA* 98: 12920–5.
135. Finch CE (1972). Enzyme activities, gene function, and aging in mammals. *Exp Gerontol.* 7: 53–67.
136. Richardson A, Roberts MS, Rutherford MS (1985). Aging and gene expression. In: Rothstein M, ed. *Review of Biological Research in Aging*. New York: Alan R. Liss, pp. 395–419.
137. Richardson A, Birchenall-Sparks MC, Staecker JL (1983). Aging and transcription. In: Rothstein M, ed. *Biological Research in Aging*. New York: Alan R. Liss, Inc., pp. 275–94.
138. Richardson A, Semsei I (1987). Effect of aging on translation and transcription In: Rothstein M, ed. *Review of Biological Research in Aging*. New York: Alan R. Liss, pp. 467–83.
139. Pahlavani MA, Haley-Zitlin V, Richardson A (1994). Influence of dietary restriction on gene expression: Changes in the transcription of specific genes. In: Yu BP, ed. *Modulation of the Aging Process by Dietary Restriction*. Boca Raton, FL: CRC Press, pp. 143–56.
140. Van Remmen H, Ward W, Sabia RV, Richardson A (1995). Effect of age on gene expression and protein degradation. In: Masoro EJ, ed. *Handbook of Physiology Volume on Aging*. New York: Oxford University Press, pp. 171–234.

141. Li Y, Huang T-T, Carlson EJ, *et al.* (1995). Dilated cardiomyopathy and neonatal lethality in mutant mice lacking manganese superoxide dismutase. *Nat Genet.* 11: 376–81.
142. Yant L, Ran Q, Rao L, *et al.* (2003). The selenoprotein GPX4 is essential for mouse development and protects from radiation and oxidative damage insults. *Free Rad Biol Med.* 34: 496–502.
143. Lebovitz RM, Zhang H, Vogel H, *et al.* (1996). Neurodegeneration, myocardial injury, and perinatal death in mitochondrial superoxide dismutase-deficient mice. *Proc Natl Acad Sci USA* 93: 9782–7.
144. Huang T-T, Carlson EJ, Gillespie AM, Epstein CJ (1998). Genetic modification of the dilated cardiomyopathy and neonatal lethality phenotype of mice lacking mangnese superoxide dismutase. *Age* 21: 83–4.
145. Van Remmen H, Salvador C, Epstein CJ, Richardson A (1999). Characterization of the antioxidant status of the heterozygous manganese superoxide dismutase knockout mouse. *Arch Biochem Biophys.* 363: 91–7.
146. Williams MD, Van Remmen H, Conrad CC, Huang T-T, Epstein CJ, Richardson A (1998). Increased oxidative damage is correlated to altered mitochondrial function in heterozygous manganese superoxide dismutase knockout mice. *J Biol Chem.* 273: 28510–15.
147. Van Remmen H, Williams MD, Guo Z, *et al.* (2001). Knockout mice heterozygous for Sod2 show alterations in cardiac mitochondrial function and apoptosis. *Am J Physiol.* 281: H1422–32.
148. Huang T-T, Yasunami M, Carlson EJ, *et al.* (1997). Superoxide-mediated cytotoxicity in superoxide dismutase-deficient fetal fibroblasts. *Arch Biochem Biophys.* 344: 424–32.
149. Ho Y-S, Gargano M, Cao J, Bronson RT, Heimler I, Hutz RJ (1998). Reduced fertility in female mice lacking CuZn superoxide dismutase. *J Biol Chem.* 273: 7765–9.
150. Reaume AG, Elliott JL, Hoffman EK, *et al.* (1996). Motor neurons in Cu/Zn superoxide dismutase-deficient mice develop normally but exhibit enhanced cell death after axonal injury. *Nat Genet.* 13: 43–7.
151. McFadden SL, Ding D, Burkard RF, *et al.* (1999). Cu/Zn SOD deficiency potentiates hearing loss and cochlear pathology in aged 129,CD-1 mice. *J Comp Neurol.* 413: 101–12.
152. Shefner JM, Reaume AG, Flood DG, *et al.* (1999). Mice lacking cytosolic copper/zinc superoxide dismutase display a distinctive motor axonopathy. *Neurology* 53: 1239–46.
153. Flood DG, Reaume AG, Gruner JA, *et al.* (1999). Hindlimb motor neurons require Cu/Zn superoxide dismutase for maintenance of neuromuscular junctions. *Am J Pathol.* 155: 663–72.
154. Migliaccio E, Giorgio M, Mele S, *et al.* (1999). The p66shc adaptor protein controls oxidative stress response and life span in mammals. *Nature* 402: 309–13.
155. Tatar M, Khazaeli AA, Curtsinger JW (1997). Chaperoning extended life. *Nature* 390: 30.
156. Clancy DJ, Gems D, Harshman LG, *et al.* (2001). Extension of life-span by loss of CHICO, a *Drosophila* insulin receptor substrate protein. *Science* 292: 104–6.
157. Fabrizio P, Pozza F, Pletcher SD, Gendron CM, Longo VD (2001). Regulation of longevity and stress resistance by Sch9 in yeast. *Science* 292: 288–90.
158. Kimura KD, Tissenbaum HA, Liu Y, Ruvkun G (1997). daf-2, an insulin receptor-like gene that regulates longevity and diapause in *Caenorhabditis elegans*. *Science* 277: 942–6.

159. Brown-Borg HM, Borg KE, Meliska CJ, Bartke A (1996). Dwarf mice and the ageing process. *Nature* 384: 33.
160. Bartke, A (1964). Histology of the anterior hypophysis, thyroid and gonads of two types of dwarf mice. *Anat Rec.* 149: 225–35.
161. Bartke A, Brown-Borg H, Kinney B, *et al.* (2000). Growth hormone and aging. *J Am Aging Assoc.* 23: 219–25.
162. Sornson MW, Wu W, Dasen JS, *et al.* (1996). Pituitary lineage determination by the Prophet of Pit-1 homeodomain factor defective in Ames dwarfism. , 384: 327–33.
163. Bartke A (2000). Delayed aging in Ames dwarf mice: relationships to endocrine function and body size. *Results Probl Cell Differ.* 29: 181–202.
164. Bartke A, Coshigano K, Kopchick J, *et al.* (2001). Genes that prolong life: Relationships of growth hormone and growth to aging and life span. *J Gerontol.* 56: B340–9.
165. Brown-Borg HM, Bode AM, Bartke A (1999). Antioxidative mechanisms and plasma growth hormone levels: potential relationship in the aging process. *Endocrinology* 11: 41–8.
166. Brown-Borg HM, Rakoczy SG (2000). Catalase expression in delayed and premature aging mouse models. *Exp Gerontol.* 35: 199–212.
167. Hauck SJ, Bartke A (2000). Effects of growth hormone on hypothalamic catalase and Cu/Zn superoxide dismutase. *Free Rad Biol Med.* 28: 970–8.
168. Hauck SJ, Aaron JM, Wright C, Kopchick JJ, Bartke A (2002). Antioxidant enzymes, free-radical damage, and response to paraquat in liver and kidney of long-living growth hormone receptor/binding protein gene-disrupted mice. *Horm Metab Res.* 34: 481–6.
169. Ikeno Y, Bronson RT, Hubbard GB, Lee S, Bartke A (2003). The delayed occurrence of fatal neoplastic diseases in Ames Dwarf mice: correlation to the extended longevity. *J Gerontol.* in press
170. Li S, Crenshaw EB III, Rawson EJ, *et al.* (1990). Dwarf locus mutants lacking three pituitary cell types result from mutations in the POU-domain gene pit-1. *Nature* 347: 528–33.
171. Miller RA (1999). Kleemeier award lecture: are there genes for aging? *J Gerontol.* 54: B297–307.
172. Zhou Y, Xu BC, Maheshwari HG, *et al.* (1997). A mammalian model for Laron syndrome produced by targeted disruption of the mouse growth hormone receptor/binding protein gene (the Laron mouse). *Proc Natl Acad Sci USA* 94: 13215–20.
173. Chandrashekar V, Bartke A, Awoniyi CA, *et al.* (2001). Testicular endocrine function in GH receptor gene disrupted mice. *Endocrinology* 142: 3443–50.
174. Coschigano KT, Clemmons D, Bellush LL, Kopchick JJ (2000). Assessment of growth parameters and life span of GHR/BP gene-disrupted mice. *Endocrinology* 141: 2608–13.

Human Aging and Longevity: Genetic Aspects

Holger Hoehn and Armin Renner
Human Genetics, University of Wurzburg, Wurzburg 97074, Germany

Introduction

In the light of evolutionary theory, human aging can be viewed as "a late-onset genetic disease that affects us all," and, like in lower organisms, there surely are both public and private mechanisms that contribute to human aging [1]. Because of the surprising similarity between major genes and metabolic pathways that control longevity in yeast, worms, drosophila, and even mice, public mechanisms of aging currently attract most of the attention of comparative aging research. The apparent degree of evolutionary conservation of key metabolic pathways related to longevity and aging is in fact impressive [2]. However, as will be pointed out with respect to some of the somatic maintenance systems that are fairly recent additions to the vertebrate genome, there undoubtedly are a number of mechanisms of aging and longevity that are private to vertebrates and mammals. Senescent human phenotypes have been properly described as "non-adaptive, non-determinative, subject to stochastic events as well as highly polygenic modulations" such that there is little justification for a simplificationist view of human aging [3]. Even though we have learned a great deal about public mechanisms of aging in lower organisms that may be partly conserved in mammals, a purely comparative and reductonist approach must surely be abandoned in view of the complexity of the extremely long-lived human organism. No other species has such a highly developed central nervous system which controls, via the hypothalamic-pituitary axis, much of our neuroendocrine homeostasis. Because of the remarkable longevity and medicalization of our species, at least in the industrialized countries, the natural history and pathogenesis of aging and longevity has been extensively documented in humans, and there is a comprehensive clinical record of age-related changes of the human phenotype [4–6]. What has not been completely resolved with respect to human aging is the question whether aging occurs, like development, via an ordered and genetically regulated, i.e., programmed process, or whether it results from (random) wear-and-tear type

H.D. Osiewacz (ed.), Aging of Organisms, 247–269.

accumulation of damage to DNA, proteins, cells and higher order structures of the human body. Clearly, a programmed process would be much more difficult to modify or reverse than a wear-and-tear type process whose course might be slowed down or even halted by preventive interventions. Most likely, the truth lies somewhere between these two alternatives, and human aging encompasses features that are both programmed and stochastic [7].

Aging and longevity in humans

A current definition describes aging as loss of homeostasis that affects all metabolic, neuroendocrine, immunologic and genomic maintenance systems of our body [8]. Typical examples are the increased mortality of old people on exceptionally hot summer days (loss of thermoregulation) or during viral and bacterial epidemics (loss of immunological homeostasis). With increasing age, humans clearly loose their adaptive capabilities by which extreme changes in environmental conditions are overcome in younger years. From a genetic point of view, the gradual loss of stability of our somatic genome is one of the most impressive examples of the age-related decline in homeostasis. As has been noted as early as 1974 by Hart and Setlow [9], there is a highly significant correlation between the maximum lifespan of a species and the capability to preserve and/or restore the integrity of its DNA. The same correlation exists for the acitivity of the poly (ADP-Ribose) polymerase (PARP) enzymatic activity as a typical longevity assurance system in mammals [10]. At present, there is little doubt that the maximum longevity of a given species is determined by its genome, and that the exceptional longevity or our species reflects the relatively high degree of stability of the human genome. Humans live much longer than mice. Compared to its human counterpart, the murine genome contains a higher proportion of retroviral sequences and mobile elements that render it less stable. The opposite holds for birds: they have a relatively small, compact and extremely stable genome, and their lifespans exceed by far those of comparable sized mammals [11].

Maximum human lifespan

Reports of 130- to 160-year-old people in the Andes or in Eastern Europe have never held up to scientific scrutiny [12]. In 1996, the male longevity record was held by a resident of California (CM) with 114 years [13]. Among women, the longevity record is currently held by Jeanne Calment who died in 1997 at the alleged age of 122 years [14]. However, three independent demographic studies raise considerable doubt about this generally accepted world-record.

(1) The paper by Wilmoth and collegues describes the increase of the maximum lifespan of Swedish women from 104 to 112 years during the past two centuries [15]. Yet, until the year 2000 there was no person older than 112 on credible record in Sweden. The authors point out that there is a remarkable acceleration of the increase in maximum lifespans for both sexes since 1970 which cannot be explained by a birth cohort effect but rather reflects decreased mortality of

people older than 80. Improved living conditions and medical interventions account for the fact that the oldest-old are the fasted growing population in wealthy countries. Wilmoth and coworkers therefore do not exclude a further rise in maximum lifespan beyond the current limit of 112 years.

(2) On the basis of extrapolations from mortality rates of 35 to 75 year olds in 27 industrialized nations Kesteloot concludes that the theoretical maximal life-span averages 107.6 ± 2.1 years for women and 104.0 ± 1.9 years for men [16]. There were only two countries each in which the extrapolated maximum age at death reached 112 years for females and 108 years for males.

(3) The Finnish demographer Kannisto published data on the mortality rates of close to 53 000 centenarian women collected from the records of 13 industrialized nations [17]. As illustrated in Figure 1, his data show a very clear exponential decline of survival until age 108. Thereafter, mortality rates appear to increase such that there were only 3 women at age 111 and a single female reaching age 112. This data appears at variance with a more recent study reporting a decrease rather than increase in the mortality of supercentenarian women [18]. However, using the data collected by Kannisto, the statistician Zelterman came to the conclusion that it would take at least 200 million women over age 100 to find (with acceptable statistical accuracy) a single women reaching age 115 [19].

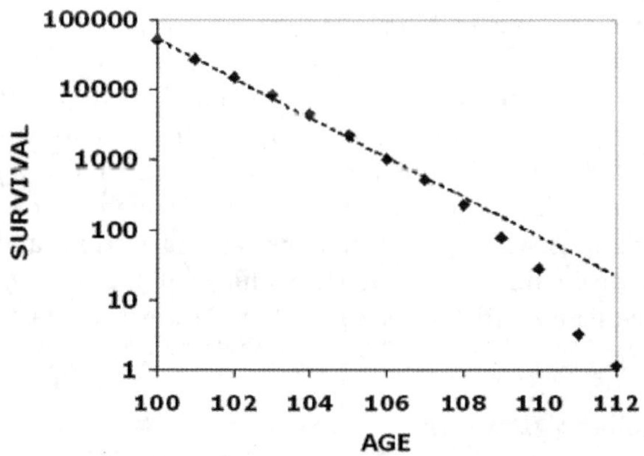

Figure 1. Mortality of 52 947 centenarian females [17].

Contrary to some rather optimistic projections concerning the possibility of extension of the human lifespan [20] these three studies suggest that the maximum human lifespan has a current limit around 112 to 115 years, and that exceptions like Jeanne Calment, if they exist at all, must be exceedingly rare. Since the maximum lifespan of a given species, including our own, appears to be genetically determined, there are no scientifically sound prospects for substantial alterations of this current biological limit.

Average human lifespan
In contrast to the genetically determined maximum lifespan, the average lifespan of human populations appears to be determined mostly by environmental factors. During the course of the 20th century, industrialized nations experienced a dramatic increase in their average lifespans [21]. Improved hygiene and nutrition, healthcare, working conditions, education, and improved socio-economic conditions together with medical progress in terms of diagnosis, prevention and therapy account for most of this gain. Extrapolations of its near linear increase since 1840 lead to the prediction that the average life expectancy for women in selected wealthy nations with small and rather homogeneous populations will increase from 85 years today to 90 years in 2020 and to 95 years in 2050 [22]. Since the number of old people in a given population is a clear function of the prevailing economic conditions, the average life expectancy will be much less in underprivileged nations. Moreover, biological reasons predict a definitive plateau of the average life expectancy around age 95 which is caused by the natural heterogeneity of unselected human populations [23]. With gene frequencies ranging from 0.1 to 1% and more for age-related diseases (e.g., familial cancers, dementias, diabetes, and cardiovascular diseases), the total burden of genetic susceptibility to late-onset diseases precludes homogeneous and promotes heterogeneous aging in human populations [1]. In addition, even if all detrimental environmental factors could be eliminated, it is unlikely that each and every member of a given population could be convinced to adapt a (frugal) lifestyle that is optimal for longevity. Looking at other organisms it seems worth mentioning that there are a number of striking examples which underline the *primacy of the environment* over genes with respect to the average lifespan. A case in point are studies in bee populations which show that the type of feeding and progeny care determine whether a given animal ends up with a lifespan of 2 or 10 months, or, as a pampered queen, lives more than 5 years. All these animals start with identical genomes [24, 25].

Why do women live longer than men?
In most industrialized nations, the average life expectancy of females exceeds that of males by 5 to 8 years [26]. Does nature or nurture account for this difference? From a genetic point of view, the mosaic status of females regarding the activity of their X-chromosomes provides an attractive explanation [27]. Mutations of X-chromosomal genes will affect 100% of cells in male individuals in which the respective gene is expressed. Due to random inactivation of either the maternal or the paternal X-chromosome in female cells, a corresponding loss of gene function will only affect

50% of the cells in females. This is the reason why females are clinically unaffected carriers for X-linked recessive diseases like hemophilia or Duchenne muscular dystrophy, whereas half of the male offspring of such carrier females are affected. In the context of aging it is relevant that the human X-chromosome appears to be enriched in genes that control various functions of our central nervous system. Mutations in these genes cause sensory deficits and/or mental retardation, but only males are affected [28]. This is the reason why there is a 30% excess of males in institutions for the mentally retarded [29]. Since maintenance of neuroendocrine homeostasis and of cognitive abilities appears to play a key role in (healthy) human aging [30], females are at a clear genetic advantage with respect to constitutional and somatic mutations of the many X-chromosomal "cognition" genes. Additional biological factors favoring a higher life-expectancy of females are hormonal differences (protective effects of estrogens) and lower mortalities due to cardiovascular and neoplastic diseases. The opposite is true for testosteron-driven males, whose mortality at age 20–25 is threefold higher than that of females [12]. Compared to females, a greater proportion of males displays high risk lifestyles (e.g., smoking, alcohol consumption, aggressive behavior) which enhance rather than delay mortality [31]. A convincing case for the dominance of biological over environmental factors as explanation for the gender difference with respect to average longevity derives from a comparison between Japan and Sweden. Inspite of major differences in the lifestyles, women outlive males in both countries by 6 to 8 years [32]. A small consolation for males is that women experience longer periods of disability and chronic conditions during aging that are in part due to both biological (muscle strength, bone density) and lifestyle factors (sedentary behavior, obesity) [33]. It should also be pointed out that there is some evidence for narrowing of the gender gap in some of the (wealthiest) industrialized countries. This trend appears to reflect significant reductions of mortality in males aged 25 to 59 through adaptation of healthier lifestyles and medical progress [34].

Heritability of life expectancy

Even though it is difficult to separate nature vs. nurture effects for highly complex and multifactorial phenotypes like human aging and longevity [35], there is the common experience that individuals with long-lived parents tend to live longer and age more successfully than individuals with short-lived parents [36, 37]. Contrary to such empirical and intuitive correlations, the analysis of factors which underly the variation in adult human lifespan yields only 25% clearly genetic factors, while 50% are due to changing factors (which include both genetic and environmental components) and the remaining 25% are due to fixed non-genetic factors [21]. Even more impressive, twin studies show that the heritability of susceptibility to death amounts to only 0.5, which means that at least half of the variance in longevity must be due to non-genetic factors [38].

Age-related disease and antagonistic pleiotropy

In the industrialized countries, currently 1 in 10 000 newborns has a life expectancy of 100 years. This means that the great majority of people will not die as centenarians but succumb to age-related frailty and morbidity at considerably lower ages [39]. The record of molecular medicine shows that many of the common age-related diseases (diabetes, cardiovascular, neoplasia, and dementia) have a strong genetic component. Why are genes that increase our susceptibility to old-age diseases so frequent in the human population? This question has been answered by evolutionary biology which posits that there is no selection against mutations that cause late-onset diseases [40]. Selection can only act on reproduction, and late-life diseases usually manifest themselves after reduction (in males) or complete cessation (in females) of reproduction. But there is a second important point: many of the genes that contribute to late-life morbidity follow the principle of antagonistic pleiotropy. This means that a given gene may have early beneficial and late detrimental effects [41, 42]. The principle of antagonistic pleiotropy of gene action can be illustrated by defects in the LDL-receptor genes which cause elevated cholesterol levels in approximately one in 300 humans. During childhood, moderate hypercholesterolemia is beneficial rather than harmful since a good supply of cholesterol is required for building cell membranes (lipid bilayer) and nerve axons. Children nutritionally or genetically deprived of cholesterol suffer various developmental defects [43]. In addition, cholesterol is the progenitor of steroid hormones, including estrogens and androgens, such that higher cholesterol levels may increase both physical and reproductive fitness. Longterm persistence of hypercholesterolemia, however, sharply increases the risk of cardiovascular disease, premature morbidity and death. Analogous arguments can be made for traits such as hypertension or adult onset diabetes. Compared to hypotension, high blood pressure increases physical fitness during childhood and adolescence, but persistence into adulthood increases the risk for cardiovascular disease and stroke. It is assumed that the evolutionary purpose of antagonistic pleiotropic gene action is to eliminate older individuals as food competitors for their (numerous) progeny [41].

The principle of antagonistic pleiotropy has been be invoked to explain the prevalence of a number of late onset diseases in humans, including Alzheimer's disease, hemochromatosis, osteoporosis, and defects of the immune and endocrine systems [42, 44–48]. A particulary instructive example with regard to the high prevalence of late onset Alzheimer's disease is the observation of gene variants in the FE65 adaptor protein which affect the tightness of binding to the cytoplasmic domain of the amyloid precursor protein. These genetic variants act antagonistically with respect to the risk of very late onset Alzheimer's disease, and selection seems to have favored the "bad" variant [48]. Due to its tight binding, the bad variant might be beneficial for the protective tasks of the amyloid precursor protein in younger years thus fulfilling the criteria for antagonistic pleiotropic gene action. Together with the mechanism of alternative splicing, the very existence of genes which follow the principle of antagonistic pleiotropy might explain the surprisingly small overall number of genes in our genome, since many of our genes seem to exert multiple functions. This also implies that we will never be able to suppress or remove gene variants from our genome that cause or contribute to late life diseases because we need these very genes during our younger and reproductive years.

Human gene variants and aging: lessons from centenarians

What is special about the oldest-old? They frequently have long-lived ancestry, good environmental conditions and a moderate lifestyle. Centenarians were thought to lack genetic variants that cause or contribute to late-life diseases [49]. The latter certainly holds for a considerable number of highly penetrant dominant mutations that underly various forms of familial cancers or familial Alzheimer and Parkinson disease. Centenarians therefore were thought to represent a highly selected population with a very rare and special genetic constitution. This idea was seemingly supported by the fact that certain variants of the apolipoprotein E gene are either over- or underrepresented in very old people [50]. There are significantly fewer carriers of the ApoE4 allele among centenarians, and ApoE4 carriers have an increased risk for a number of age-related impairments and diseases (cf. Table 1).

Table 1*. Increased risk of ApoE4 carriers for age-related impairments and diseases*

Risk category	Impairment/disease
Cardiovascular disease susceptibility	Lipoprotein metabolism; stroke
"A-beta" suceptibility phenotypes (deposition of aggregates of the gamma-secretase products of the amyloid precursor protein gene)	Senile dementia of Alzheimer type Cerebral amyloid angiopathy Lewy body dementia Down syndrome Brain trauma Cognitive decline
Other susceptibility phenotypes	Osteoporosis-correlated bone fractures Cholelithiasis (gallstones) in females Retinitis pigmentosa (ApoE4 homozygosity)

However, the search for additional susceptibility gene variants that might be underrepresented among centenarians yielded a surprising result which is refered to as "centenarian paradox." With the exception of the ApoE4 allele, centenarians were found to display the same frequencies of susceptibility gene variants as younger people [46, 51–55]. This held not only for nuclear gene variants predisposing to cardiovascular disease, stroke, thrombophilia and cancer, but also for disease-causing variants of the mitochondrial genome [56]. Figure 2 depicts an example of the centenarian paradox with respect to carrier status for the C282Y mutation in the hemochromatosis (HFE) gene. If mutation carriers have an increased risk for cardiovascular disease, one would expect fewer carriers among centenarians relative to younger age-groups. As Figure 2 shows, this is not the case.

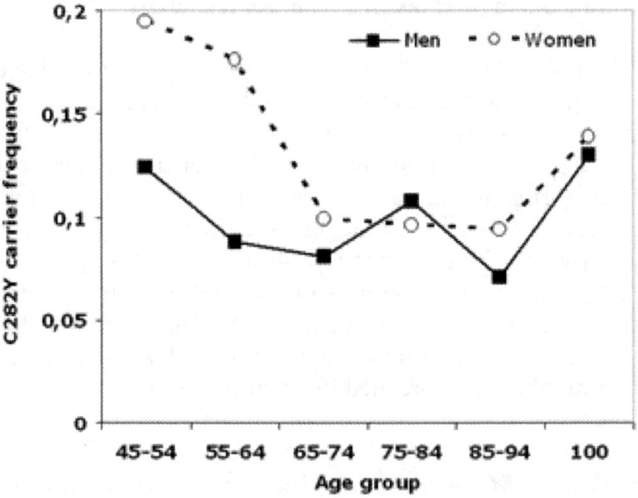

Figure 2. *Carrier frequencies for the HFE C282Y mutation as a function of age [127].*

A number of hypotheses have been invoked to explain the centenarian paradox, including effects due to genetic background, metabolic modelling, compensatory adaptation, and stochastic events [55–57]. One of the most plausible explanations comes from observations in yeast mutants which compensate deficits of their mitochondrial genome by increasing the activity of certain nuclear genes [55]. This so called "retrograde response" paradoxically leads to the prolongation of the lifespan of these yeast mutants. It is conceivable that similar compensatory mechanisms exist in humans and might be one of the explanations for the centenarian paradox.

Aging as loss of genetic homeostasis

Lessons from caretaker-genes

As shown in Table 2, there are a number of human genes whose primary function is in the maintenance of genomic stability. These so called caretaker-genes [58] are instrumental in the recognition and repair of DNA damage [59]. Recognition and elimination of different types of DNA damage appears to be highly specialized and dependent on the types of DNA lesions. For example, genes of the Xeroderma pigmentosum group safeguard us against UV-induced lesions, genes of the Ataxia telangiectasia group (which include the nibrin, Mre11 and Rad50 genes) protect us against the adverse effects of ionising radiation, and the Fanconi anemia group of genes appears to defend us against damage by alkylating agents and reactive oxygen

Table 2. *Examples of human caretaker genes [128]*

Clinical classification	Genes	Protective function	Molecular function	Neoplasia
Skin cancer group				
Xeroderma pigmentosum	XPA-XPG	UV light	Nucleotide excision repair (NER)	Skin carcinoma
Progeria group without tumors				
Cockayne syndrome	XPD, XPG	UV light	Helicase/replication-coupled repair	–
Trichothiodystrophy	XPB	UV light		–
Progeria group with tumors				
Werner syndrome	WRN	4-NQQ	Helicase	Sarcoma, etc.
Bloom syndrome	BLM	UV light	Helicase	All tumors
Rothmund-Thompson syndrome	RTS	BrdUrd, UV	Helicase	Sarcoma
Lymphoma group				
Ataxia telangiectasia	ATM	Ionizing radiation (IR)	DSB-recognition and repair (NHEJ)	Lymphoma
Nijmegen Breakage syndrome	Nibrin			Lymphoma
AT-like syndromes	Mre11, Rad50 Ligase IV, etc	Ionizing radiation (IR)	DSB-recognition and repair (NHEJ)	Lymphoma
Fanconi anemia group	FANC-A, C, D2, E, F, G, D1 (= BRCA2)	Alkylating agents and reactive oxygen species	DSB-recognition and recombinational repair (HR)	AML, squamous cell carcinomas
Breast cancer group	BRCA1, BRCA2	Reactive oxygen species (ROS)	Recombinational repair	Breast cancer Ovarian cancer
Colon carcinoma group	MSH2, MSH6, MLH1, PMS1,2	Reactive oxygen species (ROS)	Mismatch-repair	Non-polyposis Colon cancer
Li-Fraumeni group	p53	IR, chemicals (?)	Cell cycle arrest	All tumors

species [60]. Some of these caretaker genes are clearly involved in DNA repair, like the Xeroderma genes in nucleotide excision repair and the BRCA1 and 2 genes in homologous recombination repair and gene conversion [59]. The formal genetics of caretaker gene syndromes is autosomal recessive, implying that heterozygous carriers are unaffected. However, it is conceivable that heterozygous carriers are more sensitive than non-carriers under certain stress situations (including exposure towards X-rays and chemicals). Since heterozygous carriers are much more frequent in our population than homozygous affected individuals (Hardy-Weinberg rule), they may account for a large proportion of age-related morbidity and cancer.

Werner syndrome: link between genetic instability, aging, and cancer
The Werner syndrome (WS) is the most prominent example of the progeroid effects of mutations in caretaker genes [61]. WS patients experience a normal development until puberty, but there is no adolescent growth spurt. In their twenties, patients begin to show greying and loss of hair, and they develop cataracts as early manifestations of the syndrome. During adulthood, the phenotype is characterized by laryngal atrophy (altering the voice) and general atrophy of subcutaneous tissue, most prominently in the distal third of the lower extremities, with scleroderma-like skin changes and frequent malleolar ulcerations. Internally, patients suffer from premature atherosclerosis, diabetes mellitus and a strongly elevated risk of neoplasia. Typical malignancies are connective tissue sarcomas and medullary thyroid carcinomas, but other types of cancers do occur as well. Even though many of the phenotypic manifestations are reminiscent of premature aging, some organs such as the brain may be relatively spared which caused Epstein and coworkers to refer to WS as a caricature of aging [62]. The gene responsible for WS is a human homologue of the highly conserved RecQ-types of helicases. The WRN-gene has both helicase and exonuclease functions and appears to play an important role in the maintenance of genomic stability, chiefly during DNA-replication [63, 64]. Typical cytogenetic manifestation of defects in the WRN gene are multiple and clonal reciprocal chromosome exchanges ("variegated translocation mosaicism") which reflect a high rate of faulty rejoining of DNA-breaks during replication as shown in Figure 3.

Mutations in other members of the human helicase family of genes cause the Bloom and Rothmund-Thompson syndromes (see Table 2). Common to the helicase-type caretaker syndromes are progeroid phenotypic changes, genomic instability and a sharply increased risk of neoplasia. These (fortunately rare) syndromes teach us that a stable genome is a prerequisite for "normal" human aging and longevity.

Neoplasia as manifestation of genetic instability

Second only to cardiovascular disease, neoplasia is one of the most prevalent causes of morbidity and mortality in the industrialized countries. Both environmental and genetic factors play a role in the genesis of malignant disease, but the aging process *per se* with its inherent loss of genetic stability seems to be the most important trigger. This point can be nicely illustrated by plotting cancer mortality as a function of age. For Germany, such data have been provided by Lock and Funk and cover almost the

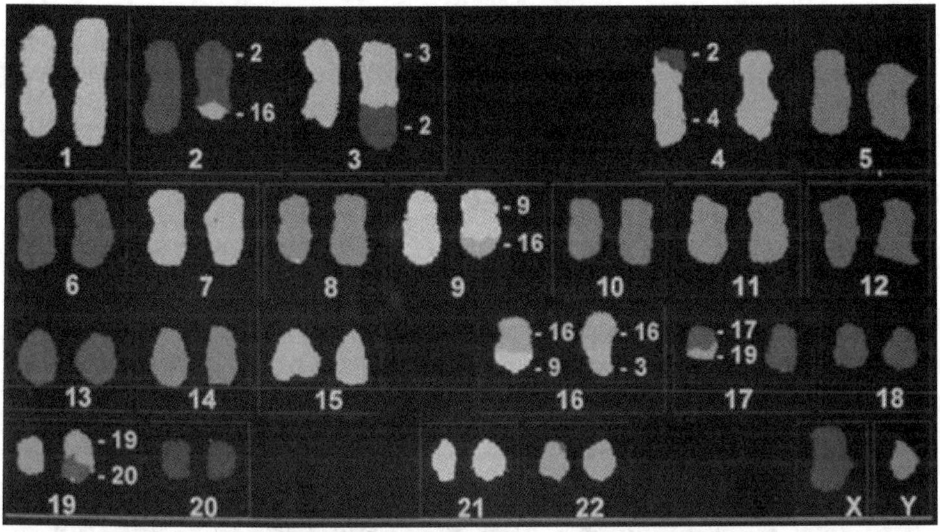

Figure 3. *Werner karotype showing multiple rearrangements [129].*

entire 20th century [65]. Their data show very clearly that the aging process per se is the most constant risk factor for death from neoplastic disease:

Surprisingly, on the basis of this data, a woman aged 60 or 65 at the end of the 20th century has the same risk to die from cancer than her counterpart at the beginning of the 20th century. This runs counter to any intuition, since there are thousands of new and potentially carcinogenic chemicals released in our environment and nutrition each year, let alone the increase in air and electronic pollution. The data of Lock and Funk clearly suggest that life in modern inustrialized societies does not seem to increase the risk for death from cancer. Rather, the increased prevalence of cancer deaths seems to reflect the dramatic increase of older people in our societies [66].

Aging and genetic instability: a fatal connection

Thermoinstability as intrinsic risk factor

In addition to various exogenous factors (ionising radiation, clastogenic chemicals, virus infections) two major endogenous factors account for the age-related increase of genetic instability in our somatic cells. One of these intrinsic factors is the thermoinstability of our DNA at 37°C body temperature. As shown in Table 3, at 37°C our somatic cells suffer a variety of potentially harmful DNA alterations within a 24 h period. These alterations include depurinations, abasic sites, strand breaks, adduct formation, and others as shown in Table 3.

Cytogenetic equivalents of these molecular alterations are increased chromosomal breakage rates in lymphocytes from old donors and accumulations of cells in the G2

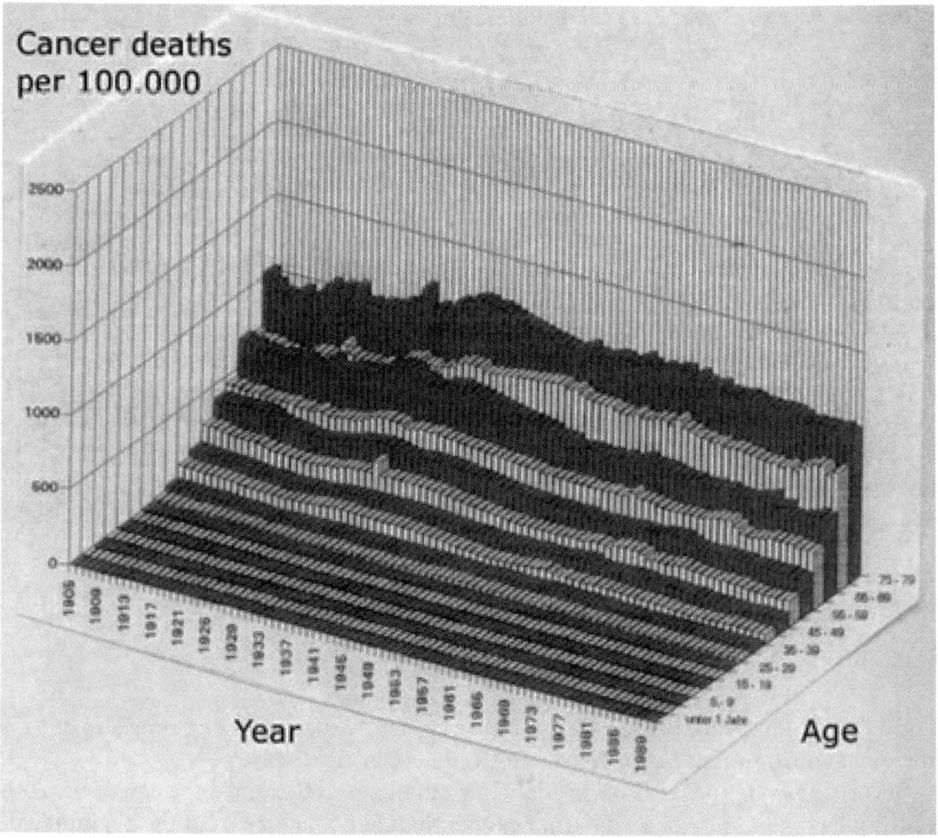

Figure 4. Age-dependent cancer mortality of women in Germany1906–1989 [65].

phase of the cell cycle [67]. Other examples of age-dependent genomic deterioration are increased non-disjunction frequencies during female gametogenesis and age-dependent mutation rates in human spermatocytes [68]. As mentioned above, our genome contains a considerable number of caretaker genes that counteract and repair such age-related genomic alterations. In his "disposable soma theory of aging," Tom Kirkwood has convincingly argued that damage recognition and damage repair cannot be 100% effective under the energetic conditions of our biological existence [69]. Deficient repair leads to genetic alterations that cause impairment of gene function and, ultimately, cell death or neoplasia. In this context it is attractive to view cell senescence as a mechanism that protects long-lived organisms from cancer [70].

***Table 3**. Types and frequencies of DNA lesions in human somatic cells at 37°C body temperature [130]*

Type of damage	Number of lesions per cell per day
Apurinic sites (AP)	12 000
Pyrimidine-loss	600
Cytosine desamination	200
Single strand breaks (SSB)	55 000
Double strand breaks (DSB)	9
Crosslinks	8
Pyrimidine dimers	?
Other (e.g., adducts)	?

The only cells of our body that need to repair any DNA damage with the highest possible effectiveness and fidelity are our germ cells. Under this premise, meiotic recombination can be viewed as a powerful mechanism of our germ cell lineage to remove and repair double strand breaks, since the intact homologous chromosome is available as template during meiotic chromosome pairing [71]. The locations of meiotic crossovers might therefore reflect sites of pre-existing DNA-damage. Under this scenario, the exchange of grandparental alleles that assures human individuality could be viewed as a welcome side-effect of the repair effort. Owing to meiotic recombination, each mature germ cell enters fertilization with a new and clean slate, assuring the immortality of our germline. In contrast, unrepaired or misrepaired damage gradually accumulates in, and thereby contributes to, the mortality of our somatic cells. Even though we know that recombinational repair of DSB occurs in somatic cells, presumably through the BRCA1-and BRCA2/Rad51-related pathways [72], pairing of homologous chromosomes as a prerequesite for error-free repair of DSB remains an exception in somatic cells of mammals.

Another fundamental difference between our somatic and our germ cell lineages involves the maintenance of telomere repeats that stabilize chromosome ends. In most of our somatic cells, there is a replication dependent loss of telomere repeats. These protective repeats are kept largely constant in germ and neoplastic cells [73]. The linear nature of the mammalian chromosome obviously causes the dilemma of replicative telomere shortening in the absence of telomerase activity, even though telomere maintenance may altenatively be achieved by somatic recombination [47]. In meiosis, telomere repeats are instrumental in enabling correct chromosome pairing, which is the prerequisite for the 1 to 1 segregation of homologous chromosomes during the reduction division [75]. In somatic cells, telomere loss is intimately linked to cell replication and might be enhanced under pathologic conditions, such as chronic inflammatory disease [76]. However, the replicative potential of the renewable stem cell compartments in our body, notably bone marrow and gut-crypt stem cells, exceeds by far a single human lifetime, and mice lacking telomerase survive surprisingly well; thus, telomere loss accompanies somatic cell senescence but may not be the one and only culprit of (cell) aging [77].

Oxidative stress as instrinsic risk factor

In addition to thermoinstability, generation of reactive oxygen and nitrogen species (ROS) is widely regarded as the second major instrinsic, and thereby unavoidable risk factor that contributes to loss of genetic homeostasis during aging [23, 78]. In additon to some other sources, ROS arise chiefly as by-products of the generation of energy through oxidative phosphorylation within the respiratory chain of our mitochondria [79]. There is an extensive body of literature that described age-related alterations of the mitochondrial genome of our somatic cell lineages [80]. Part of the vulnerability of the mitochondrial genome probably reflects the absence and/or lower efficiency of mitochondrial DNA-repair systems [81]. To counteract the continuous threat to the nuclear genome by ROS, long-lived, warm-blooded species have developed a number of scavenger systems, including superoxide dismutase, catalase and the glutathione system, but energy constraints imposed by the physico-chemical laws of our cells prohibit a 100% effectiveness of these protective systems. This leads to age-related increase of free radical damage to DNA, proteins and lipids which contributes to age-related morbidity. Degenerative diseases of advanced age such as senile dementia of the Alzheimer type, Parkinsons'disease and senile macula degeneration are believed to result, at least in part, from the lifetime detrimental action of ROS [82]. Age-related free radical-induced DNA damage has been convincingly linked to cancer [83]. Moreover, mutations in caretaker and gatekeeper genes which contribute to age-related neoplasia, may be induced by reactive oxygen and nitrogen species [84, 85]. On the other hand, inactivation of certain caretaker genes by constitutional mutations cause increased cellular sensitivity towards oxygen [86]. The resulting disease, Fanconi anemia (FA), has a high risk of bone marrow failure, myelodysplasia, neoplasia, premature aging and premature death. Like in normal aging, endocrine abnormalitics occur in more than 80% of FA patients, including hyperinsulinemia, growth hormone insufficiency and hypothyroidism [87]. The FA-family of genes, including the BRCA2 (breast cancer 2) gene [88], is a relatively recent acquisition of the vertebrate genome which may provide additional protection against oxidative damage in long-lived species [60, 89].

An impressive case for the eminent role of ROS during aging comes from the comparison of mammalian and avian species [90]. In spite of a two-fold higher metabolic rate and higher body temperature, birds outlive mammals of comparable body weight [91]. However, the rate of free radical production (and thereby oxidative damage) appears to be substantially lower in birds than in mammals. Such lower levels of ROS generation in conjunction with low levels of fatty acid unsaturation provide less opportunity for harmful synergism between ROS and advanced glycosylated (Maillard) end products (AGEs) that promote age-related cell and tissue degeneration [92]. An additional factor favoring the exceptional longevity of birds compared to mammals appears to be delayed reproduction. Long-lived bird species such as condors, ravens, eagles and seagulls start reproduction late (albatrosses as late as 10 years of age) and have smaller clutch sizes, supporting the idea that less investment in reproduction may benefit somatic maintenance [93]. Interestingly, Westendorp and Kirkwood report a similar correlation between age at first reproduction and longevity in human females [94].

Can we prevent human aging?

Theoretical and experimental approaches
Table 4 summarizes some strategies which might be successful in preventing or delaying human aging.

***Table 4**. Life-extension strategies. (See text for discussion)*

Strategy	Possible approaches
Decrease of body temperature	(theoretical possibility only)
Decrease of metabolic rate	Calorie reduction (CR) has been successful in animal experiments and, with caveats, in Biosphere 2
Reduction of reactive oxygen species	Antioxidants have positive effect *in vitro*
Improvement of damage recognition and DNA repair	Gene manipulation/gene transfer (*in vitro*; experimental stage)

Decreasing our body temperature to 34°C or less would reduce the thermoinstability of our DNA and thereby reduce the mutational load in our somatic cells. Under such hypothermic conditions, our average lifespan would be predicted to increase to 200 and our maximum lifespan to 280 years [95]. However, such a drastic reduction of body temperature would impair the function of most of our enzymatic systems which have been optimized and fine-tuned to 37°C during the course of mammalian evolution. Therefore, reducing the intrinsic thermoinstability is not a realistic option. More promising results have been achieved by caloric restriction [96] which causes complex alterations of gene expression profiles [97] and might ultimately act through the highly conserved insulin signalling pathways, including minor temperature changes [98]. However, as observed during the "Biosphere" project, longterm caloric restriction may entail untowards effects in humans if, for example cholesterol levels, blood pressure, blood sugar or leukocyte counts decrease below optimal valaues [99, 100]. Because of the impressive evidence, in lower organisms, for lifespan extension through augmentation of ROS-defense systems

[101, 102] anti-aging activists advocate the supplemental use of antioxidants and micronutrients in humans [103, 104]. Such supplements may be indicated in situations of malnutrition or disease, but appear to be largely ineffective under conditions of balanced nutrition and lifestyle [105]. With few exceptions (e.g., vitamin A) nutritional supplements can be safely consumed in large quantities, and people in wealthy countries can afford to produce expensive urine and pay for placebo effects. Much more critical is the use of hormone supplements such as estrogens, growth hormone and dehydroepiandrosterone [106]. Since genetic instability and the somatic mutational load increase with aging [107] the number of potential tumor cells increases as well. Together with the age-related decline of the immune system, hormone replacement regimens might select for genetically altered cells and thereby increase the risk for neoplasia. As a putative public mechanism of aging, much theoretical and practical attention has been given to the age-related decline of IGF-1 reflecting decline in growth hormone secretion [108]. However, there are no well-controlled and unbiased studies that would convincingly support the initial therapeutic optimism with respect to GH replacement in the aging human population [109, 110].

Practical and effective approaches

There is no doubt that balanced nutrition, stabile psyche, and a physically and mentally active lifestyle can slow down, but not prevent aging. From a genetic perspective, three major factors determine healthy aging: (a) favorable environmental conditions, (b) absence of detrimental genes, (c) moderate lifestyle. While (a) and (c) can be influenced by individual choice, (b) depends on our parents whom we cannot choose. The importance of environmental and lifestyle factors is highlighted by the fact that there are no behavioural "extremists" among centenarians, no chain smokers, no alcohol or drug addicts, and no couch-potatoes. Regular physical activity ranks top in the recipe for longevity of the Johns Hopkins University [111]. Second is a balanced, low-fat and high-fruit/vegetable diet, and third is the avoidance of excessive caloric intake and the very moderate use, if any, of alcohol, tobacco and caffeine. Nutritional supplements play only a minor, if any, role in healthy aging provided that nutrition is balanced. Medical and pharmacological interventions aimed at reducing hypertension, diabetes, thrombophilia and hyperlipidemia have greatly improved the survival of elderly people [15]. Low dose aspirin, estrogen replacement, and wide-spread use of statins may eventually reduce the risk of both cardiovascular and Alzheimer's disease [112]. However, the high number of genes with antagonistic pleiotropic gene action, in concert with intrinsic thermoinstability and oxidative stress represent irreversible barriers to the avoidance of aging, let alone mortality. In spite of optimal environmental conditions and optimal medical care, physical and mentally active lifestyle, balanced nutrition and psyche, these intrinsic "time bombs" limit the average human lifespan to under and the maximum human lifespan to somewhat over 100 years. However, the aging process *per se*, albeit irreversible, appears to be amenable to modulation by metabolic interventions and lifestyle choice [113, 114].

Conclusion and perspective

In humans, genetic individuality, environmental conditions and lifestyle choices contribute to the impressive heterogeneity of both aging and longevity [115]. If one accepts thermoinstability of our DNA and generation of reactive oxygen/nitrogen species as intrinsic and immutable factors that constantly threaten the integrity of our somatic genome [116], increasing the expression and improving the function of DNA repair and maintenance genes might counteract the aging process. Comparisons of gene expression profiles between young and old cells reveal a large number of differences that involve hundreds of genes instrumental in DNA replication, cell cycle transit and signal transduction [117]. The results of such studies at the cellular level provide strong support for the notion that multiple rather than a few critical genes play a role in the aging process [118]. To add to the complexity, one should not loose sight of the fact that epigenetic variation is likely to play a significant role in age-related diseases such as autoimmunity and cancer, but perhaps also in the aging process itself [119, 120]. This also holds for the many posttranslational modifications that are part of protein function and protein turnover in our cells [121]. Human genetics has taught us that genotype-phenotype correlations are complex rather than linear, and that environmental factors and lifestyles can easily overrule the power of genes [122]. With respect to the genetic aspects of aging this seems like a comforting prospect.

Evolutionary theory provides a satisfactory explanation why aging is both a necessity and a trade-off in complex organisms such as ours [123]. Looking back into evolution it is of course intriguing that mechanisms like insulin signalling or stress response regulation that undoubtedly are important for aging seem widely conserved among species [2]. However, the mechanisms that determine aging in yeast, *C. elegans* or drosophila may not be of immediate relevance to aging in humans. Many hundreds millions of years of evolutionary time and orders of magnitude in complexity separate us from these lower organisms. Humans do not undergo dauer formation in response to food shortage, and yeast, *C. elegans* or drosophila do not get cancer, Alzheimer's disease or osteoporosis. Extrapolations from findings in these lower organisms to human aging and longevity should therefore made with utmost caution, and prospects for modulating the aging process via molecular genetic approaches appear far less realistic in humans than in worms, flies or even mice [124].

Finally, how many genes contribute to human longevity and aging? While earlier studies suggested the interaction of many genes, a recent sib-pair study of long-lived families claims that human chromosome 4 harbors a single gene that determines exceptional longevity [125]. However, lod-scores for the chromosome 4 assignement were barely significant and may be expected by mere chance in this type of whole genome scans. Unless this study can be confirmed with a much larger number of sib-pairs, any conclusions regarding the possibility of a major "longevity" gene in our species are premature. Intuitively, a complex process like human aging is unlikely to be determined by a single or very few "master" genes, and prospects of eliminating human aging by gene manipulation therefore must be considered wishful thinking at best. There is no scientifically sound evidence to date that any type of genetic enhancement of our species could be achieved within the foreseeable future [126]. With respect to the prospects of modification of human aging and longevity, environmental and lifestyle manipulations are much more effective than any conceivable manipulations at the gene or protein levels. A good illustration of this point are studies of life expectancy among members of certain religious groups who adhere to healthy lifestyles that outlive their peer groups of comparable socio-economic and ethnic background by almost 10 years [114]. It is unlikely that such large gains in life expectancy for so many people will be ever achieved via biological manipulations.

References

1. Martin GM (1999). Genes that modulate longevity and senescence. In: Robine JM, Forette B, Franceschi C, Allard M, eds. *The Paradoxes of Longevity*. New York: Springer, pp. 11–21.
2. Partridge L, Gems D (2002). Mechanisms of ageing: public or private? *Nat Rev Genet*. 3: 165–75.
3. Martin GM (2002). Keynote: mechanisms of senescence complificationists versus simplificationists. *Mech Ageing Dev*. 123: 65–73.
4. Finch CE, Schneider EL, eds. (1985). *Handbook of the Biology of Aging*. New York: Van Nostrand Reinhold.
5. Medina JJ (1996). *The Clock of Ages*. Cambridge: Cambridge University Press.
6. Arking R (1998). *Biology of Aging*. Sunderland, MA: Sinauer Associates.
7. Papaconstantinou J (1994). Unifying model of the programmed (intrinsic) and stochastic (extrinsic) theories of aging. *Ann NY Acad Sci*. 719: 195–211.
8. Martin GM (1997). Genetics and the pathobiology of ageing. *Phil Trans R Soc London B Biol Sci*. 352: 1773–80.
9. Hart RW, Setlow RB (1974). Correlation between deoxyribonucleic acid excision-repair and life-span in a number of mammalian species. *Proc Natl Acad Sci USA* 71: 2169–73.
10. Bürkle A (2001). PARP-1: a regulator of genomic stability linked with mammalian longevity. *Chembiochemistry* 2: 725–8.
11. Holmes DJ, Austad SN (1995). Birds as animal models for the comparative biology of aging: a prospectus. *J Gerontol A Biol Med Sci*. 50: B59–66.
12. Smith DW (1993). *Human Longevity*. Baltimore: Johns Hopkins University Press.
13. Wilmoth J, Skytthe A, Friou D, Jeune B (1996). The oldest man ever? A case study of exceptional longevity. *Gerontologist* 36: 783–8.
14. Robine JM and Allard M (1998). *The oldest human*. *Science* 279: 1834–5.
15. Wilmoth JR, Deegan LJ, Lundstrom H, Horiuchi S (2000). Increase of maximum life-span in Sweden, 1861–1999. *Science* 289: 2366–8.
16. Kesteloot H (1999). On the determinants of mortality at the population level. *Acta Cardiol*. 54: 141–9.
17. Kannisto V (1988). Mortality after age 100. *Popul Stud*. 42: 389–98.
18. Robine J, Vaupel JW (2001). Supercentenarians: slower ageing in individuals or senile elderly? *Exp Gerontol*. 36: 915–30.
19. Zelterman D (1992). A statistical distribution with an unbounded hazard function and its application to a theory from demography. *Biometrics* 48: 807–18.
20. Wilmoth JR (1998). The future of human longevity: a demographer's perspective. *Science* 280: 395–7.

21. Vaupel JW, Carey JR, Christensen K, *et al.* (1998). Biodemographic trajectories of longevity. *Science* 280: 855–60.
22. Oeppen J, Vaupel JW (2002). Demography. Broken limits to life expectancy. *Science* 296: 1029–31.
23. Harman D (1998). Extending functional life span. *Exp Gerontol.* 33: 95–112.
24. Ricklefs RE, Finch CE (1995). *Aging: A Natural History.* New York: WH Freeman and Company.
25. Amdam GV, Omholt SW (2002). The regulatory anatomy of honeybee lifespan. *J Theor Biol.* 216: 209–28.
26. Guralnik JM, Balfour JL, Volpato S (2000). The ratio of older women to men: historical perspectives and cross-national comparisons. *Aging (Milano)* 12: 321–2.
27. Clarke CA, Mittwoch U (1994). Puzzles in longevity. *Perspect Biol Med.* 37: 327–36.
28. Chelly J and Mandel JL (2001). Monogenic causes of X-linked mental retardation. *Nat Rev Genet.* 2: 669–80.
29. Zechner U, Wilda M, Kehrer-Sawatzki H, Vogel W, Fundele R, Hameister H (2001). A high density of X-linked genes for general cognitive ability: a run-away process shaping evolution? *Trends Genet.* 17: 697–701.
30. Lindenberger U, Mayr U, Kliegl R (1993). Speed and intelligence in old age. *Psychol Aging* 8: 207–20.
31. Rogers RG, Powell-Griner E (1991). Life expectancies of cigarette smokers and nonsmokers in the United States. *Soc Sci Med.* 32: 1151–9.
32. Harman D (1991). The aging process: major risk factor for disease and health. *Proc Natl Acad Sci USA* 88: 5360–3.
33. Leveille SG, Resnick HE, Balfour J (2000). Gender differences in disability: evidence and underlying reasons. *Aging (Milano)* 12: 106–12.
34. Trovato F, Lalu NM (1996). Narrowing sex differentials in life expectancy in the industrialized world: early 1970's to early 1990's. *Soc Biol.* 43: 20–37.
35. Begun A, Desjardins N, Iachine I, Yashin A (2000). Multivariate frailty model with a major gene: application to genealogical data. *Stud Health Technol Inform.* 77: 412–6.
36. Cournil A, Kirkwood TBL (2001). If you would live long, choose your parents well. *Trends Genet* 17: 233 5.
37. Frederiksen H, McGue M, Jeune B, *et al.* (2002). Do children of long-lived parents age more successfully? *Epidemiology* 13: 334–9.
38. Iachine IA, Holm NV, Harris JR, *et al.* (1998). How heritable is individual susceptibility to death? The results of an analysis of survival data on Danish, Swedish and Finnish twins. *Twin Res.* 1: 196–205.
39. Kirkwood T (1999). *Time of our Lives. The Science of Human Aging*. London: Weidenfeld and Nicolson.
40. Williams GC (1957). Pleiotropy, natural selection, and the evolution of senescence. *Evolution* 11: 398–411.
41. Rose MR (1991). *Evolutionary Biology of Aging*. Oxford, New York: Oxford University Press.
42. Albin RL (1993). Antagonistic pleiotropy, mutation accumulation, and human genetic disease. *Genetica* 91: 279–86.
43. Opitz JM, Gilbert-Barness E, Ackerman J, Lowichik A (2002). Cholesterol and development: the RSH ("Smith-Lemli-Opitz") syndrome and related conditions. *Pediatr Pathol Mol Med.* 21: 153–81.
44. Toupance B, Godelle B, Gouyon PH, Schachter F (1998). A model for antagonistic pleiotropic gene action for mortality and advanced age. *Am J Hum Genet.* 62: 1525–34.

45. Stine WA (1998). Calcium homeostasis and human evolution. *Coll Antropol.* 22: 411–25.
46. Franceschi C, Valensin, S, Fagnoni F, *et al.* (1999). Biomarkers of immunosenescence within an evolutionary perspective: the challenge of heterogeneity and the role of antigenic load. *Exp Gerontol.* 34: 911–9.
47. Waters DJ, Shen S, Glickman LT (2000). Life expectancy, antagonistic pleiotropy, and the testis of dogs and men. *The Prostate* 43: 272–7.
48. Hu Q, Cool BH, Wang B, Hearn MG, Martin GM (2002). A candidate molecular mechanism for the association of an intronic polymorphism of FE65 with resistance to very late onset dementia of the Alzheimer type. *Hum Mol Genet.* 11: 465–75.
49. Perls T, Kunkel LM, Puca AA (2002). The genetics of exceptional human longevity. *J Mol Neurosc*i 19: 233–8.
50. Schachter F, Faure-Delanef L, Guenot F, *et al.* (1994). Genetic associations with human longevity at the APOE and ACE loci. *Nat Genet.* 6: 29–32.
51. Muiras ML, Verasdonck P, Cottet F, Schachter F (1998). Lack of association between human longevity and genetic polymorphisms in drug-metabolizing enzymes at the NAT2, GSTM1 and CYP2D6 loci. *Hum Genet.* 102: 526–32.
52. Bonafe M, Olivieri F, Mari D, *et al.* (1999). p53 variants predisposing to cancer are present in healthy centenarians. *Am J Hum Genet.* 64: 292–5.
53. Yashin AI, De Benedictis G, Vaupel JW, *et al.* (2000). Genes and longevity: lessons from studies of centenarians. *J Gerontol A Biol Med Sci.* 55: B319–28.
54. Heijmans BT, Westendorp RG, Slagboom PE (2000). Common gene variants, mortality and extreme longevity in humans. *Exp Gerontol.* 35: 865–77.
55. De Benedictis G, Carrieri G, Garasto S, *et al.* (2000). Does a retrograde response in human aging and longevity exist? *Exp Gerontol.* 35: 795–801.
56. Rose G, Passarino G, Carrieri G, *et al.* (2001). Paradoxes in longevity: sequence of mt DNA haplogroup J in centenarians. *Eur J Hum Genet.* 9: 701–7.
57. Paolisso G, Barbieri M, Bonafe M, Franceschi C (2000). Metabolic age modelling: the lesson from centenarians. *Eur J Clin Invest.* 30: 888–94.
58. Kinzler KW, Vogelstein B (1997). Cancer susceptibility genes: gatekeepers and caretakers. *Nature* 386: 761–3.
59. Hoeijmakers JH (2001). Genome maintenance mechanisms for preventing cancer. *Nature* 411: 366–74.
60. Joenje H, Patel KJ (2001). The emerging genetic and molecular basis of Fanconi anaemia. *Nat Rev Genet.* 2: 446–57.
61. Martin GM, Oshima J, Gray M, Poot M (1999). What geriatricians should know about the Werner syndrome. *J Am Ger Assoc.* 47: 1136–44.
62. Epstein CJ, Martin GM, Schultz AL, Motulsky AG (1966). Werner's syndrome: a review of its symptomatology, natural history, pathologic features, genetics and relationship to the natural aging process. *Medicine* 45: 177–221.
63. Oshima J (2000). The Werner syndrome protein: an update. *Bioessays* 22: 894–901.
64. Fry M (2002). The Werner syndrome helicase-nuclease- one protein, many mysteries. <http: / /sageke.sciencemag.org/cgi/content/full/sageke;2002/13/re2>.
65. Lock W, Funk W (1995). Aging and cancer. *Versicherungsmedizin* 47: 157–165.
66. Ames B (1997). Environmental pollution, pesticides, and the prevention of cancer: misconceptions. *FASEB J.* 11: 1041–52.
67. Maurer B, Guttenbach M, Schmid M (2002). Chromosome instability in normative aging. In: Hasima FM, Martin GM, Weissman SM, eds. *Chromosomal Instability and Aging: Basic Science and Clinical Implications.* New York: Marcel Dekker.
68. Vogel F, Motulsky AG (1997). *Human Genetics.* Berlin Heidelberg New York: Springer.

69. Kirkwood TL, Kapahi P, Shanley DP (2000). Evolution, stress, and longevity. *J Anat.* 197: 587–90.
70. Greider CW (2000–2001). Cellular responses to telomere shortening: cellular senescence as a tumor suppressor mechanism. *Harvey Lect.* 96: 33–50.
71. Gensler HL, Bernstein H (1981). Evolutionary significance of meiosis. *Q Rev Biol.* 56: 279–92.
72. Thompson LH, Schild D (2002). Recombinational DNA repair and human disease. *Mutat Res.* 509: 49–78.
73. Shay JW, Wright WE (2001). Ageing and cancer: the telomere and telomerase connection. *Novartis Found Symp.* 235: 116–25.
74. Aragona M, Maisano R, Panetta S, *et al.* (2000). Telomere maintenance in aging and carcinogenesis. *Int J Oncol.* 17: 981–9.
75. Takahashi Y, Kuro M, Ishikawa F (2000). Aging mechanisms. *Proc Natl Acad Sci USA* 97: 12407–8.
76. O'Sullivan JN, Bronner MP, Brentnall TA, *et al.* (2002). Chromosomal instability in ulcerative colitis is related to telomere shortening. *Nat Genet.* 32: 280–4.
77. Goyns MH, Lavery WL (2000). Telomerase and mammalian ageing: a critical appraisal. *Mech Ageing Dev.* 114: 69–77.
78. Martin GM, Austadt SN, Johnson TE (1996). Genetic analysis of ageing: role of oxidative damage and environmental stresses. *Nat Genet.* 13: 25–34.
79. Richter C (1992). Reactive oxygen and DNA damage in mitochondria. *Mutat Res.* 275: 249–55.
80. Wei YH, Lee HS (2002). Oxidative stress, mitochondrial DNA mutation, and impairment of antioxidant enzymes in aging. *Proc Soc Exp Biol Med.* 227: 671–82.
81. Croteau DL, Rhys CMJ, Hudson EK, Dianov GL, Hansford RG, Bohr VA (1997). An oxidative damage-specific endonuclease from rat liver mitochondria. *J Biol Chem.* 272: 27338–44.
82. Sayre LM, Smith MA, Perry G (2001). Chemistry and biochemistry of oxidative stress in neurodegenerative disease. *Curr Med Chem.* 8: 721–38.
83. Malins DC, Johnson PM, Wheeler TM, Barker EA, Polissar NL, Vinson MA (2001). Age-related radical-induced DNA damage is linked to prostate cancer. *Cancer Res.* 61: 6025–8.
84. Grollman AP, Moriya M (1993). Mutagenesis by 8-oxoguanine: an enemy within. *Trends Genet.* 9: 246–9.
85. Forsberg L, de Faire U, Morgenstern R (2001). Oxidative stress, human genetic variation, and disease. *Arch Biochem Biophys.* 389: 84–93.
86. Schindler D, Hoehn H (1988). Fanconi anemia mutations causes cellular susceptibility to ambient oxygen. *Am J Hum Genet.* 43: 429–35.
87. Wajnrajch MP, Gertner JM, Huma Z, *et al.* (2001). Evaluation of growth and hormonal status in patients referred to the International Fanconi Anemia registry. *Pediatrics* 107: 744–54.
88. Howlett NG, Taniguchi T, Olson S, *et al.* (2002). Biallelic inactivation of BRCA2 in Fanconi anemia. *Science* 297: 606–9.
89. Futaki M, Liu JM (2001). Chromosomal breakage syndromes and the BRCA1 genome surveillance complex. *Trends Mol Med.* 7: 560–5.
90. Pamplona R, Barja G, Portero-Otin M (2002). Membrane fatty acid unsaturation, protection against oxidative stress, and maximum life span. *Ann NY Acad Sci.* 959: 475–90.

91. Calder WA (1990). Avian longevity and aging. In: Harrison DE, ed. *Genetic Effects on Aging. II.* Caldwell: Telford Press
92. Kristal BS, Yu BP (1992). An emerging hypothesis: synergistic inductions of aging of free radicals and Maillard reaction. *J Geront Biol Sci.* 47: B107–14.
93. Lithgow GJ, Kirkwood TB (1996). Mechanisms and evolution of aging. *Science* 273: 80.
94. Westendorp RG, Kirkwood TB (1998). Human longevity at the cost of reproductive success. *Nature* 396: 743–6.
95. Rosenberg B, Kemeny G, Smith LG, *et al.* (1973). The kinetics and thermodynamics of death in multicellular organisms. *Mech Ageing Dev.* 2: 275–93.
96. Weindruch R, Walford RL (1988). *The Retardation of Aging and Disease by Dietary Restriction.* Springfield, Illinois: Charles Thomas.
97. Lee CK, Klopp RG, Weindruch R, Prolla TA (1999). Gene expression profile of aging and its retardation by caloric restriction. *Science* 285: 1390–3.
98. Roth GS, Lane MA, Ingram DK, *et al.* (2002). Biomarkers of caloric restriction may predict longevity in humans. *Science* 297: 811.
99. Walford RL, Weber L, Panov S (1995). Caloric restriction and aging as viewed from Biosphere 2. *Receptor* 5: 29–33.
100. Walford RL, Mock D, MacCallum T, Laseter JL (1999). Physiologic changes in humans subjected to severe selective caloric restriction for two years in biosphere 2: health, aging, and toxicological perspectives. *Toxicol Sci.* 52: 61–5.
101. Parkes TL, Elia AJ, Dickinson D, Hilliker AJ, Philliphs JP, Boulianne GL (1998). Extension of *Drosophila* lifespan by overexpression of human SOD1 in motorneurons. *Nat Genet.* 19: 171–4.
102. Sohal RS, Mockett RJ, Orr WC (2000). Current issues concerning the role of oxidative stress in aging: a perspective. *Results Probl Cell Differ.* 29: 45–66.
103. Ames BN (1998). Micronutrients prevent cancer, delay aging. *Toxicol Lett.* 102/103: 5–18.
104. Reiter RJ, Tan DX, Burkhardt S (2002). Reactive oxygen and nitrogen species and cellular and organismal decline: amelioration with melatonin. *Mech Ageing Dev.* 123: 1007–19.
105. Muntwyler J, Hennekens CH, Manson JE, Gaziano JM (2002). Vitamin supplement use in a low-risk population of US male physicians and subsequent cardiovascular mortality. *Arch Intern Med.* 162: 1472–6.
106. Johnson MD, Benn RA, Sirrs SM (2002). Uses of DHEA in aging and other disease states. *Ageing Res Rev.* 1: 29–41.
107. Vijg J (2000). Somatic mutations and aging: a re-evaluation. *Mutat Res.* 447: 117–35.
108. Carter CS, Ramsey MM, Sonntag WE (2002). A critical analysis of the role of growth hormone and IGF-1 in aging and lifespan. *Trends Genet.* 18: 295–301.
109. Taaffe DR, Jin IH, Vu TH, Hoffman AR, Marcus R (1996). Lack of effect of recombinant human growth hormone (GH) on muscle morphology and GH-insulin-like growth factor expression in resistance-trained elderly men. *J Clin Endocrinol Metab.* 81: 421–5.
110. Ruiz-Torres A, de Melo Kirzner MS (2002). Ageing and longevity are related to growth hormone/insulin-like growth factor-1 secretion. *Gerontology* 48: 401–7.
111. Johns Hopkins Prescription for Longevity. In: *The Johns Hopkins Medical Letter.* December 1998, pp 4–6.
112. Fassbender K, Simon M, Bergman C, *et al.* (2001). Simvastatin strongly reduces levels of Alzheimer's disease b-amyloid peptides Ab42 and Ab40 *in vitro* and *in vivo*. *Proc Natl Acad Sci USA* 98: 5856–61.

113. Weindruch R, Keenan KP, Carney JM, Fernandes G, *et al.* (2001). Caloric restriction mimetics: metabolic interventions. J *Gerontol A Biol Sci Med Sci.* 56: 20–3.
114. Fraser GE, Shavlik DJ (2001). Ten years of life: is it a matter of choice? *Arch Intern Med.* 161: 1645–52.
115. Stadtman ER (2002). Importance of individuality on oxidative stress and aging. *Free Rad Biol Med.* 33: 597–604.
116. Von Zglinicki T, Bürkle A, Kirkwood TB (2001). Stress, DNA damage and ageing an integrative approach. *Exp Gerontol.* 36: 1049–62.
117. Shelton DN, Chang E, Whittier PS, *et al.* (1999). Microarray analysis of replicative senescence. *Curr Biol.* 9: 939–45.
118. Lee CK, Weindruch R, Prolla TA (2000). Gene-expression profiling of the ageing brain in mice. *Nat Genet.* 25: 294–7.
119. Issa JP (2002). Epigenetic variation and human disease. *J Nutr.* 132: 2388S–92S.
120. Richardson BC (2002). Role of DNA methylation in the regulation of cell function: autoimmunity, aging and cancer. *J Nutr.* 132: 2401S–5S.
121. Stadtman ER (2001). Protein oxidation in aging and age-related diseases. *Ann NY Acad Sci.* 928: 22–38.
122. Strohman R (2002). Maneuvering in the complex path from genotype to phenotype. *Science* 296: 701–3.
123. Kirkwood TB, Austad SN (2000). Why do we age? *Nature* 408: 233–8.
124. Helfand SL, Inouye SK (2002). Rejuvenating views of the aging process. *Nat Genet Rev.* 3: 149–53.
125. Puca AA, Daly MJ, Brewster SJ, *et al.* (2001). A genome-wide scan for linkage to human exceptional longevity identifies a locus on chromosome 4. *Proc Natl Acad USA* 98: 10505–8.
126. Gordon JW (1999). Genetic enhancement in humans. *Science* 283: 2023–6.
127. Bathum L, Christiansen L, Nybo H, *et al.* (2001). Association of mutations in the hemochromatosis gene with shorter life expectancy. *Arch Intern Med..* 162: 1196–7.
128. Levitt NC, Hickson ID (2002). Caretaker tumor supressor genes that defend genome integrity. *Trends Mol Med.* 8: 179–86.
129. Melcher R, von Golitschek R, Steinlein C, *et al.* (2000). Spectral karotyping of Werner syndrome fibroblast cultures. *Cytogenet Cell Genet.* 91: 180–5.
130. Kulozik AE, Hentze MW, Hagemeier C, Bartram CR (2000). *Molecular Medicine.* Berlin, New York: de Gruyter, p. 129.

Index